ELEMENTS
OF LINEAR ALGEBRA

Prentice-Hall, Inc., Englewood Cliffs, New Jersey

ELEMENTS

OF LINEAR ALGEBRA

ANTHONY J. PETTOFREZZO
Professor of Mathematical Sciences
Florida Technological University
Orlando, Florida

TO ROBBY

*Elements of Linear Algebra
by Anthony J. Pettofrezzo*

© *1970 by Prentice-Hall, Inc.,
Englewood Cliffs, New Jersey*

*All rights reserved. No part of this book
may be reproduced in any form
or by any means without permission
in writing from the publisher.*

Current printing (first digit): 1 2 3 4 5 6 7 8 9 10
13-266155-1
*Library of Congress catalog card number 71-95704
Printed in the United States of America*

*Prentice-Hall International, Inc., London
Prentice-Hall of Australia, Pty. Ltd., Sydney
Prentice-Hall of Canada, Ltd., Toronto
Prentice-Hall of India Private Ltd., New Delhi
Prentice-Hall of Japan, Inc., Tokyo*

PREFACE

Linear algebra is a significant topic in contemporary mathematics curricula. The purpose of this book is to present to the reader at an introductory level the fundamental concepts of linear algebra. It is my belief that most linear algebra texts require, as a prerequisite on the part of the reader, a level of mathematical sophistication which he does not usually reach before he completes a course in modern abstract algebra. No such prerequisite is assumed here. Those elements of vector algebra and matrix algebra readily accessible to the reader uninitiated in abstract algebra are presented first in an intuitive framework and then in a more formal manner. Such a presentation of linear algebra can contribute much toward the development of the reader's mathematical maturity.

In Chapter 1 vector algebra is developed as an algebra of directed line segments. Coordinate position vectors are treated. The study of geometric vectors of two-dimensional and three-dimensional space is an aid toward the understanding of abstract vector spaces, matrices, tensors, and other multi-component algebras. The key concepts necessary for the study of generalized vector spaces are presented and developed in Chapters 1 and 2. The abstract concept of a real vector space is included and a number of concrete examples are discussed. Applications of the products of vectors to elementary plane geometry, plane and spherical trigonometry, and coordinate geometry are also illustrated in Chapter 2. The use of vector algebra as a mathematical vehicle in the study of coordinate geometry of three-dimensional space is demonstrated in Chapter 3. Chapters 4 and 5 include those basic concepts of matrix algebra which are important in the study of physics, statistics, economics, engineering, and mathematics. Matrices are considered as elements of an algebra. The abstract concept of a linear algebra is presented and a number of concrete examples are discussed. The concept of a linear transformation of the plane and the use of matrices in discussing such transformations are illustrated in Chapter 6. Some aspects of the algebra of transformations and its relation to the algebra of matrices are included here. Chapter 7 contains material on the application of the properties of eigenvalues and eigenvectors to the study of the conics.

The motivations of the concepts presented are included wherever appropriate. Considerable attention has been paid to the formulation of precise definitions and statements of theorems. Detailed proofs of most of the theorems stated are included in this book.

Many more illustrative examples than usual have been included as aids to the reader in his mastery of the concepts and methods presented. There are a sufficient number of exercises, which either supplement the theory presented or give the reader an opportunity to reinforce his understanding of the applications. Answers are provided to the odd-numbered exercises.

This book contains enough material for a one-semester course at the college level. The many detailed illustrative examples make the book quite useful for individual study, summer institutes for teachers of mathematics, and in-service programs.

I am indebted to Mr. James Walsh of Prentice-Hall, Inc., for suggesting that this book be written, and for his encouragement and interest. I wish to express my appreciation to my wife, Betty, and my son, Paul, for their assistance in the preparation of the manuscript. Finally, a special note of appreciation is due the editorial-production staff of Prentice-Hall, Inc., for their kind cooperation in the production of this book.

Anthony J. Pettofrezzo

CONTENTS

ONE

ELEMENTARY
VECTOR OPERATIONS   1

    1-1 / Scalars and vectors   1
    1-2 / Equality of vectors   4
    1-3 / Vector addition and subtraction   6
    1-4 / Multiplication of a vector by a scalar   9
    1-5 / Linear dependence of vectors   14
    1-6 / Applications of linear dependence   19
    1-7 / Position vectors   26

TWO

PRODUCTS OF VECTORS 32

2-1 / The scalar product 32
2-2 / Applications of the scalar product 39
2-3 / Circles and lines on a coordinate plane 45
2-4 / Orthogonal bases 48
2-5 / The vector product 51
2-6 / Applications of the vector product 58
2-7 / The scalar triple product 64
2-8 / The vector triple product 69
2-9 / Quadruple products 70
2-10 / Quaternions 74
2-11 / Real vector spaces 77

THREE

PLANES AND LINES IN SPACE 82

3-1 / Direction cosines and numbers 82
3-2 / Equation of a plane 86
3-3 / Equation of a sphere 92
3-4 / Angle between two planes 96
3-5 / Distance between a point and a plane 99
3-6 / Equation of a line 102

FOUR

MATRICES 111

4-1 / Definitions and elementary properties 111
4-2 / Matrix multiplication 117
4-3 / Diagonal matrices 125
4-4 / Special real matrices 128
4-5 / Special complex matrices 133

FIVE

INVERSES AND
SYSTEMS OF MATRICES   137

     5–1 / Determinants   137
     5–2 / Inverse of a matrix   146
     5–3 / Systems of matrices   154
     5–4 / Linear algebras   161
     5–5 / Rank of a matrix   163
     5–6 / Systems of linear equations   169

SIX

TRANSFORMATIONS
OF THE PLANE   175

     6–1 / Mappings   175
     6–2 / Rotations   178
     6–3 / Reflections, dilations, and magnifications   183
     6–4 / Other transformations   189
     6–5 / Linear homogeneous transformations   193
     6–6 / Orthogonal matrices   195
     6–7 / Translations   198
     6–8 / Rigid motion transformations   205

SEVEN

EIGENVALUES
AND EIGENVECTORS   213

     7–1 / Characteristic functions   213
     7–2 / A geometric interpretation of eigenvectors   218
     7–3 / Some theorems   221
     7–4 / Diagonalization of matrices   225
     7–5 / The Hamilton-Cayley theorem   231
     7–6 / Quadratic forms   236
     7–7 / Classification of the conics   238
     7–8 / Invariants for conics   245

ANSWERS

TO ODD-NUMBERED
EXERCISES 249

INDEX 287

ELEMENTS
OF LINEAR ALGEBRA

ONE

ELEMENTARY VECTOR OPERATIONS

1-1 / SCALARS AND VECTORS

In discussing physical space it is necessary to consider several types of physical quantities. One class of quantities consists of those quantities which have associated with them some measure of undirected magnitude. Such quantities are called *scalar quantities* or simply *scalars*. Each scalar quantity can be represented by a real number which indicates the magnitude of the quantity according to some arbitrarily chosen convenient scale or unit of measure. Since scalars are real numbers, scalars enter into combinations according to the rules of the algebra of real numbers. Mass, density, area, volume, time, work, electrical charge, potential, temperature, and population are examples of scalar quantities.

A second class of physical quantities consists of those quantities which have associated with them both the property of magnitude and the property of direction. Such quantities are called *vector quantities* or simply *vectors*. Force, velocity, acceleration, and momentum are examples of vector quantities.

The following example illustrates the need to distinguish between scalars and vectors. Consider a plane which flies from point A to point B, 300 miles east of A, and then proceeds to fly north to a third point C, 400 miles north of B as shown in Figure 1-1. The distances that the plane has flown are scalars and may be added in the usual manner to determine the total distance covered by the flight; that is, $300 + 400$, or 700, miles. The flight may also be considered in terms of the displacement of the plane from point A to point C. This displacement may be considered as the sum of two displacements: one 300 miles east from A to B and the second 400 miles north from B to C; the sum of these displacements is the displacement 500 miles with a bearing of approximately $36°52'$ east of north from A to C. Notice that direction as well as magnitude is considered in describing displacements. Displacements are examples of vector quantities. Furthermore, the sum of two displacements is calculated in a rather different manner than is the sum of two scalar quantities; in other words, vector addition is quite different from scalar addition.

Notice that in Figure 1-1 the displacements were denoted by means of directed line segments or arrows. In mathematics it is convenient to construct a geometric model of the physical concept of a vector and of situations involving vector quantities.

DEFINITION 1.1 / *A **geometric vector** or simply a **vector** is a directed line segment* (*Figure 1-2*).

In Figure 1-2 the length of the directed line segment with reference to some conveniently chosen unit of length is associated with the magnitude of the vector; thus, lengths of directed line segments represent scalars. Notice that the magnitude of a vector is a nonnegative real number. Unless otherwise restricted, we shall consider a geometric vector as a vector in three-dimensional space.

Symbolically the vector represented in Figure 1-2 is denoted by $\overrightarrow{AB}$, where A is the **initial point** (sometimes called the **origin** or **origin point**) of the directed line segment and B is the **terminal point**. Symbolically the magnitude of $\overrightarrow{AB}$ is denoted by $|\overrightarrow{AB}|$. Whenever convenient, a second notation for vectors will be used which consists of single small letters beneath a half arrow such as $\vec{a}, \vec{b}, \vec{c}, \ldots$. Then $|\vec{a}|, |\vec{b}|, |\vec{c}|, \ldots$ will represent the magnitudes of such vectors, respectively.

In our study of vectors we shall often "associate" a vector with a line

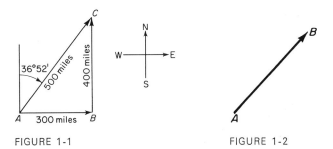

FIGURE 1-1   FIGURE 1-2

segment. It is possible to associate either $\overrightarrow{AB}$ or $\overrightarrow{BA}$ with the line segment whose end points are A and B, where in each case the magnitude of the vector is equal to the length of the line segment. However, the vectors $\overrightarrow{AB}$ and $\overrightarrow{BA}$ are not equal. When we use the second notation of a single small letter to represent a vector associated with a line segment, we will find it convenient to adopt the following convention for establishing the association we desire: "*associate $\vec{a}$ with line segment AB*" shall mean $\vec{a} = \overrightarrow{AB}$; "*associate $\vec{a}$ with line segment BA*" shall mean $\vec{a} = \overrightarrow{BA}$.

A vector of particular interest is the *null vector* or *zero vector*, which will be denoted by $\vec{0}$.

DEFINITION 1.2 / *A **null vector** or **zero vector** is a vector whose magnitude is zero.*

By Definition 1.2, a geometric vector is a null vector if its initial and terminal points coincide. We choose to consider the null vector as a vector without a unique direction and, specifically, with a direction that is indeterminate. The null vector is the only vector whose direction is indeterminate. Some mathematicians choose to consider the null vector as having any arbitrary direction; that is, the null vector could be considered as the limit of any one of infinitely many finite vectors as its magnitude approaches zero. The wording of many theorems in the subsequent development of vector algebra must be carefully changed if the direction of the null vector is considered arbitrary.

EXERCISES

Identify each quantity as either a scalar quantity or a vector quantity.

1. Distance between New York and Boston.
2. Displacement from Chicago to St. Louis.
3. Temperature of 97° Fahrenheit.
4. Weight of 100 pounds.
5. Pressure of 18 pounds per square inch.
6. 250 horsepower.

1-2 / EQUALITY OF VECTORS

In choosing an appropriate definition for the equality of two vectors, one is usually guided by the applications that will be made of the vectors. For example, in the study of the theory of mechanics of rigid bodies, vectors $\vec{a}$ and $\vec{b}$ (denoting forces) have the same mechanical effect in that the "line of action" of these two vectors is the same and the vectors have the same magnitude. However, as in Figure 1-3, the mechanical effect of $\vec{c}$, a vector of equal magnitude to $\vec{a}$ and $\vec{b}$, would be to rotate the shaded object acted upon. In this type of problem $\vec{a}$ and $\vec{b}$ would be considered equal since they have equal magnitudes and lie along the same line with the same orientation. Consideration of this type of problem requires a definition of equality for a class of vectors commonly called **line vectors.**

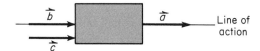

FIGURE 1-3

In the theory of mechanics of deformable bodies, one needs a more restrictive definition for the equality of vectors. For example, consider $\vec{a}$ and $\vec{b}$, both having equal magnitudes and directed along the same line with the same orientation, acting upon an elastic material as indicated in Figure 1-4. Each vector would deform the material in a different way: $\vec{a}$ would tend to compress it, whereas $\vec{b}$ would tend to stretch it. A consideration of this type of problem leads to another definition for the equality of vectors. In order for two vectors to be equal they must have equal magnitudes, act along the same line of action with the same orientation, and be applied at the same point in space. A study of vectors under this definition of equality is a study of **bound vectors.**

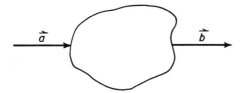

FIGURE 1-4

However, those properties of geometry and trigonometry in which one is generally interested in mathematics allow mathematicians to use a less restrictive definition for the equality of vectors.

DEFINITION 1.3 / *Two vectors $\vec{a}$ and $\vec{b}$ are **equal** if, and only if, $|\vec{a}| = |\vec{b}|$ and $\vec{a}$ is parallel to $\vec{b}$ with the same orientation; that is, $\vec{a} = \vec{b}$ if, and only if, $\vec{a}$ and $\vec{b}$ have the same magnitude and direction.*

In Definition 1.3 the word "parallel" is used in a generalized sense to mean that the vectors are on the same or parallel lines; "orientation" refers to the "sense" of the vector along the line; "direction" refers to both parallelism and orientation. From the definition it follows that any vector may be subjected to a parallel displacement without considering its magnitude or direction as being changed. The vectors in Figure 1-5 are all equal to one another. Since the vectors are equal, any one of the vectors may be considered to represent a whole class of equal vectors of which the others are members.

Under Definition 1.3 for the equality of vectors, the vectors discussed are called **free vectors**. Unless otherwise specified, we shall assume that all vectors are free vectors.

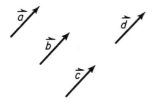

FIGURE 1-5

1-2 / EQUALITY OF VECTORS

EXERCISES

In Exercises 1 through 6 state whether or not the two vectors appear to be equal.

In Exercises 7 through 9 copy the given figure and draw a vector with the given point P as its initial point and equal to the given vector.

10–12. Copy the given figure in Exercises 7 through 9 and draw a vector with the given point P as its terminal point and equal to the given vector.

1-3 / VECTOR ADDITION AND SUBTRACTION

Consideration of the displacement problem of §1-1 and similar physical problems concerning vectors motivates the following definition, called the **law of vector addition.**

DEFINITION 1.4 / *Given two vectors $\vec{a}$ and $\vec{b}$, if $\vec{b}$ is translated so that its initial point coincides with the terminal point of $\vec{a}$, then a third vector $\vec{c}$ with the same initial point as $\vec{a}$ and the same terminal point as $\vec{b}$ is equal to $\vec{a} + \vec{b}$ (Figure 1-6).*

The sum of two vectors is a uniquely determined vector; that is, if $\vec{c} = \vec{a} + \vec{b}$ and $\vec{d} = \vec{a} + \vec{b}$, then $\vec{c} = \vec{d}$.

Consider any parallelogram $ABCD$. Associate vectors $\vec{a}$ and $\vec{b}$ with sides AB and BC, respectively, as in Figure 1-7. Then $\vec{a} + \vec{b}$ may be associated with the diagonal AC. Since the opposite sides of any parallelogram are equal in length and parallel, $\vec{b}$ and $\vec{a}$ may be associated with sides AD

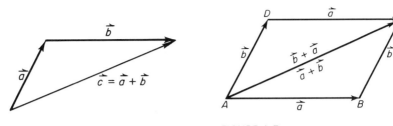

FIGURE 1-6   FIGURE 1-7

and DC, respectively. By Definition 1.4, $\vec{b} + \vec{a}$ may also be associated with the diagonal AC. Hence, one of the fundamental properties of the addition of real numbers is also valid for the addition of vectors.

THEOREM 1.1 / *Vector addition is commutative; that is,*

$$\vec{a} + \vec{b} = \vec{b} + \vec{a}. \tag{1-1}$$

The law of vector addition is sometimes called the **triangle law of addition** or the **parallelogram law of addition**. A consideration of Figures 1-6 and 1-7 justifies the use of these terms.

Associativity is another familiar property of the addition of real numbers that also holds for the addition of vectors.

THEOREM 1.2 / *Vector addition is associative; that is,*

$$\vec{a} + (\vec{b} + \vec{c}) = (\vec{a} + \vec{b}) + \vec{c}. \tag{1-2}$$

PROOF / Consider a parallelepiped $ABCDEFGH$. Associate vectors $\vec{a}$, $\vec{b}$, and $\vec{c}$ with sides AB, AD, and AE, respectively, as in Figure 1-8. Then

$$\overrightarrow{AG} = \overrightarrow{AH} + \overrightarrow{HG} \quad \text{and} \quad \overrightarrow{AG} = \overrightarrow{AC} + \overrightarrow{CG}. \quad \text{(Def. 1.4)}$$

Therefore,

$$\overrightarrow{HG} + \overrightarrow{AH} = \overrightarrow{AC} + \overrightarrow{CG}, \quad \text{(Th. 1.1)}$$
$$\overrightarrow{HG} + (\overrightarrow{AD} + \overrightarrow{DH}) = (\overrightarrow{AB} + \overrightarrow{BC}) + \overrightarrow{CG}, \quad \text{(Def. 1.4)}$$
$$\overrightarrow{AB} + (\overrightarrow{AD} + \overrightarrow{AE}) = (\overrightarrow{AB} + \overrightarrow{AD}) + \overrightarrow{AE}. \quad \text{(Def. 1.3)}$$

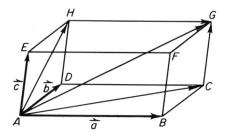

FIGURE 1-8

Hence,
$$\vec{a} + (\vec{b} + \vec{c}) = (\vec{a} + \vec{b}) + \vec{c}.$$

The associative property of vector addition may be extended to find the sum of more than three vectors. If the initial point of each succeeding vector is placed at the terminal point of the preceding one, then that vector with the same initial point as the first vector and the same terminal point as the last vector represents the sum of the vectors.

DEFINITION 1.5 / *If two vectors $\vec{a}$ and $\vec{b}$ have a common initial point, then their difference $\vec{a} - \vec{b}$ is the vector $\vec{c}$ extending from the terminal point of $\vec{b}$ to the terminal point of $\vec{a}$ (Figure 1-9).*

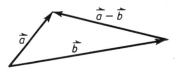

FIGURE 1-9

If $\vec{b} = \vec{a}$, then $\vec{a} - \vec{a} = \vec{0}$, the null vector. Furthermore, if $-\vec{b}$ is used to represent a vector equal in magnitude to $\vec{b}$ but having the opposite direction from $\vec{b}$, then $\vec{a} - \vec{b} = \vec{a} + (-\vec{b})$ as shown in Figure 1-10.

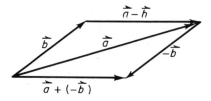

FIGURE 1-10

EXERCISES

1. Simplify (a) $\overrightarrow{AB} + \overrightarrow{BC} + \overrightarrow{CD}$; (b) $\overrightarrow{RS} + \overrightarrow{ST} + \overrightarrow{TU} + \overrightarrow{UR}$.

2. Simplify (a) $\overrightarrow{AB} - \overrightarrow{CB}$; (b) $\overrightarrow{MN} - \overrightarrow{RP} - \overrightarrow{PN}$.

3. Use geometric constructions to copy the coplanar vectors in the figure, and show that the sum of the vectors appears to be the null vector.

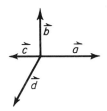

4. If $\vec{a}$ and $\vec{b}$ are vectors associated with adjacent sides AB and BC of a regular hexagon $ABCDEF$, determine vectors that may be associated with the other four sides CD, DE, EF, and FA.

5. Prove that $(\vec{a} + \vec{b}) + \vec{c} = (\vec{c} + \vec{a}) + \vec{b}$ by using Theorems 1.1 and 1.2.

6. Use the properties of a triangle to prove that (a) $|\vec{a} + \vec{b}| \leq |\vec{a}| + |\vec{b}|$; (b) $|\vec{a} - \vec{b}| \geq |\vec{a}| - |\vec{b}|$.

7. If $ABCD$ is a quadrilateral where $\overrightarrow{OB} - \overrightarrow{OA} = \overrightarrow{OC} - \overrightarrow{OD}$, prove that $ABCD$ is a parallelogram.

8. If $\vec{a}$, $\vec{b}$, and $\vec{c}$ have a common initial point and are associated with the edges of a parallelepiped, determine vectors representing the diagonals of the parallelepiped.

9. Let $\vec{a}$, $\vec{b}$, and $\vec{c}$ have a common initial point and be associated with the edges of a cube. (a) Determine the vectors drawn from the common initial point to each of the other seven vertices. (b) Show that the sum of three of these vectors is equal to the sum of the other four.

10. Prove that the sum of the vectors from the center to the vertices of a regular hexagon is the null vector.

1-4 / MULTIPLICATION OF A VECTOR BY A SCALAR

DEFINITION 1.6 / *Given any real number (scalar) k and any vector $\vec{a}$, the product $k\vec{a}$ is a vector such that $|k\vec{a}| = |k||\vec{a}|$. If $\vec{a}$ is a nonzero vector and k is* positive, then $k\vec{a}$ has the same direction as $\vec{a}$; if $\vec{a}$ is a nonzero vector and k is negative, then $k\vec{a}$ has an opposite direction to $\vec{a}$.

In Figure 1-11 several geometric examples of the scalar multiplication of $\vec{a}$ are given. Note that $-\vec{a}$ need not be considered as a special vector, but rather as the product of the scalar -1 and $\vec{a}$. Also note that $\vec{a}$ may be considered as being transformed when multiplied by a scalar k. If $k > 1$, the transformation is a *stretch* or *dilation*; if $0 < k < 1$, the transformation is a *contraction*; if $k < 0$, the transformation is a stretch or a contraction followed by a reversal of direction.

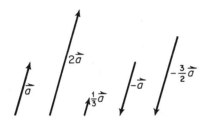

FIGURE 1-11

Definition 1.6 for any positive integer k is a logical consequence of a consideration of the way in which the sum of k addends $\vec{a} + \vec{a} + \cdots + \vec{a}$ may be represented.

The multiplication of a vector by a scalar satisfies the following "distributive" and "associative" laws:

$$(m + n)\vec{a} = m\vec{a} + n\vec{a}; \tag{1-3}$$
$$m(\vec{a} + \vec{b}) = m\vec{a} + m\vec{b}; \tag{1-4}$$
$$m(n\vec{a}) = (mn)\vec{a}. \tag{1-5}$$

The distributive property (1-4) may be illustrated geometrically, as in Figure 1-12, by using the properties of similar triangles.

If two nonzero vectors $\vec{a}$ and $\vec{b}$ are parallel, then it is possible to find a nonzero scalar m that will transform $\vec{b}$ into $\vec{a}$; that is, $\vec{a} = m\vec{b}$. Similarly, a nonzero scalar n exists that will transform $\vec{a}$ into $\vec{b}$; that is, $\vec{b} = n\vec{a}$. For example, if $\vec{a}$ is a vector parallel to $\vec{b}$ and in the same sense, such that $|\vec{a}| = 2|\vec{b}|$, then $\vec{a} = 2\vec{b}$ and $\vec{b} = \frac{1}{2}\vec{a}$. If $\vec{a}$ is a vector parallel to $\vec{b}$ and in the opposite sense, such that $|\vec{a}| = 2|\vec{b}|$, then $\vec{a} = -2\vec{b}$ and $\vec{b} = -\frac{1}{2}\vec{a}$. Now, with regard to Definition 1.6, it should be evident that two nonzero vectors

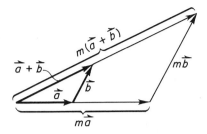

FIGURE 1-12

are parallel if, and only if, either one may be expressed as a nonzero scalar multiple of the other. We summarize this as a theorem.

THEOREM 1.3 / *If $\vec{a}$ and $\vec{b}$ are nonzero vectors, then $\vec{a}$ is parallel to $\vec{b}$ if, and only if, there exists a nonzero scalar m or n such that $\vec{a} = m\vec{b}$ or $\vec{b} = n\vec{a}$.*

Since the direction of the null vector is indeterminate, the null vector cannot be parallel to any vector. However, note that the null vector is a zero multiple of every other vector.

Many theorems of plane geometry which involve parallel line segments with related magnitudes may be proved by using Theorem 1.3.

EXAMPLE 1 / The line segment joining the midpoints of any two sides of a triangle is parallel to the third side and equal to one-half the length of the third side (Figure 1-13).

Let N and M be the midpoints respectively of sides OB and AB in any triangle OAB. Let $\vec{a}$, $\vec{b}$, and $\vec{c}$ be associated with line segments OA, OB, and NM, respectively. Then

$$\overrightarrow{OM} = \overrightarrow{ON} + \overrightarrow{NM} = \overrightarrow{OA} + \overrightarrow{AM}$$

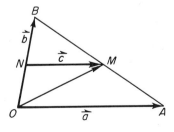

FIGURE 1-13

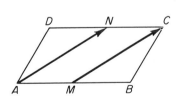

FIGURE 1-14

$$= \tfrac{1}{2}\vec{b} + \vec{c} = \vec{a} + \tfrac{1}{2}(\vec{b} - \vec{a})$$
$$= \tfrac{1}{2}\vec{b} + \vec{c} = \tfrac{1}{2}\vec{b} + \tfrac{1}{2}\vec{a}.$$

Hence $\vec{c} = \tfrac{1}{2}\vec{a}$; that is, line segment NM is parallel to side OA, and the length of NM is equal to one-half the length of OA.

EXAMPLE 2 / In parallelogram $ABCD$, if M and N are the midpoints of AB and CD, respectively, then $AMCN$ is a parallelogram (Figure 1-14).

Now, $\overrightarrow{AN} = \overrightarrow{AD} + \overrightarrow{DN} = \overrightarrow{AD} + \tfrac{1}{2}\overrightarrow{DC}$, and $\overrightarrow{MC} = \overrightarrow{MB} + \overrightarrow{BC} = \tfrac{1}{2}\overrightarrow{AB} + \overrightarrow{BC}$. Since $\overrightarrow{AB} = \overrightarrow{DC}$ and $\overrightarrow{BC} = \overrightarrow{AD}$, then $\overrightarrow{MC} = \overrightarrow{AN}$. Hence, $AMCN$ is a parallelogram since two opposite sides are equal in length and parallel.

EXAMPLE 3 / The diagonals of a parallelogram bisect each other (Figure 1-15).

Let $OABC$ be any parallelogram. Let $\vec{a}$ and $\vec{c}$ be associated with sides OA and OC, respectively. Then $\vec{c} - \vec{a}$ and $\vec{c} + \vec{a}$ are the vectors associated with diagonals AC and OB, respectively. If D and E are the midpoints of diagonals OB and AC, respectively, then $\overrightarrow{OD} = \tfrac{1}{2}(\vec{c} + \vec{a})$, and $\overrightarrow{OE} = \vec{a} + \tfrac{1}{2}(\vec{c} - \vec{a}) = \tfrac{1}{2}(\vec{c} + \vec{a})$. Since $\overrightarrow{OD}$ and $\overrightarrow{OE}$ are equal to the same vector, $\overrightarrow{OD} = \overrightarrow{OE}$ and D and E coincide; that is, the diagonals of a parallelogram bisect each other.

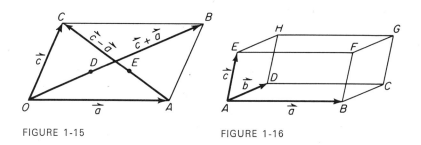

FIGURE 1-15   FIGURE 1-16

EXAMPLE 4 / The diagonals of a parallelepiped bisect each other (Figure 1-16).

Let $\vec{a}$, $\vec{b}$, and $\vec{c}$ be three vectors associated with the edges AB, AD, and AE of any parallelepiped $ABCDEFGH$. Let M, N, R, and S be the midpoints of diagonals *AG*, *EC*, *BH*, and *DF*, respectively. Then

$$\overrightarrow{AM} = \tfrac{1}{2}\overrightarrow{AG} = \tfrac{1}{2}(\overrightarrow{AB} + \overrightarrow{BC} + \overrightarrow{CG}) = \tfrac{1}{2}(\vec{a} + \vec{b} + \vec{c});$$
$$\overrightarrow{AN} = \tfrac{1}{2}\overrightarrow{AC} + \tfrac{1}{2}\overrightarrow{AE} = \tfrac{1}{2}(\overrightarrow{AB} + \overrightarrow{BC} + \overrightarrow{AE}) = \tfrac{1}{2}(\vec{a} + \vec{b} + \vec{c});$$
$$\overrightarrow{AR} = \tfrac{1}{2}\overrightarrow{AB} + \tfrac{1}{2}\overrightarrow{AH} = \tfrac{1}{2}(\overrightarrow{AB} + \overrightarrow{AD} + \overrightarrow{DH}) = \tfrac{1}{2}(\vec{a} + \vec{b} + \vec{c});$$
$$\overrightarrow{AS} = \tfrac{1}{2}\overrightarrow{AF} + \tfrac{1}{2}\overrightarrow{AD} = \tfrac{1}{2}(\overrightarrow{AB} + \overrightarrow{BF} + \overrightarrow{AD}) = \tfrac{1}{2}(\vec{a} + \vec{b} + \vec{c}).$$

Hence, points *M*, *N*, *R*, and *S* coincide.

Consider the multiplication of any nonzero vector $\vec{a}$ by a scalar equal to the reciprocal of the magnitude of $\vec{a}$; that is, $(1/|\vec{a}|)\vec{a}$. Let $(1/|\vec{a}|)\vec{a} = \vec{b}$. Then $\vec{a} = |\vec{a}|\vec{b}$. By Definition 1.6, $|\vec{a}| = |\vec{a}||\vec{b}|$. Therefore $|\vec{b}| = 1$. Each vector whose magnitude is one is called a **unit vector**. By Theorem 1.3 and Definition 1.6, $\vec{b}$ is parallel to $\vec{a}$ in the same sense. Hence $\vec{b}$, or $(1/|\vec{a}|)\vec{a}$, is a unit vector in the direction $\vec{a}$. We shall make frequent reference to unit vectors.

EXERCISES

1. Simplify (a) $\overrightarrow{AB} + \overrightarrow{AB} + \overrightarrow{AB}$; (b) $(\overrightarrow{OA} - \overrightarrow{OB}) - (\overrightarrow{OB} - \overrightarrow{OA})$.

2. Select two vectors $\vec{r}$ and $\vec{s}$ of equal magnitude with a common initial point and making an angle whose measure is 60°. Then construct the vectors (a) $2\vec{r} + \vec{s}$; (b) $\vec{r} - 2\vec{s}$; (c) $\tfrac{3}{2}\vec{s} - \vec{r}$; (d) $2(\vec{r} + \vec{s})$.

3. Illustrate geometrically that $-(\vec{a} - \vec{b}) = -\vec{a} + \vec{b}$.

4. Interpret $\overrightarrow{AB}/|\overrightarrow{AB}|$ for any nonzero vector $\overrightarrow{AB}$.

5. Describe how to construct a parallelogram if two vectors $\vec{a}$ and $\vec{b}$ are given as the diagonals of the parallelogram.

In Exercises 6 through 10 prove the stated theorem.

6. If a line divides two sides of a triangle proportionally, then it is parallel to the third side.

7. The median of a trapezoid is parallel to the bases and equal to one-half their sum.

8. The line segment joining the midpoints of the diagonals of a trapezoid is parallel to the bases and equal to one-half their difference.

9. If *M*, *N*, *R*, and *S* divide sides *AB*, *BC*, *CD*, and *DA* of a parallelogram in the same ratio, then *MNRS* is a parallelogram.

10. If the diagonals of a quadrilateral bisect each other, then the quadrilateral is a parallelogram.

1-5 / LINEAR DEPENDENCE OF VECTORS

In Examples 3 and 4 of §1-4 we considered two geometric problems: one on a plane, and a second in space. In Example 3 we associated two nonzero, nonparallel vectors $\vec{a}$ and $\vec{c}$ with line segments. Then every vector that we considered on the plane was expressed in the form $m\vec{a} + n\vec{c}$; that is, as a function of $\vec{a}$ and $\vec{c}$. Similarly, in Example 4 we associated three nonzero, noncoplanar vectors $\vec{a}$, $\vec{b}$, and $\vec{c}$ with line segments. Then every vector that we considered in space was expressed in the form $m\vec{a} + n\vec{b} + p\vec{c}$; that is, as a function of $\vec{a}$, $\vec{b}$, and $\vec{c}$.

DEFINITION 1.7 / *An expression such as* $m_1\vec{a}_1 + m_2\vec{a}_2 + \cdots + m_n\vec{a}_n$ *which represents the sum of a set of n scalar multiples of n vectors is called a **linear function** of the n vectors.*

Consider a nonzero vector $\vec{a}$. Then every vector on a line parallel to $\vec{a}$ is a linear function of $\vec{a}$; that is, of the form $m\vec{a}$ where m is a real number. The set of vectors $m\vec{a}$ is an example of a **one-dimensional vector space,** and the nonzero vector $\vec{a}$ is called a **basis** for that vector space. Essentially, $\vec{a}$ serves as a reference vector in that every other vector in the one-dimensional vector space may be expressed in terms of $\vec{a}$ or is *dependent* upon $\vec{a}$. Furthermore, any vector of the form $m\vec{a}$ where $m \neq 0$ may be considered a basis of the same one-dimensional vector space for which $\vec{a}$ is a basis.

DEFINITION 1.8 / *A set of n vectors* $\vec{a}_1, \vec{a}_2, \ldots, \vec{a}_n$ *is called a set of **linearly independent** vectors if* $m_1\vec{a}_1 + m_2\vec{a}_2 + \cdots + m_n\vec{a}_n = \vec{0}$ *implies* $m_1 = m_2 = \cdots = m_n = 0$. *A set of n vectors is called a set of **linearly dependent** vectors if the vectors are not linearly independent.*

The set of vectors $\vec{a}$ and $\vec{c}$ in Example 3 of §1-4 is a set of linearly independent vectors since $m_1\vec{a} + m_2\vec{c} = \vec{0}$ if, and only if, $m_1 = m_2 = 0$. The set of vectors $\vec{a}, \vec{c}, \vec{c} + \vec{a}$, and $\vec{c} - \vec{a}$ in that same example is a set of linearly dependent vectors since $m_1\vec{a} + m_2\vec{c} + m_3(\vec{c} + \vec{a}) + m_4(\vec{c} - \vec{a}) = \vec{0}$ is true for $m_1 = m_2 = -1, m_3 = 1$, and $m_4 = 0$; that is, for at least one $m_i \neq 0$.

In general, note that by Definition 1.8 n vectors are linearly dependent if at least one of the m_i's in a vector equation of the form $m_1\vec{a_1} + m_2\vec{a_2} + \cdots + m_n\vec{a_n} = \vec{0}$ is not zero. If n vectors are linearly dependent, then at least one of the vectors may be written as a linear function of the other $n - 1$ vectors. For example, if $m_3 \neq 0$, then

$$\vec{a_3} = -\frac{m_1}{m_3}\vec{a_1} - \frac{m_2}{m_3}\vec{a_2} - \frac{m_4}{m_3}\vec{a_4} - \cdots - \frac{m_n}{m_3}\vec{a_n}.$$

Conversely, if one vector may be expressed as a linear function of $n - 1$ other vectors, then the n vectors are linearly dependent. It follows from our discussion of a one-dimensional vector space that any two vectors in a one-dimensional space are linearly dependent.

THEOREM 1.4 / *If $\vec{a}$ and $\vec{b}$ are nonzero, nonparallel vectors, then $x\vec{a} + y\vec{b} = \vec{0}$ implies $x = y = 0$.*

PROOF / Suppose that $x \neq 0$. Then $x\vec{a} = -y\vec{b}$. Therefore $\vec{a} = -(y/x)\vec{b}$; that is, $\vec{a}$ is parallel to $\vec{b}$, contradicting the hypothesis. Hence, $x = 0$ and $y\vec{b} = \vec{0}$ implies that $y = 0$ from Definition 1.6.

Any two nonzero, nonparallel vectors $\vec{a}$ and $\vec{b}$ must by Theorem 1.4 be linearly independent. The set of vectors of the form $m\vec{a} + n\vec{b}$, where $\vec{a}$ and $\vec{b}$ are linearly independent vectors and m and n are real numbers, is an example of a **two-dimensional vector space.** The two vectors $\vec{a}$ and $\vec{b}$ form a **basis** for that vector space.

THEOREM 1.5 / *If $\vec{a}$ and $\vec{b}$ are nonzero, nonparallel vectors with the same initial point, and if $\vec{c}$ is any vector on the plane determined by $\vec{a}$ and $\vec{b}$, then $\vec{c}$ can be expressed as a linear function of $\vec{a}$ and $\vec{b}$.*

PROOF / If $\vec{c}$ is parallel to $\vec{a}$, then $\vec{c} = m\vec{a}$; $n = 0$. If $\vec{c}$ is parallel to $\vec{b}$, then $\vec{c} = n\vec{b}$; $m = 0$. If $\vec{c}$ is the null vector, then $m = n = 0$. If $\vec{c}$ is not the null vector and is not parallel to $\vec{a}$ or $\vec{b}$, then there exists a parallelogram $ABCD$ with edges parallel to $\vec{a}$ and $\vec{b}$ and with a diagonal $\vec{c}$. Any vectors parallel to $\vec{a}$ and $\vec{b}$ can be expressed as linear functions of $\vec{a}$ and $\vec{b}$, respectively. Hence, $\vec{c} = m\vec{a} + n\vec{b}$ (Figure 1-17).

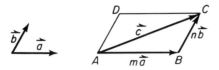

FIGURE 1-17

The converse of Theorem 1.5 is also true; that is, if $\vec{c} = m\vec{a} + n\vec{b}$, where $\vec{a}$ and $\vec{b}$ are nonzero, nonparallel vectors with the same initial point, then $\vec{c}$ may be considered to be on the plane determined by $\vec{a}$ and $\vec{b}$.

Theorem 1.5 implies that any three vectors in a two-dimensional space are linearly dependent.

THEOREM 1.6 / *If $\vec{a}$, $\vec{b}$, and $\vec{c}$ are nonzero, noncoplanar vectors with the same initial point, then $x\vec{a} + y\vec{b} + z\vec{c} = \vec{0}$ implies $x = y = z = 0$.*

PROOF / Suppose that $x \neq 0$. Then $x\vec{a} = -y\vec{b} - z\vec{c}$ and $\vec{a} = -(y/x)\vec{b} - (z/x)\vec{c}$; that is, $\vec{a}$ lies on the plane of $\vec{b}$ and $\vec{c}$, which contradicts the hypothesis that $\vec{a}, \vec{b}$, and $\vec{c}$ are noncoplanar. Hence $x = 0$, $y\vec{b} + z\vec{c} = \vec{0}$, and, by Theorem 1.4, $x = y = z = 0$.

Any three nonzero, noncoplanar vectors $\vec{a}$, $\vec{b}$, and $\vec{c}$ with the same initial point must by Theorem 1.6 be linearly independent. The set of vectors of the form $m\vec{a} + n\vec{b} + p\vec{c}$, where $\vec{a}, \vec{b}$, and $\vec{c}$ are linearly independent vectors and m, n, and p are real numbers, is an example of a **three-dimensional vector space**. The vectors $\vec{a}, \vec{b}$, and $\vec{c}$ form a **basis** for that vector space.

THEOREM 1.7 / *If $\vec{a}$, $\vec{b}$, and $\vec{c}$ are nonzero, noncoplanar vectors with the same initial point, then any vector in space can be expressed as a linear function of $\vec{a}, \vec{b}$, and $\vec{c}$.*

The proof of Theorem 1.7 is analogous to the proof of Theorem 1.5 and is left as an exercise (Exercise 14). Theorem 1.7 implies that any four vectors in space are linearly dependent.

The concept of a set of linearly independent or dependent vectors plays a key role in proving numerous theorems of geometry by vector methods. Two useful theorems concerning the linear dependence of vectors will now be considered. Careful attention should be paid to these theorems.

THEOREM 1.8 / *If $\vec{a}$, $\vec{b}$, and $\vec{c}$ are linearly independent vectors, then any vector in space can be expressed as a linear function of $\vec{a}$, $\vec{b}$, and $\vec{c}$ in only one way; that is, every vector in space is a unique linear function of $\vec{a}$, $\vec{b}$, and $\vec{c}$.*

PROOF / By Theorem 1.7 every vector $\vec{d}$ in space is a linear function of $\vec{a}$, $\vec{b}$, and $\vec{c}$. Suppose that some vector $\vec{d}$ can be represented in terms of $\vec{a}$, $\vec{b}$, and $\vec{c}$ in two ways; that is,

$$\vec{d} = x\vec{a} + y\vec{b} + z\vec{c} = m\vec{a} + n\vec{b} + p\vec{c}.$$

Then

$$(x - m)\vec{a} + (y - n)\vec{b} + (z - p)\vec{c} = \vec{0}.$$

Since $\vec{a}$, $\vec{b}$, and $\vec{c}$ are linearly independent vectors, then

$$x - m = 0; \quad y - n = 0; \quad z - p = 0.$$

Hence, $x = m$; $y = n$; $z = p$; and the two representations of $\vec{d}$ are identical.

In any geometry of n dimensions, Theorem 1.8 may be generalized for a set of n linearly independent vectors $\vec{a_1}, \vec{a_2}, \ldots, \vec{a_n}$. Any vector that is a linear function of $\vec{a_1}, \vec{a_2}, \ldots, \vec{a_n}$ can be expressed as a linear function of the n vectors in only one way. The proof of the generalized theorem is left as an exercise (Exercise 15).

THEOREM 1.9 / *The points A, B, and C are collinear if, and only if, for any point O in space, $\overrightarrow{OC} = (1 - n)\overrightarrow{OA} + n\overrightarrow{OB}$.*

PROOF / Let O be any reference point in space and C be any point on the line AB. Then $\overrightarrow{OC}$ is a vector whose terminal point C lies on the line through the terminal points A and B of $\overrightarrow{OA}$ and $\overrightarrow{OB}$, respectively. Furthermore, consider C dividing line segment BA in the ratio m to n where $m + n = 1$ (Figure 1-18). Now,

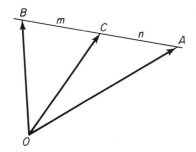

FIGURE 1-18

$$\overrightarrow{OC} = \overrightarrow{OA} + \overrightarrow{AC} = \overrightarrow{OA} + n\overrightarrow{AB}$$
$$= \overrightarrow{OA} + n(\overrightarrow{OB} - \overrightarrow{OA})$$
$$= (1-n)\overrightarrow{OA} + n\overrightarrow{OB}.$$

Conversely, if $\overrightarrow{OC} = (1-n)\overrightarrow{OA} + n\overrightarrow{OB}$, then

$$\overrightarrow{OC} = \overrightarrow{OA} - n\overrightarrow{OA} + n\overrightarrow{OB},$$
$$\overrightarrow{OC} - \overrightarrow{OA} = n(\overrightarrow{OB} - \overrightarrow{OA}),$$
$$\overrightarrow{AC} = n\overrightarrow{AB}.$$

Therefore, $\overrightarrow{AC}$ is parallel to $\overrightarrow{AB}$; line segment AC lies along line segment AB; and A, B, and C are collinear.

EXERCISES

In Exercises 1 through 8 consider the vectors given in the figure and express each vector as a linear function of (a) $\overrightarrow{OA}$ and $\overrightarrow{OB}$; (b) $\overrightarrow{OB}$ and $\overrightarrow{OC}$; (c) $\overrightarrow{OA}$ and $\overrightarrow{OD}$.

1. $\overrightarrow{OA}$.
2. $\overrightarrow{OB}$.
3. $\overrightarrow{OC}$.
4. $\overrightarrow{OD}$.
5. $\overrightarrow{OE}$.
6. $\overrightarrow{CA}$.
7. $\overrightarrow{AB}$.
8. $\overrightarrow{ED}$.

B: (0, 2)
D: (3, 1)
C: (−2, 0)
A: (1, 0)
E: (−1, −2)

9. Prove that M is the midpoint of line segment AB if $\overrightarrow{OM} + \overrightarrow{OM} = \overrightarrow{OA} + \overrightarrow{OB}$.

10. Prove that $4\overrightarrow{MN} = \overrightarrow{AB} + \overrightarrow{AD} + \overrightarrow{CB} + \overrightarrow{CD}$ if M and N are the midpoints of line segments AC and BD, respectively.

11. Prove that the terminal points of vectors $\overrightarrow{AB}, \overrightarrow{AC}$, and $\overrightarrow{AD}$ are collinear if $\overrightarrow{AD} = \overrightarrow{AB} + \frac{2}{5}(\overrightarrow{OC} - \overrightarrow{OB})$.

12. Prove that the vectors $\vec{a} + 2\vec{b} + \vec{c}, \vec{a} + 3\vec{b} - 2\vec{c}, \vec{a} + \vec{b} + 4\vec{c}$ are linearly dependent.

13. Find another set of vectors which may be a basis for the two-dimensional vector space for which the set $\vec{a}$ and $\vec{b}$ is a basis.

14. Prove Theorem 1.7.

15. Prove that any vector which is a linear function of n linearly independent vectors $\vec{a}_1, \vec{a}_2, \ldots, \vec{a}_n$ is a unique linear function of the n vectors.

16. Prove that a necessary and sufficient condition for the terminal points of any three nonzero vectors $\vec{a}, \vec{b}$, and $\vec{c}$ with common initial point to be collinear is that equation $x\vec{a} + y\vec{b} + z\vec{c} = \vec{0}$ imply $x + y + z = 0$, where $x^2 + y^2 + z^2 \neq 0$.

1-6 / APPLICATIONS OF LINEAR DEPENDENCE

In this section several theorems of plane synthetic geometry will be proved by using the concept of linear dependence of a set of vectors. Most vector proofs involving the linear dependence concept depend upon Theorems 1.8 and 1.9.

EXAMPLE 1 / The line segment joining a vertex of a parallelogram to the midpoint of a nonadjacent side intersects the diagonal at a trisection point (Figure 1-19).

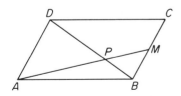

FIGURE 1-19

Let $ABCD$ be any parallelogram. Let M be the midpoint of side BC

and P be the point of intersection of line segment AM and diagonal BD. Now,

$$\overrightarrow{AM} = \tfrac{1}{2}\overrightarrow{AB} + \tfrac{1}{2}\overrightarrow{AC} = \tfrac{1}{2}\overrightarrow{AB} + \tfrac{1}{2}(\overrightarrow{AD} + \overrightarrow{DC})$$
$$= \tfrac{1}{2}\overrightarrow{AB} + \tfrac{1}{2}(\overrightarrow{AD} + \overrightarrow{AB}) = \overrightarrow{AB} + \tfrac{1}{2}\overrightarrow{AD},$$

and

$$\overrightarrow{AP} = k\overrightarrow{AM} = k\overrightarrow{AB} + \frac{k}{2}\overrightarrow{AD}.$$

Since B, P, and D are collinear, $k + k/2 = 1$ or $k = \tfrac{2}{3}$. Hence, $\overrightarrow{AP} = \tfrac{2}{3}\overrightarrow{AB} + \tfrac{1}{3}\overrightarrow{AD}$; that is, P divides diagonal BD in the ratio 1 to 2.

EXAMPLE 2 / The medians of a triangle are concurrent at a point called the *centroid* of the triangle. This point is two-thirds of the distance from each vertex to the midpoint of the opposite side (Figure 1-20).

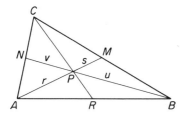

FIGURE 1-20

Let M, N, and R be the midpoints of sides BC, CA, and AB, respectively, of any triangle ABC. Let P be the point of intersection of the medians AM and BN. Consider P dividing median AM in the ratio r to s, where $r + s = 1$, and dividing median BN in the ratio u to v, where $u + v = 1$. Let O be any reference point in space not in the plane of triangle ABC. Now,

$$\overrightarrow{OP} = r\overrightarrow{OM} + s\overrightarrow{OA} = u\overrightarrow{ON} + v\overrightarrow{OB}.$$

Since M and N are midpoints of sides BC and CA, respectively,

$$\overrightarrow{OM} = \tfrac{1}{2}\overrightarrow{OB} + \tfrac{1}{2}\overrightarrow{OC} \quad \text{and} \quad \overrightarrow{ON} = \tfrac{1}{2}\overrightarrow{OA} + \tfrac{1}{2}\overrightarrow{OC}.$$

By substitution,

$$\overrightarrow{OP} = \frac{r}{2}(\overrightarrow{OB} + \overrightarrow{OC}) + s\overrightarrow{OA} = \frac{u}{2}(\overrightarrow{OA} + \overrightarrow{OC}) + v\overrightarrow{OB}.$$

Since $\overrightarrow{OA}$, $\overrightarrow{OB}$, and $\overrightarrow{OC}$ are linearly independent vectors,

$$s = \frac{u}{2}; \qquad \frac{r}{2} = v; \qquad \frac{r}{2} = \frac{u}{2}.$$

Therefore, $r = 2s$ and $u = 2v$. Hence, P divides both medians AM and BN in the ratio 2 to 1. Consider the medians CR and AM. Since their point of intersection must divide medians AM and CR in the ratio 2 to 1 by the same argument, median CR passes through point P.

EXAMPLE 3 / In any quadrilateral the line segments joining the midpoints of opposite sides bisect each other (Figure 1-21).

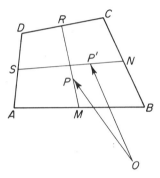

FIGURE 1-21

Let M, N, R, and S be the midpoints of sides AB, BC, CD, and DA, respectively, of any quadrilateral. Let P and P' be the midpoints of line segments MR and NS, respectively. If O is any reference point, then

$$\overrightarrow{OP} = \tfrac{1}{2}\overrightarrow{OM} + \tfrac{1}{2}\overrightarrow{OR} = \tfrac{1}{4}(\overrightarrow{OA} + \overrightarrow{OB} + \overrightarrow{OC} + \overrightarrow{OD});$$
$$\overrightarrow{OP'} = \tfrac{1}{2}\overrightarrow{ON} + \tfrac{1}{2}\overrightarrow{OS} = \tfrac{1}{4}(\overrightarrow{OB} + \overrightarrow{OC} + \overrightarrow{OA} + \overrightarrow{OD}).$$

Since $\overrightarrow{OP} = \overrightarrow{OP'}$, the points P and P' coincide.

The remaining examples of this section illustrate the use of the concept of linear dependence of vectors in proving two important theorems of higher geometry.

EXAMPLE 4 / THEOREM OF MENELAUS Given triangle ABC, three distinct points P_1, P_2, and P_3 on lines along sides AB, BC, and CA, respectively, are collinear if, and only if,

$$\left(\frac{\overrightarrow{AP_1}}{\overrightarrow{P_1B}}\right)\left(\frac{\overrightarrow{BP_2}}{\overrightarrow{P_2C}}\right)\left(\frac{\overrightarrow{CP_3}}{\overrightarrow{P_3A}}\right) = -1$$

where each quotient represents the real number that may be obtained when the two collinear vectors are expressed as multiples of the same vector (Figure 1-22).

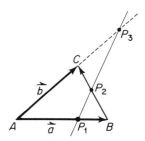

FIGURE 1-22

Let $\overrightarrow{AB} = \vec{a}$ and $\overrightarrow{AC} = \vec{b}$. Then $\overrightarrow{BC} = \vec{b} - \vec{a}$. Let $\overrightarrow{AP_1} = n\vec{a}$, $\overrightarrow{P_1B} = (1-n)\vec{a}$, $\overrightarrow{BP_2} = k(\vec{b} - \vec{a})$, $\overrightarrow{P_2C} = (1-k)(\vec{b} - \vec{a})$, $\overrightarrow{CP_3} = (m-1)\vec{b}$, and $\overrightarrow{P_3A} = -m\vec{b}$.

We first prove the necessary condition for P_1, P_2, and P_3 to be collinear. If P_1, P_2, and P_3 are collinear, then

$$\overrightarrow{AP_2} = t\overrightarrow{AP_1} + (1-t)\overrightarrow{AP_3}$$
$$= tn\vec{a} + (1-t)m\vec{b}.$$

Since $\overrightarrow{AP_2} = (1-k)\vec{a} + k\vec{b}$, and $\vec{a}$ and $\vec{b}$ are linearly independent vectors, then $tn = 1-k$ and $(1-t)m = k$; that is,

$$n = \frac{1-k}{t} \quad \text{and} \quad m = \frac{k}{1-t}.$$

Therefore,

$$\left(\frac{\overrightarrow{AP_1}}{\overrightarrow{P_1B}}\right)\left(\frac{\overrightarrow{BP_2}}{\overrightarrow{P_2C}}\right)\left(\frac{\overrightarrow{CP_3}}{\overrightarrow{P_3A}}\right) = \frac{n\vec{a}}{(1-n)\vec{a}} \cdot \frac{k(\vec{b}-\vec{a})}{(1-k)(\vec{b}-\vec{a})} \cdot \frac{(m-1)\vec{b}}{-m\vec{b}}$$

$$= \frac{n}{1-n} \cdot \frac{k}{1-k} \cdot \frac{m-1}{-m}$$

$$= \frac{1-k}{t-1+k} \cdot \frac{k}{1-k} \cdot \frac{k-1+t}{-k} = -1.$$

Next we assume that P_1, P_2, and P_3 divide sides AB, BC, and CA in the ratios $n/(1-n)$, $k/(1-k)$, and $(1-n)(k-1)/nk$, respectively; that is, the product of the ratios is -1. Let O be any point in the plane of triangle ABC. Then

$$\overrightarrow{OP_1} = n\overrightarrow{OB} + (1-n)\overrightarrow{OA}; \qquad \overrightarrow{OP_2} = k\overrightarrow{OC} + (1-k)\overrightarrow{OB};$$

and since $(1-n)(k-1) + nk = n + k - 1$,

$$\overrightarrow{OP_3} = \frac{(1-n)(k-1)}{n+k-1}\overrightarrow{OA} + \frac{nk}{n+k-1}\overrightarrow{OC}.$$

Substituting for $(1-n)\overrightarrow{OA}$ and $k\overrightarrow{OC}$, we obtain

$$\overrightarrow{OP_3} = \frac{k-1}{n+k-1}(\overrightarrow{OP_1} - n\overrightarrow{OB}) + \frac{n}{n+k-1}[\overrightarrow{OP_2} - (1-k)\overrightarrow{OB}]$$

$$= \frac{k-1}{n+k-1}\overrightarrow{OP_1} + \frac{n}{n+k-1}\overrightarrow{OP_2}.$$

Since

$$\frac{k-1}{n+k-1} + \frac{n}{n+k-1} = 1,$$

P_1, P_2, and P_3 are collinear (Theorem 1.9).

EXAMPLE 5 / DESARGUES' THEOREM If two triangles ABC and $A'B'C'$ are such that lines joining corresponding vertices are concurrent and no two corresponding sides are parallel, then the lines determined by the corresponding sides intersect in three collinear points.

Figure 1-23 represents Desargues' configuration for coplanar triangles, and Figure 1-24 represents Desargues' configuration for triangles that are not coplanar.

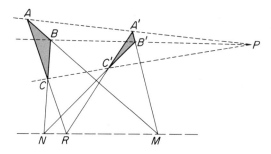

FIGURE 1-23

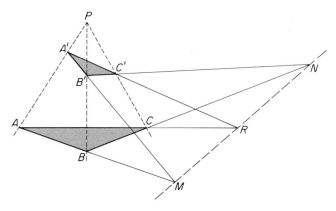

FIGURE 1-24

Let O be any reference point in space. Now,

$$\overrightarrow{OP} = x\overrightarrow{OA} + (1-x)\overrightarrow{OA'}$$
$$= y\overrightarrow{OB} + (1-y)\overrightarrow{OB'}$$
$$= z\overrightarrow{OC} + (1-z)\overrightarrow{OC'}.$$

Then $x \neq y$ since line AB is not parallel to line $A'B'$. Therefore,

$$\frac{x\overrightarrow{OA} - y\overrightarrow{OB}}{x - y} = \frac{(1-y)\overrightarrow{OB'} - (1-x)\overrightarrow{OA'}}{x - y};$$

that is, there exists a point common to lines AB and $A'B'$. Call the point M. Similarly, $y \neq z$, $x \neq z$, and points N and R exist common to the lines BC and $B'C'$ and the lines AC and $A'C'$, respectively, since

$$\frac{y\overrightarrow{OB} - z\overrightarrow{OC}}{y - z} = \frac{(1-z)\overrightarrow{OC'} - (1-y)\overrightarrow{OB'}}{y - z}$$

and

$$\frac{z\overrightarrow{OC} - x\overrightarrow{OA}}{z - x} = \frac{(1-x)\overrightarrow{OA'} - (1-z)\overrightarrow{OC'}}{z - x}.$$

Now,

$$(x - y)\overrightarrow{OM} + (y - z)\overrightarrow{ON} + (z - x)\overrightarrow{OR}$$
$$= x\overrightarrow{OA} - y\overrightarrow{OB} + y\overrightarrow{OB} - z\overrightarrow{OC} + z\overrightarrow{OC} - x\overrightarrow{OA} = \vec{0},$$

and the sum of the multiples of $\overrightarrow{OM}$, $\overrightarrow{ON}$, and $\overrightarrow{OR}$ is zero. Hence M, N, and R are collinear.

EXERCISES

Use the concept of linear dependence of vectors to prove the theorems stated in Exercises 1 through 10.

1. A line segment joining a vertex of a parallelogram to a point that divides an opposite side in the ratio 1 to n divides the diagonal in the ratio 1 to $n + 1$ or n to $n + 1$.

2. The sum of the vectors associated with the medians of a triangle is equal to the null vector.

3. A vector $\overrightarrow{OP}$ from any reference point O to the centroid P of any triangle ABC may be expressed by

$$\overrightarrow{OP} = \tfrac{1}{3}(\overrightarrow{OA} + \overrightarrow{OB} + \overrightarrow{OC}).$$

4. If line segments from two vertices of a triangle trisect the opposite sides as shown in the figure, then the point P divides both line segments AM and BN in the ratio 3 to 2.

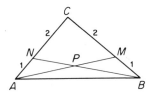

5. The line segments joining the midpoints of the adjacent sides of any quadrilateral form a parallelogram.

6. Let $ABCD$ be any quadrilateral. If O is any reference point, then $\overrightarrow{OP} = \frac{1}{4}(\overrightarrow{OA} + \overrightarrow{OB} + \overrightarrow{OC} + \overrightarrow{OD})$ where P is the midpoint of the line segment joining the midpoints of diagonals AC and BD.

7. The line segments joining the midpoints of the opposite edges of a tetrahedron bisect each other.

8. If a parallelogram is defined as a quadrilateral whose opposite sides are parallel, then the opposite sides of a parallelogram are equal in length.

9. The sum of the vectors from the center to the vertices of a regular pentagon is the null vector.

10. THEOREM OF CEVA In any triangle ABC, lines joining three points P_1, P_2, and P_3 on sides AB, BC, and CA, respectively, to the opposite vertices are concurrent if

$$\left(\frac{\overrightarrow{AP_1}}{\overrightarrow{P_1B}}\right)\left(\frac{\overrightarrow{BP_2}}{\overrightarrow{P_2C}}\right)\left(\frac{\overrightarrow{CP_3}}{\overrightarrow{P_3A}}\right) = 1.$$

1-7 / POSITION VECTORS

It is sometimes convenient to consider a rectangular Cartesian coordinate system in discussing vectors. In three-dimensional space the position of a point P may be determined by its directed perpendicular distances from three mutually perpendicular reference planes. These planes intersect in three mutually perpendicular lines called the **coordinate axes** which are usually designated as the x-axis, y-axis, and z-axis. The three **coordinate planes** are usually designated by the pair of axes which determine the planes; that is, the xy-plane, yz-plane, and zx-plane contain the x- and y-axes, the y- and z-axes, and the z- and x-axes, respectively. An ordered triple of real numbers (a, b, c) may be associated with each point P in space such that a, b, and c represent the directed distances from the yz-, zx-, and xy-planes, respectively. The scalars a, b, and c of the ordered triple (a, b, c) are called the x-, y-, and z-coordinates, respectively, of P. We shall often speak of the ordered triple (a, b, c) as the **coordinates** of P. The point of intersection of the coordinate planes has coordinates $(0, 0, 0)$ and is called the **origin.**

It is possible to assign two different orientations to a rectangular Cartesian coordinate system in space. The coordinate system is said to have a right-handed orientation if, when the positive half of the x-axis is rotated 90° onto the positive half of the y-axis, then a right-hand screw rotated in the same manner would advance along the positive half of the z-axis. A right-handed coordinate system may also be described by placing the back of a person's right hand on a table top and extending the thumb, index finger, and middle finger in three mutually perpendicular directions such that the middle finger points upward from the table top. Let the thumb, index finger, and middle finger be associated with the positive rays of the x-, y-, and z-axes, respectively. It is possible to describe a left-handed coordinate system in a similar manner with the use of a person's left hand. While both orientated coordinate systems may be used in discussing vectors in space, it is necessary to choose and use one, and only one, of the two systems since certain fundamental considerations change with orientation (see §2-5). A right-handed coordinate system will be assumed throughout our discussion of vectors.

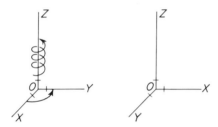

A right-handed coordinate system   A left-handed coordinate system

FIGURE 1-25

It is convenient to choose three unit vectors, designated as $\vec{i}, \vec{j}$, and $\vec{k}$, with a common initial point at the origin and terminal points at (1, 0, 0), (0, 1, 0), and (0, 0, 1), respectively, as a basis for vectors in three-dimensional coordinate space. Note that $\vec{i}, \vec{j}$, and $\vec{k}$ constitute a basis since they are linearly independent. Now, with each point $P: (x, y, z)$ in space we may associate a **position vector** $\overrightarrow{OP}$ whose initial point is the origin and whose terminal point is P. Then position vector $\overrightarrow{OP}$ may be expressed as $x\vec{i} + y\vec{j} + z\vec{k}$. Note that in Figure 1-26 the position vector $\overrightarrow{OP}$ may be associated with a diagonal of a rectangular parallelepiped whose edges are equal to $|x|$, $|y|$, and $|z|$. Therefore, $|\overrightarrow{OP}|$ can be determined by using the Pythagorean theorem twice; that is,

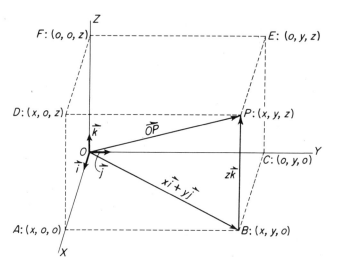

FIGURE 1-26

$$|\overrightarrow{OP}|^2 = |\overrightarrow{OB}|^2 + |\overrightarrow{BP}|^2$$
$$= |\overrightarrow{OA}|^2 + |\overrightarrow{OC}|^2 + |\overrightarrow{OF}|^2;$$
$$|\overrightarrow{OP}| = \sqrt{x^2 + y^2 + z^2}.$$

The real numbers x, y, and z are sometimes called the **components** of $\overrightarrow{OP}$. Since $\overrightarrow{OP}$ is a unique linear function of $\vec{i}$, $\vec{j}$, and $\vec{k}$, then *two position vectors are equal if, and only if, their corresponding components are equal.* The following theorems may be proved by using the previous definitions and theorems about vectors.

THEOREM 1.10 / *The components of the sum (difference) of two position vectors are equal to the sum (difference) of the corresponding components of the vectors;* that is, if $\vec{a} = x_1\vec{i} + y_1\vec{j} + z_1\vec{k}$ and $\vec{b} = x_2\vec{i} + y_2\vec{j} + z_2\vec{k}$, then

$$\vec{a} \pm \vec{b} = (x_1 \pm x_2)\vec{i} + (y_1 \pm y_2)\vec{j} + (z_1 \pm z_2)\vec{k}.$$

THEOREM 1.11 / *The components of a scalar multiple of a position vector are equal to the scalar multiple of the corresponding components of the vector;* that is, if $\vec{a} = x\vec{i} + y\vec{j} + z\vec{k}$ and m is a scalar, then $m\vec{a} = mx\vec{i} + my\vec{j} + mz\vec{k}$.

EXAMPLE 1 / Determine the position vector of the point P: $(3, -4, 12)$. Find its magnitude. Determine the unit position vector in the direction of $\overrightarrow{OP}$.

The position vector of the point $P: (3, -4, 12)$ is

$$\overrightarrow{OP} = 3\vec{i} - 4\vec{j} + 12\vec{k}.$$

Its magnitude is given as

$$|\overrightarrow{OP}| = \sqrt{3^2 + (-4)^2 + 12^2} = 13.$$

The unit position vector in the direction of $\overrightarrow{OP}$ is given as

$$\frac{\overrightarrow{OP}}{|\overrightarrow{OP}|} = \frac{3}{13}\vec{i} - \frac{4}{13}\vec{j} + \frac{12}{13}\vec{k}.$$

EXAMPLE 2 / Determine the coordinates of the point that divides the line segment from $A: (1, -2, 0)$ to $B: (6, 8, -10)$ in the ratio 2 to 3.

Let P be the desired point. Using Theorem 1.9, we have $\overrightarrow{OP} = \frac{2}{5}\overrightarrow{OB} + \frac{3}{5}\overrightarrow{OA}$. Now, $\overrightarrow{OB} = 6\vec{i} + 8\vec{j} - 10\vec{k}$ and $\overrightarrow{OA} = \vec{i} - 2\vec{j}$. Then

$$\tfrac{2}{5}\overrightarrow{OB} = \tfrac{12}{5}\vec{i} + \tfrac{16}{5}\vec{j} - \tfrac{20}{5}\vec{k} \quad \text{and} \quad \tfrac{3}{5}\overrightarrow{OA} = \tfrac{3}{5}\vec{i} - \tfrac{6}{5}\vec{j}.$$

Therefore,

$$\overrightarrow{OP} = (\tfrac{12}{5} + \tfrac{3}{5})\vec{i} + (\tfrac{16}{5} - \tfrac{6}{5})\vec{j} + (-\tfrac{20}{5} + 0)\vec{k}$$
$$= 3\vec{i} + 2\vec{j} - 4\vec{k}.$$

Hence, the coordinates of P are $(3, 2, -4)$.

EXAMPLE 3 / Determine the two-point form of the equation of a line on a plane (Figure 1-27).

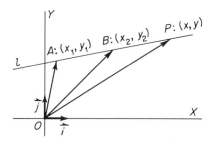

FIGURE 1-27

Let $A: (x_1, y_1)$ and $B: (x_2, y_2)$ be any two given points on line l. Let $P: (x, y)$ be any point on line l. Since A, B, and P are collinear, then $\overrightarrow{AP} = t\overrightarrow{AB}$ for any real number t; $\overrightarrow{OP} - \overrightarrow{OA} = t(\overrightarrow{OB} - \overrightarrow{OA})$. Now, $\overrightarrow{OP} = x\vec{i} + y\vec{j}$; $\overrightarrow{OA} = x_1\vec{i} + y_1\vec{j}$; $\overrightarrow{OB} = x_2\vec{i} + y_2\vec{j}$. Therefore,

$$(x - x_1)\vec{i} + (y - y_1)\vec{j} = t(x_2 - x_1)\vec{i} + t(y_2 - y_1)\vec{j}.$$

Since $\vec{i}$ and $\vec{j}$ are linearly independent vectors, then

$$(x - x_1) = t(x_2 - x_1) \quad \text{and} \quad (y - y_1) = t(y_2 - y_1)$$

represents a *parametric form* of the equation of the line with parameter t. Equating the t's in the two equations, we have

$$\frac{x - x_1}{x_2 - x_1} = \frac{y - y_1}{y_2 - y_1},$$

which represents the two-point form of the equation of a line through (x_1, y_1) and (x_2, y_2).

EXERCISES

In Exercises 1 through 4 find (a) $\overrightarrow{OP}$; (b) $|\overrightarrow{OP}|$; (c) *a unit position vector in the direction of* $\overrightarrow{OP}$.

1. $P: (2, 2, 1)$.
2. $P: (-3, 4, 5)$.
3. $P: (3, 0, 0)$.
4. $P: (1, 1, 1)$.

In Exercises 5 through 10 describe the locus of the points (x, y, z) such that:

5. $x = 0$.
6. $x = y = 0$.
7. $x = 3$.
8. $|z| < 1$.
9. $x^2 + y^2 + z^2 = 1$.
10. $|x| < \frac{1}{2}$, $|y| < \frac{1}{2}$, and $|z| < \frac{1}{2}$.

11. Determine the distance between the point $P: (x, y, z)$ and (a) the xy-plane; (b) the x-axis.

12. Determine if the three points $M: (2, 5, 9)$, $N: (0, 1, 1)$, and $P: (3, 7, 13)$ are collinear.

13. Determine if the three points $M: (1, 2, 3)$, $N: (5, 4, -2)$, and $P: (-2, -1, 5)$ are collinear.

14. Given $A: (3, 2, 4)$ and $B: (5, 3, 4)$, show that $\overrightarrow{AB}$ is parallel to the xy-plane.

15. Given $A: (3, -1, 17)$ and $B: (8, 9, 2)$, find the coordinates of the point that divides the line segment AB in the ratio (a) 1 to 1; (b) 2 to 3; (c) 2 to -1.

16. Find the coordinates of the centroid of the triangle with vertices $A: (x_1, y_1, z_1)$, $B: (x_2, y_2, z_2)$, and $C: (x_3, y_3, z_3)$.

TWO

PRODUCTS OF VECTORS

2-1 / THE SCALAR PRODUCT

The concept of a vector as a physical quantity having magnitude and direction, or as a directed line segment, does not imply how the product of two vectors shall be defined. No *a priori* definition exists. We may arbitrarily define the multiplication of vectors in any one of several ways. The interpretation of the results of our definition may differ in some instances from the interpretation of ordinary multiplication of scalars; that is, certain algebraic properties of the algebra of real numbers may not be valid for a particular definition of the product of two vectors. The admittedly arbitrary choice of one definition over another has been motivated by heuristic considerations. That is, we choose the definition that is most suitable for the applications of the product. By studying the ways in which vector quantities are combined in physical situations, we are motivated to define two types of products. The first type of product of two vectors results in a scalar and is called the *scalar product*.

When an object is moved from a point A to a point B along a straight line, by being acted upon by a constant force $\vec{f}$ as in Figure 2-1, we are often concerned with the amount of work that is done by the force. The force $\vec{f}$ may be considered as the sum of two vectors $\vec{f}_1$ and $\vec{f}_2$ with magnitudes $|\vec{f}|\cos\theta$ and $|\vec{f}|\sin\theta$, respectively. Only the vector component whose magnitude is $|\vec{f}|\cos\theta$ results in the displacement $\vec{d}$ of the object from A to B, a distance equal to $|\vec{d}|$. The amount of work accomplished in displacing the object from A to B may be represented by the product $(|\vec{f}|\cos\theta)|\vec{d}|$. This scalar may be considered as one type of product of $\vec{f}$ and $\vec{d}$. The consideration of a number of similar physical problems suggests to us a definition for the product of two vectors.

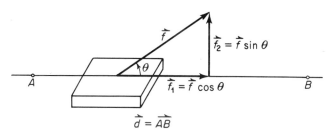

FIGURE 2-1

DEFINITION 2.1 / The **scalar product** or **dot product** of two given vectors $\vec{a}$ and $\vec{b}$, designated $\vec{a}\cdot\vec{b}$ and read "a dot b," is defined by the identity

$$\vec{a}\cdot\vec{b} = |\vec{a}||\vec{b}|\cos(\vec{a},\vec{b}), \tag{2-1}$$

where $(\vec{a},\vec{b})$ is the smallest angle between the two vectors measured (counterclockwise or clockwise) from $\vec{a}$ to $\vec{b}$ when the vectors have a common initial point; that is, $0° \leq |(\vec{a},\vec{b})| \leq 180°$.

The right-hand member of the identity (2-1) is a scalar quantity; hence the term *scalar product* of two vectors. The dot used to indicate the type of multiplication between $\vec{a}$ and $\vec{b}$ leads to the term *dot product*. Another name frequently used for the scalar product of two vectors is the **inner product**.

In Figure 2-2 it can be seen that, geometrically, $\vec{a}\cdot\vec{b}$ is equal to the product of the directed magnitude (signed length) of the projection of $\vec{a}$ onto $\vec{b}$ and the magnitude (length) of $\vec{b}$.

From the definition, the scalar product $\vec{b}\cdot\vec{a}$ is given by the expression

$$\vec{b}\cdot\vec{a} = |\vec{b}||\vec{a}|\cos(\vec{b},\vec{a}). \tag{2-2}$$

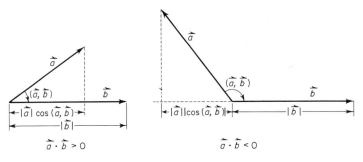

FIGURE 2-2

Now, since angle $(\vec{b}, \vec{a})$ is the negative of angle $(\vec{a}, \vec{b})$ and $\cos(\vec{a}, \vec{b})$ equals $\cos[-(\vec{a}, \vec{b})]$, then $\cos(\vec{a}, \vec{b})$ equals $\cos(\vec{b}, \vec{a})$. Since $|\vec{a}||\vec{b}| = |\vec{b}||\vec{a}|$, the following theorem is proved.

THEOREM 2.1 / *The scalar product of two vectors is commutative; that is,*

$$\vec{a} \cdot \vec{b} = \vec{b} \cdot \vec{a}. \qquad (2\text{-}3)$$

By the definition of the scalar product and the fact that $\cos 0° = 1$ and $\cos 90° = 0$, we may obtain the next two theorems.

THEOREM 2.2 / *The scalar product of a vector with itself equals the square of its magnitude; that is,*

$$\vec{a} \cdot \vec{a} = |\vec{a}|^2. \qquad (2\text{-}4)$$

THEOREM 2.3 / *If two nonzero vectors $\vec{a}$ and $\vec{b}$ are perpendicular, then $\vec{a} \cdot \vec{b} = 0$.*

Notice that the converse of Theorem 2.3 is not necessarily true. If $\vec{a} \cdot \vec{b} = 0$, then either at least one of the vectors is the null vector or the two nonzero vectors are perpendicular. A modified converse of Theorem 2.3 is valid, however.

THEOREM 2.4 / *If the scalar product of two nonzero vectors is zero, then the vectors are perpendicular.*

Since the scalar product of two vectors may be zero when neither factor is the null vector, division of a scalar product by a vector as an inverse process to finding the scalar product cannot be performed. Consider $\vec{a}\cdot\vec{b} = \vec{a}\cdot\vec{c}$; then $\vec{b} = \vec{c}$ may or may not be true. For example, $\vec{a}$ and $\vec{b}$ may be two nonzero perpendicular vectors and $\vec{c}$ may be the null vector. Then $\vec{a}\cdot\vec{b} = \vec{a}\cdot\vec{c}$, but $\vec{b} \neq \vec{c}$.

THEOREM 2.5 / *The scalar product is distributive with respect to the addition of vectors; that is,*

$$\vec{a}\cdot(\vec{b} + \vec{c}) = \vec{a}\cdot\vec{b} + \vec{a}\cdot\vec{c}. \tag{2-5}$$

PROOF / Let b', c' be the signed magnitudes of the projections of $\vec{b}$ and $\vec{c}$, respectively, along $\vec{a}$, as in Figure 2-3. Then $b' + c'$ is the signed magnitude of the projection of $\vec{b} + \vec{c}$ along $\vec{a}$. The equation

$$|\vec{a}|(b' + c') = |\vec{a}|b' + |\vec{a}|c'$$

is then equivalent to $\vec{a}\cdot(\vec{b} + \vec{c}) = \vec{a}\cdot\vec{b} + \vec{a}\cdot\vec{c}$.

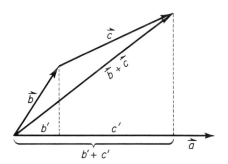

FIGURE 2-3

THEOREM 2.6 / *A real multiple of the scalar product of two vectors is equal to the scalar product of one of the vectors and the real multiple of the other; that is,*

$$m(\vec{a}\cdot\vec{b}) = (m\vec{a})\cdot\vec{b} = \vec{a}\cdot(m\vec{b}). \tag{2-6}$$

PROOF / $m(\vec{a}\cdot\vec{b}) = m|\vec{a}||\vec{b}|\cos(\vec{a}, \vec{b})$
$= |m\vec{a}||\vec{b}|\cos(m\vec{a}, \vec{b}) = (m\vec{a})\cdot\vec{b}$
$= |\vec{a}||m\vec{b}|\cos(\vec{a}, m\vec{b}) = \vec{a}\cdot(m\vec{b})$.

2-1 / THE SCALAR PRODUCT

Any vector in space may be represented in the form $x\vec{i} + y\vec{j} + z\vec{k}$ in terms of the unit vectors $\vec{i}$ along the x-axis, $\vec{j}$ along the y-axis, and $\vec{k}$ along the z-axis. By Theorem 2.3,

$$\vec{i}\cdot\vec{j} = \vec{j}\cdot\vec{i} = 0; \quad \vec{j}\cdot\vec{k} = \vec{k}\cdot\vec{j} = 0; \quad \vec{k}\cdot\vec{i} = \vec{i}\cdot\vec{k} = 0. \tag{2-7}$$

By Theorem 2.2,

$$\vec{i}\cdot\vec{i} = 1; \quad \vec{j}\cdot\vec{j} = 1; \quad \vec{k}\cdot\vec{k} = 1. \tag{2-8}$$

The scalar product of two position vectors may now be determined as shown in Theorem 2.7.

THEOREM 2.7 / *The scalar product of two position vectors is equal to the sum of the products of their corresponding components; that is, if $\vec{a} = x_1\vec{i} + y_1\vec{j} + z_1\vec{k}$ and $\vec{b} = x_2\vec{i} + y_2\vec{j} + z_2\vec{k}$, then*

$$\vec{a}\cdot\vec{b} = x_1 x_2 + y_1 y_2 + z_1 z_2. \tag{2-9}$$

PROOF / Making use of Theorems 2.1, 2.5, and 2.6, we have

$$\vec{a}\cdot\vec{b} = (x_1\vec{i} + y_1\vec{j} + z_1\vec{k})\cdot(x_2\vec{i} + y_2\vec{j} + z_2\vec{k})$$
$$= x_1 x_2(\vec{i}\cdot\vec{i}) + x_1 y_2(\vec{i}\cdot\vec{j}) + x_1 z_2(\vec{i}\cdot\vec{k})$$
$$+ y_1 x_2(\vec{j}\cdot\vec{i}) + y_1 y_2(\vec{j}\cdot\vec{j}) + y_1 z_2(\vec{j}\cdot\vec{k})$$
$$+ z_1 x_2(\vec{k}\cdot\vec{i}) + z_1 y_2(\vec{k}\cdot\vec{j}) + z_1 z_2(\vec{k}\cdot\vec{k}).$$

But

$$\vec{i}\cdot\vec{i} = \vec{j}\cdot\vec{j} = \vec{k}\cdot\vec{k} = 1$$

and

$$\vec{i}\cdot\vec{j} = \vec{j}\cdot\vec{i} = \vec{j}\cdot\vec{k} = \vec{k}\cdot\vec{j} = \vec{k}\cdot\vec{i} = \vec{i}\cdot\vec{k} = 0.$$

Therefore,

$$\vec{a}\cdot\vec{b} = x_1 x_2 + y_1 y_2 + z_1 z_2.$$

EXAMPLE 1 / Determine the magnitude of the position vector $\vec{a} = 4\vec{i} + 3\vec{j} + 12\vec{k}$. Then determine a unit vector in the same direction as $\vec{a}$.

By Theorem 2.2, $|\vec{a}|^2 = \vec{a}\cdot\vec{a}$. Therefore,

$$|\vec{a}| = \sqrt{\vec{a}\cdot\vec{a}} = \sqrt{(4)^2 + (3)^2 + (12)^2} = 13.$$

The unit vector in the same direction as $\vec{a}$ is that vector $\vec{u} = x\vec{i} + y\vec{j} + z\vec{k}$ such that $13\vec{u} = \vec{a}$; that is, $13(x\vec{i} + y\vec{j} + z\vec{k}) = 4\vec{i} + 3\vec{j} + 12\vec{k}$. Hence, $\vec{u} = \frac{4}{13}\vec{i} + \frac{3}{13}\vec{j} + \frac{12}{13}\vec{k}$.

EXAMPLE 2 / Determine the angle between the position vectors

$$\vec{a} = 3\vec{i} - 2\vec{j} + 6\vec{k} \quad \text{and} \quad \vec{b} = -3\vec{i} - 5\vec{j} + 8\vec{k}.$$

Let θ denote the angle between $\vec{a}$ and $\vec{b}$. By Definition 2.1, $\vec{a} \cdot \vec{b} = |\vec{a}||\vec{b}| \cos \theta$. Since neither $\vec{a}$ nor $\vec{b}$ is the null vector,

$$\cos \theta = \frac{\vec{a} \cdot \vec{b}}{|\vec{a}||\vec{b}|} = \frac{(3)(-3) + (-2)(-5) + (6)(8)}{|\vec{a}||\vec{b}|} = \frac{49}{|\vec{a}||\vec{b}|},$$

where $|\vec{a}|^2 = \vec{a} \cdot \vec{a}$ and $|\vec{b}|^2 = \vec{b} \cdot \vec{b}$.

Now,

$$|\vec{a}| = \sqrt{(3)^2 + (-2)^2 + (6)^2} = 7$$

and

$$|\vec{b}| = \sqrt{(-3)^2 + (-5)^2 + (8)^2} = 7\sqrt{2}.$$

Hence,

$$\cos \theta = \frac{49}{49\sqrt{2}} = \frac{1}{\sqrt{2}} \quad \text{and} \quad \theta = 45°.$$

EXAMPLE 3 / Find the signed magnitude of the projection of

$$\vec{a} = 3\vec{i} - \vec{j} - 2\vec{k} \quad \text{on} \quad \vec{b} = \vec{i} + 2\vec{j} - 3\vec{k}.$$

The signed magnitude of the projection of $\vec{a}$ on $\vec{b}$ is equal to $|\vec{a}| \cos \theta$ and, since $|\vec{b}| \neq 0$,

$$|\vec{a}| \cos \theta = \frac{\vec{a} \cdot \vec{b}}{|\vec{b}|}.$$

Therefore,

$$|\vec{a}| \cos \theta = \frac{(3)(1) + (-1)(2) + (-2)(-3)}{\sqrt{(1)^2 + (2)^2 + (-3)^2}} = \frac{7}{\sqrt{14}} = \frac{\sqrt{14}}{2}.$$

EXAMPLE 4 / Verify Theorem 2.5 (the scalar product is distributive with respect to the addition of vectors) when

$$\vec{a} = 2\vec{i} - 3\vec{j} + 4\vec{k}, \quad \vec{b} = \vec{i} - \vec{j} + 2\vec{k}, \quad \text{and} \quad \vec{c} = 3\vec{i} + 2\vec{j} + \vec{k}.$$

We evaluate and compare the two members of equation (2-5) for the given vectors $\vec{a}$, $\vec{b}$, and $\vec{c}$:

$$\begin{aligned}
\vec{a} \cdot (\vec{b} + \vec{c}) &= (2\vec{i} - 3\vec{j} + 4\vec{k}) \cdot [(\vec{i} - \vec{j} + 2\vec{k}) + (3\vec{i} + 2\vec{j} + \vec{k})] \\
&= (2i - 3\vec{j} + 4\vec{k}) \cdot (4\vec{i} + \vec{j} + 3\vec{k}) \\
&= (2)(4) + (-3)(1) + (4)(3) = 17; \\
\vec{a} \cdot \vec{b} + \vec{a} \cdot \vec{c} &= (2\vec{i} - 3\vec{j} + 4\vec{k}) \cdot (\vec{i} - \vec{j} + 2\vec{k}) \\
&\quad + (2\vec{i} - 3\vec{j} + 4\vec{k}) \cdot (3\vec{i} + 2\vec{j} + \vec{k}) \\
&= [(2)(1) + (-3)(-1) + (4)(2)] \\
&\quad + [(2)(3) + (-3)(2) + (4)(1)] \\
&= 13 + 4 = 17.
\end{aligned}$$

Hence,

$$\vec{a} \cdot (\vec{b} + \vec{c}) = \vec{a} \cdot \vec{b} + \vec{a} \cdot \vec{c}.$$

EXERCISES

1. Determine the magnitude of the position vector $\vec{a} = 2\vec{i} + 2\vec{j} + \vec{k}$. Determine a unit vector in the same direction as $\vec{a}$.

2. Determine the angle between the position vectors $\vec{a} = 3\vec{i} - \vec{j} - 2\vec{k}$ and $\vec{b} = \vec{i} + 2\vec{j} - 3\vec{k}$.

3. Find the magnitude of the projection of the vector $\vec{a} = \vec{i} - 3\vec{j} + 2\vec{k}$ on $\vec{b} = 4\vec{i} - 3\vec{j}$.

4. Prove that if $\vec{a} = \vec{i} + 3\vec{j} - 2\vec{k}$ and $\vec{b} = \vec{i} - \vec{j} - \vec{k}$, then $\vec{a}$ and $\vec{b}$ are perpendicular.

5. Prove that (a) $|\vec{r_1} \cdot \vec{r_2}| \leq |\vec{r_1}||\vec{r_2}|$. State the conditions for (b) $\vec{r_1} \cdot \vec{r_2} = |\vec{r_1}||\vec{r_2}|$; (c) $\vec{r_1} \cdot \vec{r_2} = -|\vec{r_1}||\vec{r_2}|$.

6. Find the magnitude of the vector $x\vec{i} + y\vec{j} + z\vec{k}$.

7. Find the cosine of the angle between the diagonal of a cube and one of its edges.

8. Find (a) the magnitude of $\vec{RS}$ where $R: (-1, 2, 0)$ and $S: (5, 5, 6)$; (b) a unit vector in the direction of $\vec{RS}$.

9. Use scalar products to prove that the triangle whose vertices are $A: (1, 0, 1)$, $B: (1, 1, 1)$, and $C: (1, 1, 0)$ is a right isosceles triangle.

10. Explain why there is not an associative law for the scalar product.

11. Prove that $\vec{a} \cdot \vec{a} = 0$ is a necessary and sufficient condition for $\vec{a}$ to be the null vector.

12. Prove that if the scalar product of a vector $\vec{a}$ with each of the three linearly independent vectors $\vec{i}$, $\vec{j}$, and $\vec{k}$ is zero, then the vector is the null vector.

13. If A, B, and C are points whose coordinates are $(1, 0, 0)$, $(0, 1, 0)$, and $(0, 0, 1)$, respectively, then find the magnitude of the projection of $\vec{AB}$ on $\vec{AC}$.

2-2 / APPLICATIONS OF THE SCALAR PRODUCT

Many theorems in elementary plane geometry, trigonometry, and analytic geometry have exceedingly simple vector proofs. In this section the proofs of several such theorems shall be illustrated. Vector proofs involving the scalar product usually involve one of two special cases: the first, when the vectors are nonzero perpendicular vectors, in which case the scalar product is zero (Theorem 2.3); the second, when the vectors are equal, in which case the scalar product is equal to the square of the magnitude of the vector (Theorem 2.2).

EXAMPLE 1 / If the diagonals of a parallelogram are perpendicular, then the parallelogram is a rhombus (Figure 2-4).

Let $ABCD$ be a parallelogram. Let $\vec{a}$ and $\vec{b}$ be associated with the adjacent sides AB and BC, respectively. Then $\vec{a} + \vec{b}$ and $\vec{a} - \vec{b}$ are vectors associated with the diagonals. If the diagonals are perpendicular, then $(\vec{a} + \vec{b}) \cdot (\vec{a} - \vec{b}) = 0$. However,

$$(\vec{a} + \vec{b}) \cdot (\vec{a} - \vec{b}) = (\vec{a} + \vec{b}) \cdot \vec{a} - (\vec{a} + \vec{b}) \cdot \vec{b}$$
$$= \vec{a} \cdot \vec{a} + \vec{b} \cdot \vec{a} - \vec{a} \cdot \vec{b} - \vec{b} \cdot \vec{b}$$
$$= \vec{a} \cdot \vec{a} - \vec{b} \cdot \vec{b} \quad \text{(since } \vec{b} \cdot \vec{a} = \vec{a} \cdot \vec{b}\text{)}$$
$$= |\vec{a}|^2 - |\vec{b}|^2.$$

Therefore, $|\vec{a}|^2 - |\vec{b}|^2 = 0$, or $|\vec{a}| = |\vec{b}|$. Hence, two adjacent sides of the parallelogram are equal in length, and the parallelogram is a rhombus.

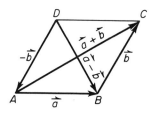
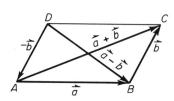

FIGURE 2-4   FIGURE 2-5

EXAMPLE 2 / The sum of the squares of the diagonals of any parallelogram is equal to the sum of the squares of the sides (Figure 2-5).

Let $ABCD$ be any parallelogram. Let $\vec{a}$ and $\vec{b}$ be associated with the adjacent sides AB and BC, respectively. Then $\overrightarrow{AC}$ and $\overrightarrow{DB}$ are vectors associated with the diagonals where $\overrightarrow{AC} = \vec{a} + \vec{b}$ and $\overrightarrow{DB} = \vec{a} - \vec{b}$. Now,

$$\overrightarrow{AC} \cdot \overrightarrow{AC} = (\vec{a} + \vec{b}) \cdot (\vec{a} + \vec{b})$$
$$|\overrightarrow{AC}|^2 = |\vec{a}|^2 + |\vec{b}|^2 + 2\vec{a} \cdot \vec{b}$$

and

$$\overrightarrow{DB} \cdot \overrightarrow{DB} = (\vec{a} - \vec{b}) \cdot (\vec{a} - \vec{b})$$
$$|\overrightarrow{DB}|^2 = |\vec{a}|^2 + |\vec{b}|^2 - 2\vec{a} \cdot \vec{b}.$$

Therefore,

$$|\overrightarrow{AC}|^2 + |\overrightarrow{DB}|^2 = 2|\vec{a}|^2 + 2|\vec{b}|^2.$$

Since

$$|\vec{a}| = |\overrightarrow{AB}| = |\overrightarrow{CD}| \quad \text{and} \quad |\vec{b}| = |\overrightarrow{BC}| = |\overrightarrow{DA}|,$$

then

$$|\overrightarrow{AC}|^2 + |\overrightarrow{DB}|^2 = |\overrightarrow{AB}|^2 + |\overrightarrow{CD}|^2 + |\overrightarrow{BC}|^2 + |\overrightarrow{DA}|^2;$$

that is, the sum of the squares of the diagonals of any parallelogram is equal to the sum of the squares of the sides.

EXAMPLE 3 / An angle inscribed in a semicircle is a right angle (Figure 2-6).

Consider the semicircle ACB with center at O where the points A, B, C, and O do not pairwise coincide. Associate vectors $\vec{a}$ and $\vec{c}$ with radii OA and OC, respectively. Then the vectors $-\vec{a}$, $\vec{a} - \vec{c}$, and $-\vec{a} - \vec{c}$ may be associated with the line segments OB, CA, and CB, respectively. Now,

$$(\vec{a} - \vec{c}) \cdot (-\vec{a} - \vec{c}) = -\vec{a} \cdot \vec{a} - \vec{a} \cdot \vec{c} + \vec{c} \cdot \vec{a} + \vec{c} \cdot \vec{c} = |\vec{c}|^2 - |\vec{a}|^2.$$

But $|\vec{c}| = |\vec{a}|$ since the radii of the same circle are equal in length. Hence $(\vec{a} - \vec{c}) \cdot (-\vec{a} - \vec{c}) = 0$. It follows that the line segments CA and CB associated with these vectors are perpendicular, or C coincides with A or B. Therefore, the angle inscribed in a semicircle is a right angle.

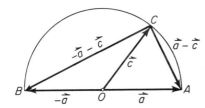

FIGURE 2-6   FIGURE 2-7

EXAMPLE 4 / Derive the law of cosines (Figure 2-7).

Let ABC be any triangle. Associate $\vec{a}$, $\vec{b}$, and $\vec{c}$ with sides CB, CA, and BA, respectively. Denote angle ACB by θ. Now,

$$\vec{c} = \vec{b} - \vec{a}$$

and

$$\begin{aligned}\vec{c} \cdot \vec{c} &= (\vec{b} - \vec{a}) \cdot (\vec{b} - \vec{a}) \\ &= (\vec{b} - \vec{a}) \cdot \vec{b} - (\vec{b} - \vec{a}) \cdot \vec{a} \\ &= \vec{b} \cdot \vec{b} - \vec{a} \cdot \vec{b} - \vec{b} \cdot \vec{a} + \vec{a} \cdot \vec{a}.\end{aligned}$$

Then

$$\begin{aligned}|\vec{c}|^2 &= |\vec{b}|^2 + |\vec{a}|^2 - 2\vec{a} \cdot \vec{b} \quad \text{(since } \vec{b} \cdot \vec{a} = \vec{a} \cdot \vec{b}\text{)} \\ &= |\vec{b}|^2 + |\vec{a}|^2 - 2|\vec{a}||\vec{b}| \cos \theta.\end{aligned}$$

EXAMPLE 5 / The median to the base of an isosceles triangle is perpendicular to the base (Figure 2-8).

Consider an isosceles triangle *ABC* with *M* the midpoint of the base *AB*. Associate $\vec{a}$, $\vec{b}$, and $\vec{m}$ with sides *CA*, *CB*, and median *CM*, respectively. Side *AB* may now be associated with $\vec{b} - \vec{a}$. Now,

$$\vec{m} = \tfrac{1}{2}(\vec{b} + \vec{a}).$$

Then

$$\vec{m} \cdot (\vec{b} - \vec{a}) = \tfrac{1}{2}(\vec{b} + \vec{a}) \cdot (\vec{b} - \vec{a})$$
$$= \tfrac{1}{2}(|\vec{b}|^2 - |\vec{a}|^2).$$

Since sides *CA* and *CB* are equal in length, $|\vec{a}| = |\vec{b}|$, $|\vec{b}|^2 - |\vec{a}|^2 = 0$, and $\vec{m} \cdot (\vec{b} - \vec{a}) = 0$. Therefore, $\vec{m}$ is perpendicular to $\vec{b} - \vec{a}$ since $|\vec{m}| \neq 0$ and $|\vec{b} - \vec{a}| \neq 0$; that is, the median *CM* is perpendicular to side *AB*.

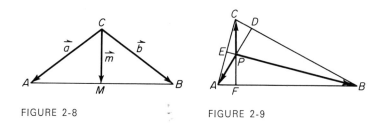

FIGURE 2-8   FIGURE 2-9

EXAMPLE 6 / The altitudes of any triangle are concurrent (Figure 2-9).

If *ABC* is a right triangle, the theorem is obviously true. Let *ABC* be any triangle, other than a right triangle, with altitudes *AD* and *BE* intersecting at point *P*. Let *F* be the point of intersection of line *CP* and side *AB*. Then

$$\vec{PB} \cdot (\vec{PC} - \vec{PA}) = 0 \quad \text{and} \quad \vec{PA} \cdot (\vec{PB} - \vec{PC}) = 0.$$

Therefore,

$$\vec{PB} \cdot (\vec{PC} - \vec{PA}) + \vec{PA} \cdot (\vec{PB} - \vec{PC}) = 0,$$
$$\vec{PB} \cdot \vec{PC} - \vec{PA} \cdot \vec{PC} = 0,$$
$$(\vec{PB} - \vec{PA}) \cdot \vec{PC} = 0,$$
$$\vec{AB} \cdot \vec{PC} = 0.$$

Since $|\overrightarrow{AB}| \neq 0$ and $|\overrightarrow{PC}| \neq 0$, the line PC is perpendicular to side AB; that is, line segment CF is the altitude to side AB and contains the point P, the point of intersection of the other two altitudes. Hence, the altitudes of any triangle are concurrent.

Notice that the proofs of the theorems in Examples 1 through 6 involved the use of vectors that were not expressed with reference to a coordinate system. The following examples make use of the concept of the scalar product of two position vectors.

EXAMPLE 7 / Prove that $\cos(\theta - \phi) = \cos\theta\cos\phi + \sin\theta\sin\phi$ (Figure 2-10).

Let $\vec{a}$ and $\vec{b}$ be unit position vectors on a rectangular Cartesian coordinate plane, such that the vectors $\vec{a}$ and $\vec{b}$ form angles θ and ϕ, respectively, with the positive half of the x-axis. Then $\vec{a} = \cos\theta\,\vec{i} + \sin\theta\,\vec{j}$ and $\vec{b} = \cos\phi\,\vec{i} + \sin\phi\,\vec{j}$. Now, by Definition 2.1,

$$\vec{a}\cdot\vec{b} = |\vec{a}||\vec{b}|\cos(\vec{a},\vec{b}) = |\vec{a}||\vec{b}|\cos(\phi - \theta)$$
$$= |\vec{a}||\vec{b}|\cos(\theta - \phi) = \cos(\theta - \phi).$$

Note that we may consider $\vec{a}$ and $\vec{b}$ as vectors in space whose third components are zero. By Theorem 2.7, the scalar product of the unit position vectors $\vec{a}$ and $\vec{b}$ may be expressed as

$$\vec{a}\cdot\vec{b} = \cos\theta\cos\phi + \sin\theta\sin\phi.$$

Hence,

$$\cos(\theta - \phi) = \cos\theta\cos\phi + \sin\theta\sin\phi.$$

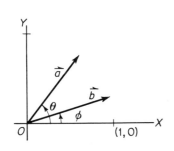

FIGURE 2-10

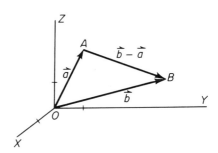

FIGURE 2-11

EXAMPLE 8 / Determine a formula for the distance between two points in space (Figure 2-11).

Let $A: (x_1, y_1, z_1)$ and $B: (x_2, y_2, z_2)$ be any two points in space with position vectors $\vec{a}$ and $\vec{b}$. Then $\vec{a} = x_1\vec{i} + y_1\vec{j} + z_1\vec{k}$, $\vec{b} = x_2\vec{i} + y_2\vec{j} + z_2\vec{k}$, and $\vec{AB} = \vec{b} - \vec{a}$. The distance between A and B equals $|\vec{b} - \vec{a}|$. Now,

$$\vec{b} - \vec{a} = (x_2 - x_1)\vec{i} + (y_2 - y_1)\vec{j} + (z_2 - z_1)\vec{k}.$$

According to Theorem 2.2,

$$|\vec{b} - \vec{a}| = \sqrt{(\vec{b} - \vec{a}) \cdot (\vec{b} - \vec{a})} \qquad (2\text{-}10)$$

represents the vector formula for the distance between two points in space. Since

$$(\vec{b} - \vec{a}) \cdot (\vec{b} - \vec{a}) = (x_2 - x_1)^2 + (y_2 - y_1)^2 + (z_2 - z_1)^2,$$

$$|\vec{b} - \vec{a}| = \sqrt{(x_2 - x_1)^2 + (y_2 - y_1)^2 + (z_2 - z_1)^2} \qquad (2\text{-}11)$$

represents the Cartesian coordinate formula for the distance between two points (x_1, y_1, z_1) and (x_2, y_2, z_2).

Other applications of the scalar product will be considered in subsequent sections of this chapter. In Chapter 3 the use of the scalar product, along with other concepts to be developed in the present chapter, will be considered as they apply to the study of the coordinate geometry of planes and lines in three-dimensional space.

EXERCISES

Make use of the concept of the scalar product of two vectors to prove the theorems stated in Exercises 1 through 10.

1. The diagonals of a rhombus are perpendicular.

2. In any right triangle the square of the hypotenuse is equal to the sum of the squares of the other two sides (PYTHAGOREAN THEOREM).

3. In any right triangle the median to the hypotenuse is equal to one-half the hypotenuse.

4. The line segments joining consecutive midpoints of the sides of any square form a square.

5. The sum of the squares of the diagonals of any quadrilateral is equal to twice the sum of the squares of the line segments joining the midpoints of the opposite sides.

6. If, from a point outside any circle, a tangent and a secant through the diameter are drawn, then the tangent is the mean proportional between the secant and its external segment.

7. $\cos(\theta + \phi) = \cos\theta \cos\phi - \sin\theta \sin\phi$.

8. If a line is perpendicular to two intersecting lines at their point of intersection, then the line is perpendicular to the plane determined by the intersecting lines.

9. For any triangle ABC, $|\overrightarrow{AB}| = |\overrightarrow{AC}|\cos A + |\overrightarrow{BC}|\cos B$.

10. The perpendicular bisectors of the sides of a triangle are concurrent.

2-3 / CIRCLES AND LINES ON A COORDINATE PLANE

In analytic geometry many properties and equations related to the circle and the line may be derived by vector methods. In this section a selection of those properties and equations whose derivations may be obtained by use of the scalar product are illustrated through a sequence of examples.

EXAMPLE 1 / Determine the equation of a circle with center at $C: (x_1, y_1)$ and radius a (Figure 2-12).

Let $P: (x, y)$ be any point on the circle with center at $C: (x_1, y_1)$ and radius a. Then $|\overrightarrow{CP}| = a$, and

$$\overrightarrow{CP} \cdot \overrightarrow{CP} = a^2 \qquad (2\text{-}12)$$

represents a vector form of the equation of the circle. Since $\overrightarrow{CP} = \overrightarrow{OP} - \overrightarrow{OC}$,

$$(\overrightarrow{OP} - \overrightarrow{OC}) \cdot (\overrightarrow{OP} - \overrightarrow{OC}) = a^2 \qquad (2\text{-}13)$$

also represents a vector form of the equation of the circle. Now, $\overrightarrow{OP} = x\vec{i} + y\vec{j}$, $\overrightarrow{OC} = x_1\vec{i} + y_1\vec{j}$, and $\overrightarrow{CP} = \overrightarrow{OP} - \overrightarrow{OC} = (x - x_1)\vec{i} + (y - y_1)\vec{j}$. Hence by (2-12),

$$(x - x_1)^2 + (y - y_1)^2 = a^2 \qquad (2\text{-}14)$$

represents a rectangular Cartesian coordinate form of the equation of a circle with center at $C: (x_1, y_1)$ and radius a.

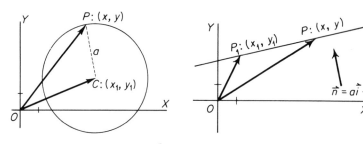

FIGURE 2-12   FIGURE 2-13

EXAMPLE 2 / Determine the equation of the line l passing through a given point $P_1: (x_1, y_1)$ and perpendicular to a given vector $\vec{n} = a\vec{i} + b\vec{j}$ (Figure 2-13).

Let $P: (x, y)$ be a general point on the line l. Then $\overrightarrow{OP} - \overrightarrow{OP_1}$ is perpendicular to $\vec{n}$, and

$$(\overrightarrow{OP} - \overrightarrow{OP_1}) \cdot \vec{n} = 0 \qquad (2\text{-}15)$$

represents a vector form of the equation of line l. Since $\overrightarrow{OP} = x\vec{i} + y\vec{j}$, $\overrightarrow{OP_1} = x_1\vec{i} + y_1\vec{j}$, and $\vec{n} = a\vec{i} + b\vec{j}$, then

$$\overrightarrow{OP} - \overrightarrow{OP_1} = (x - x_1)\vec{i} + (y - y_1)\vec{j}$$

and

$$(\overrightarrow{OP} - \overrightarrow{OP_1}) \cdot \vec{n} = a(x - x_1) + b(y - y_1).$$

Therefore,

$$a(x - x_1) + b(y - y_1) = 0 \qquad (2\text{-}16)$$

represents a rectangular Cartesian coordinate form of the equation of the line l. Note that the coefficients of x and y are the horizontal and vertical components of the vector perpendicular to the line. We shall make use of this fact in the next two examples.

EXAMPLE 3 / Determine a formula for the distance r between a point $P_1: (x_1, y_1)$ and a line $ax + by + c = 0$ (Figure 2-14).

Note that the distance between a point P_1 and a given line is defined to be the shortest distance, and is measured along the line through P_1 perpendicular to the given line. Let $P_0: (x_0, y_0)$ be any point on the given line; then $ax_0 + by_0 + c = 0$. Let $\vec{n}$ be any vector perpendicular to the line, such as $a\vec{i} + b\vec{j}$. Then the distance r is the magnitude of the projection of $\overrightarrow{P_0P_1}$ on $\vec{n}$ and, as in Example 3 of §2-1,

$$r = \frac{|\overrightarrow{P_0P_1} \cdot \vec{n}|}{|\vec{n}|}. \tag{2-17}$$

Since $\overrightarrow{P_0P_1} = (x_1 - x_0)\vec{i} + (y_1 - y_0)\vec{j}$ and $\vec{n} = a\vec{i} + b\vec{j}$, then

$$r = \frac{|a(x_1 - x_0) + b(y_1 - y_0)|}{\sqrt{a^2 + b^2}}.$$

Now, $-ax_0 - by_0 = c$. Therefore,

$$r = \frac{|ax_1 + by_1 + c|}{\sqrt{a^2 + b^2}}. \tag{2-18}$$

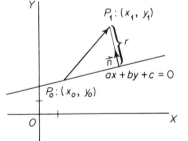

FIGURE 2-14   FIGURE 2-15

EXAMPLE 4 / Determine the equation of the line l tangent at $P_1: (x_1, y_1)$ to the circle whose center is $P_0: (x_0, y_0)$ (Figure 2-15).

Let $P: (x, y)$ be a general point on the tangent line l. Line l is tangent to the circle if, and only if, $\overrightarrow{P_1P}$ is perpendicular to $\overrightarrow{P_0P_1}$, the radius vector to the tangent. Therefore,

$$\overrightarrow{P_1P} \cdot \overrightarrow{P_0P_1} = 0 \tag{2-19}$$

and
$$(\vec{OP} - \vec{OP_1}) \cdot (\vec{OP_1} - \vec{OP_0}) = 0 \qquad (2\text{-}20)$$

represent vector forms of the tangent line l. Since $\vec{P_1P} = (x - x_1)\vec{i} + (y - y_1)\vec{j}$ and $\vec{P_0P_1} = (x_1 - x_0)\vec{i} + (y_1 - y_0)\vec{j}$, then

$$(x - x_1)(x_1 - x_0) + (y - y_1)(y_1 - y_0) = 0 \qquad (2\text{-}21)$$

represents a rectangular Cartesian coordinate form of the equation of the line l tangent at $P_1: (x_1, y_1)$ to the circle whose center is $P_0: (x_0, y_0)$.

EXERCISES

1. Determine the equation of a circle with center at $C: (1, -3)$ and radius 4.

2. Determine the equation of the line passing through $P_1: (2, 1)$ and perpendicular to $\vec{n} = 3\vec{i} - \vec{j}$.

3. Find the distance from $P_1: (3, 2)$ to the line $3x + 4y - 7 = 0$.

4. Determine the equation of the line tangent at $P_1: (5, 1)$ to the circle whose center is $P_0: (3, -2)$.

5. State a vector form of the equation of the straight line through the origin parallel to $\vec{b}$.

6. Use the result of Exercise 5 to determine a rectangular Cartesian coordinate form of the equation of the line through the origin parallel to $\vec{b} = b_1\vec{i} + b_2\vec{j}$.

7. State a vector form of the equation of the straight line through the terminal point of $\vec{a}$ and parallel to $\vec{b}$.

8. Use the result of Exercise 7 to determine a rectangular Cartesian coordinate form of the equation of the line through (a_1, a_2) parallel to $\vec{b} = b_1\vec{i} + b_2\vec{j}$.

9. Find a formula for the distance r from the origin to the line $ax + by + c = 0$.

2-4 / ORTHOGONAL BASES

Two nonzero vectors $\vec{x_1}$ and $\vec{x_2}$ such that $\vec{x_1} \cdot \vec{x_2} = 0$ are called **orthogonal vectors**. A set of two such vectors constitutes an **orthogonal basis** for a vector space of two dimensions. Three nonzero vectors $\vec{x_1}$, $\vec{x_2}$, and $\vec{x_3}$, such that $\vec{x}_m \cdot \vec{x}_n = 0$ for all pairs m, n where $m \neq n$, form a set of three orthogonal vectors which constitutes an **orthogonal basis** for a vector space of three dimensions. The set of vectors $\vec{i}$, $\vec{j}$, and $\vec{k}$ is an example of a set of three orthogonal vectors. Since each of these vectors is a unit vector, the set constitutes a **normal orthogonal basis**, sometimes called an **orthonormal basis**.

It is often necessary to determine an orthogonal or orthonormal basis for a vector space of two dimensions embedded in a vector space of three dimensions. If $\vec{y}_1$ and $\vec{y}_2$ are two independent three-dimensional vectors, it is possible to determine two orthogonal vectors $\vec{x}_1$ and $\vec{x}_2$ in the two-dimensional vector space *spanned* by $\vec{y}_1$ and $\vec{y}_2$ (that is, for which the set $\vec{y}_1$ and $\vec{y}_2$ is a basis) by means of the following process. First, let $\vec{x}_1 = \vec{y}_1$ and $\vec{x}_2 = \vec{y}_1 + t\vec{y}_2$. Then t is determined such that $\vec{x}_1 \cdot \vec{x}_2 = 0$:

$$\vec{y}_1 \cdot (\vec{y}_1 + t\vec{y}_2) = 0,$$

$$(\vec{y}_1 \cdot \vec{y}_1) + t(\vec{y}_1 \cdot \vec{y}_2) = 0,$$

$$t = -\frac{\vec{y}_1 \cdot \vec{y}_1}{\vec{y}_1 \cdot \vec{y}_2},$$

and

$$\vec{x}_2 = \vec{y}_1 - \frac{\vec{y}_1 \cdot \vec{y}_1}{\vec{y}_1 \cdot \vec{y}_2}\vec{y}_2. \tag{2-22}$$

Therefore,

$$\vec{y}_1 \quad \text{and} \quad \vec{y}_1 - \frac{\vec{y}_1 \cdot \vec{y}_1}{\vec{y}_1 \cdot \vec{y}_2}\vec{y}_2$$

constitute an orthogonal basis for the vector space spanned by $\vec{y}_1$ and $\vec{y}_2$. An orthonormal basis may be obtained by multiplying each vector by the reciprocal of its magnitude. The process that has been used is an elementary example of a more general process, called the **Gram-Schmidt process**, for finding orthogonal bases; this process finds application in the study of n-dimensional vector spaces.

EXAMPLE 1 / Find an orthogonal basis for the two-dimensional vector space spanned by the vectors $\vec{i} - \vec{j}$ and $\vec{i} + 2\vec{k}$.

Let $\vec{y}_1 = \vec{i} - \vec{j}$ and $\vec{y}_2 = \vec{i} + 2\vec{k}$. Then a set of vectors $\vec{x}_1$ and $\vec{x}_2$

which constitutes an orthogonal basis for the two-dimensional vector space spanned by $\vec{y_1}$ and $\vec{y_2}$ may be expressed as

$$\vec{x_1} = \vec{i} - \vec{j}$$

and

$$\begin{aligned}\vec{x_2} &= \vec{i} - \vec{j} - \frac{(\vec{i} - \vec{j})\cdot(\vec{i} - \vec{j})}{(\vec{i} - \vec{j})\cdot(\vec{i} + 2\vec{k})}(\vec{i} + 2\vec{k}) \\ &= \vec{i} - \vec{j} - 2(\vec{i} + 2\vec{k}) \\ &= -\vec{i} - \vec{j} - 4\vec{k}.\end{aligned}$$

Note that $\vec{x_1} \cdot \vec{x_2} = 0$.

EXAMPLE 2 / Find an orthonormal basis for the two-dimensional vector space spanned by the vectors given in Example 1.

Now, $\vec{x_1}/|\vec{x_1}|$ and $\vec{x_2}/|\vec{x_2}|$ constitutes an orthonormal basis for the two-dimensional vector space of Example 1. Hence, the vectors

$$\frac{1}{\sqrt{2}}\vec{i} - \frac{1}{\sqrt{2}}\vec{j} \quad \text{and} \quad -\frac{1}{3\sqrt{2}}\vec{i} - \frac{1}{3\sqrt{2}}\vec{j} - \frac{4}{3\sqrt{2}}\vec{k}$$

represent one orthonormal basis.

EXAMPLE 3 / Determine a second orthogonal basis for the vector space of Example 1.

The Gram-Schmidt process may be used again; this time we let $\vec{y_1} = \vec{i} + 2\vec{k}$ and $\vec{y_2} = \vec{i} - \vec{j}$. Then $\vec{x_1} = \vec{i} + 2\vec{k}$ and

$$\begin{aligned}\vec{x_2} &= \vec{i} + 2\vec{k} - \frac{(\vec{i} + 2\vec{k})\cdot(\vec{i} + 2\vec{k})}{(\vec{i} + 2\vec{k})\cdot(\vec{i} - \vec{j})}(\vec{i} - \vec{j}) \\ &= \vec{i} + 2\vec{k} - 5(\vec{i} - \vec{j}) \\ &= -4\vec{i} + 5\vec{j} + 2\vec{k}.\end{aligned}$$

EXERCISES

1. Determine the values of x for which the vectors $(x - 1)\vec{i} + x\vec{j} + (x + 4)\vec{k}$ and $2\vec{i} + x\vec{j} + \vec{k}$ are orthogonal.

2. Find an orthogonal basis for the two-dimensional vector space spanned by the vectors $\vec{i} + \vec{k}$ and $\vec{i} + \vec{j} + 3\vec{k}$.

3. Find an orthonormal basis for the vector space of Exercise 2.

4. Find an orthogonal basis for the two-dimensional vector space spanned by the vectors $2\vec{i} - \vec{j} + 2\vec{k}$ and $3\vec{i} + \vec{j} + 2\vec{k}$.

5. Find an orthonormal basis for the vector space of Exercise 4.

2-5 / THE VECTOR PRODUCT

A second type of product of two vectors has for its result a vector, and is called the *vector product*. Consider two vectors $\vec{a}$ and $\vec{b}$ subjected to parallel displacement, if necessary, to bring their initial points into coincidence. Let the smallest angle from $\vec{a}$ to $\vec{b}$ be designated by the symbol $(\vec{a}, \vec{b})$; that is, $0° \leq |(\vec{a}, \vec{b})| \leq 180°$ (Figure 2-16).

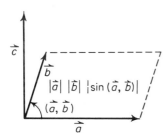

FIGURE 2-16

DEFINITION 2.2 / *The **vector product** or **cross product** of $\vec{a}$ and $\vec{b}$, designated $\vec{a} \times \vec{b}$ and read "$\vec{a}$ cross $\vec{b}$," is defined in general as a third vector $\vec{c}$ such that:*

(i) *the magnitude of $\vec{c}$ is equal to the measure of the area of a parallelogram with adjacent sides $\vec{a}$ and $\vec{b}$; that is,*

$$|\vec{c}| = |\vec{a}||\vec{b}||\sin(\vec{a}, \vec{b})|; \tag{2-23}$$

(ii) *the direction of $\vec{c}$ is perpendicular to the plane determined by $\vec{a}$ and $\vec{b}$ and follows the advances of a right-hand screw when $\vec{a}$ is rotated into $\vec{b}$.*

Another name frequently used for the vector product of two vectors is the **outer product**.

From Definition 2.2 it is evident that $|\vec{b} \times \vec{a}| = |\vec{a} \times \vec{b}|$ and the direction of $\vec{b} \times \vec{a}$ is opposite the direction of $\vec{a} \times \vec{b}$. Hence, the vector product of two vectors is generally not commutative.

THEOREM 2.8 / *The vector product of two vectors is **skew-commutative**; that is,*

$$\vec{a} \times \vec{b} = -\vec{b} \times \vec{a}. \tag{2-24}$$

Note that when $\vec{a}$ and $\vec{b}$ are parallel vectors the parallelogram of Definition 2.2 may be thought of as a parallelogram with area equal to zero. We sometimes call such a parallelogram a **null parallelogram**.

THEOREM 2.9 / *If $\vec{a}$ and $\vec{b}$ are parallel, then $\vec{a} \times \vec{b} = \vec{0}$.*

THEOREM 2.10 / *For any vector $\vec{a}$,*

$$\vec{a} \times \vec{a} = \vec{0}. \tag{2-25}$$

Notice that if it is given that $\vec{a} \times \vec{b} = \vec{0}$, then $\vec{a}$ and $\vec{b}$ are not necessarily parallel; either $\vec{a}$ and $\vec{b}$ are parallel, or $\vec{a} = \vec{0}$, or $\vec{b} = \vec{0}$.

THEOREM 2.11 / *If the vector product of two nonzero vectors $\vec{a}$ and $\vec{b}$ is the null vector, then the vectors are parallel.*

Since the vector product of two vectors may be the null vector when neither vector is the null vector, division of a vector product by a vector, as an inverse process to finding the vector product, cannot be defined. If we consider $\vec{a} \times \vec{b} = \vec{a} \times \vec{c}$, then $\vec{b} = \vec{c}$ may or may not be true. For example, consider $\vec{a}$ and $\vec{b}$ to be two nonzero parallel vectors and $\vec{c}$ to be the null vector. Then $\vec{a} \times \vec{b} = \vec{a} \times \vec{c}$, but $\vec{b} \neq \vec{c}$.

The next theorem will be useful in proving the distributive property of the vector product with respect to the addition of vectors.

THEOREM 2.12 / *If $\vec{b}'$ is the vector component of $\vec{b}$ on a line perpendicular to $\vec{a}$ in the plane of $\vec{a}$ and $\vec{b}$, then $\vec{a} \times \vec{b} = \vec{a} \times \vec{b}'$ (Figure 2-17).*

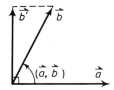

FIGURE 2-17

PROOF / Now,

$$|\vec{a} \times \vec{b}| = |\vec{a}||\vec{b}||\sin(\vec{a}, \vec{b})|$$

and

$$|\vec{a} \times \vec{b}'| = |\vec{a}||\vec{b}'|.$$

Since

$$|\vec{b}'| = |\vec{b}| \cos[90° - (\vec{a}, \vec{b})]$$
$$= |\vec{b}||\sin(\vec{a}, \vec{b})|,$$

it follows that $|\vec{a} \times \vec{b}| = |\vec{a} \times \vec{b}'|$. Furthermore, the direction of $\vec{a} \times \vec{b}$ is the same as the direction of $\vec{a} \times \vec{b}'$. Hence, $\vec{a} \times \vec{b} = \vec{a} \times \vec{b}'$.

THEOREM 2.13 / *The vector product is distributive with respect to the addition of vectors; that is,*

$$\vec{a} \times (\vec{b} + \vec{c}) = \vec{a} \times \vec{b} + \vec{a} \times \vec{c}. \tag{2-26}$$

PROOF / Let $\vec{a}$ be a vector perpendicular to the plane of the paper in the direction of the reader. Let $\vec{b}'$ and $\vec{c}'$ be the vector components of $\vec{b}$ and $\vec{c}$, respectively, in a plane perpendicular to $\vec{a}$. Then $\vec{a} \times \vec{b}'$ and $\vec{a} \times \vec{c}'$ are in the same plane (the plane of the paper) and are perpendicular to $\vec{b}'$ and $\vec{c}'$, respectively, as shown in Figure 2-18. Then

$$\frac{|\vec{a} \times \vec{c}'|}{|\vec{a} \times \vec{b}'|} = \frac{|\vec{a}||\vec{c}'|\sin 90°}{|\vec{a}||\vec{b}'|\sin 90°} = \frac{|\vec{c}'|}{|\vec{b}'|}$$

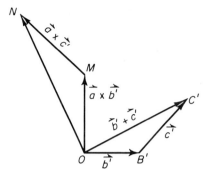

FIGURE 2-18

and, since $\angle OMN = \angle OB'C'$, $\triangle OMN \sim \triangle OB'C'$. Hence, $\overrightarrow{ON} \perp \overrightarrow{OC'}$ and

$$\frac{|\overrightarrow{ON}|}{|\vec{b}' + \vec{c}'|} = \frac{|\vec{a} \times \vec{b}'|}{|\vec{b}'|};$$

$|\overrightarrow{ON}| = |\vec{a}||\vec{b}' + \vec{c}'| \sin 90°$; that is, $\overrightarrow{ON} = \vec{a} \times (\vec{b}' + \vec{c}')$. From $\triangle OMN$,

$$\overrightarrow{ON} = \overrightarrow{OM} + \overrightarrow{MN} = \vec{a} \times \vec{b}' + \vec{a} \times \vec{c}'.$$

Therefore,

$$\vec{a} \times (\vec{b}' + \vec{c}') = \vec{a} \times \vec{b}' + \vec{a} \times \vec{c}'.$$

Using Theorem 2.12, we may replace $\vec{b}'$ and $\vec{c}'$ by $\vec{b}$ and $\vec{c}$, respectively, so that

$$\vec{a} \times (\vec{b} + \vec{c}) = \vec{a} \times \vec{b} + \vec{a} \times \vec{c}.$$

THEOREM 2.14 / *A real multiple of the vector product of two vectors is equal to the vector product of one of the vectors and the real multiple of the other;* that is,

$$m(\vec{a} \times \vec{b}) = (m\vec{a}) \times \vec{b} = \vec{a} \times (m\vec{b}). \qquad (2\text{-}27)$$

PROOF / From Definition 2.2 we may write

$$\vec{a} \times \vec{b} = |\vec{a}||\vec{b}||\sin(\vec{a}, \vec{b})|\vec{n},$$

where $\vec{n}$ is a unit vector perpendicular to the plane of $\vec{a}$ and $\vec{b}$, having the direction of the advance of a right-hand screw when the first vector $\vec{a}$ is rotated into the second vector $\vec{b}$. Now, if $m \geq 0$, then

$$m(\vec{a} \times \vec{b}) = m|\vec{a}||\vec{b}||\sin(\vec{a}, \vec{b})|\vec{n}$$
$$= |m\vec{a}||\vec{b}||\sin(m\vec{a}, \vec{b})|\vec{n} = (m\vec{a}) \times \vec{b};$$
$$= |\vec{a}||m\vec{b}||\sin(\vec{a}, m\vec{b})|\vec{n} = \vec{a} \times (m\vec{b}).$$

If $m < 0$, then

$$m(\vec{a} \times \vec{b}) = m|\vec{a}||\vec{b}||\sin(\vec{a}, \vec{b})|\vec{n}$$
$$= |m\vec{a}||\vec{b}||\sin(m\vec{a}, \vec{b})|(-\vec{n}) = (m\vec{a}) \times \vec{b};$$
$$= |\vec{a}||m\vec{b}||\sin(\vec{a}, m\vec{b})|(-\vec{n}) = \vec{a} \times (m\vec{b}).$$

Consider the unit vectors $\vec{i}$, $\vec{j}$, and $\vec{k}$ for the rectangular Cartesian coordinate system. By Definition 2.2,

$$\vec{i} \times \vec{j} = -\vec{j} \times \vec{i} = \vec{k};$$
$$\vec{j} \times \vec{k} = -\vec{k} \times \vec{j} = \vec{i}; \quad (2\text{-}28)$$
$$\vec{k} \times \vec{i} = -\vec{i} \times \vec{k} = \vec{j}.$$

By Theorem 2.10,

$$\vec{i} \times \vec{i} = \vec{0};$$
$$\vec{j} \times \vec{j} = \vec{0}; \quad (2\text{-}29)$$
$$\vec{k} \times \vec{k} = \vec{0};$$

The components of the vector product of two position vectors may now be determined as shown in Theorem 2.15.

THEOREM 2.15 / If $\vec{a} = x_1\vec{i} + y_1\vec{j} + z_1\vec{k}$ and $\vec{b} = x_2\vec{i} + y_2\vec{j} + z_2\vec{k}$, then

$$\vec{a} \times \vec{b} = (y_1z_2 - y_2z_1)\vec{i} + (z_1x_2 - z_2x_1)\vec{j} + (x_1y_2 - x_2y_1)\vec{k}. \quad (2\text{-}30)$$

PROOF /

$$\vec{a} \times \vec{b} = (x_1\vec{i} + y_1\vec{j} + z_1\vec{k}) \times (x_2\vec{i} + y_2\vec{j} + z_2\vec{k})$$
$$= x_1x_2(\vec{i} \times \vec{i}) + x_1y_2(\vec{i} \times \vec{j}) + x_1z_2(\vec{i} \times \vec{k})$$
$$+ y_1x_2(\vec{j} \times \vec{i}) + y_1y_2(\vec{j} \times \vec{j}) + y_1z_2(\vec{j} \times \vec{k})$$
$$+ z_1x_2(\vec{k} \times \vec{i}) + z_1y_2(\vec{k} \times \vec{j}) + z_1z_2(\vec{k} \times \vec{k}).$$

By the formulas (2-28) and (2-29),

$$\vec{a} \times \vec{b} = \vec{0} + x_1 y_2 \vec{k} - x_1 z_2 \vec{j} - y_1 x_2 \vec{k} + \vec{0} + y_1 z_2 \vec{i} + z_1 x_2 \vec{j} - z_1 y_2 \vec{i} + \vec{0};$$

$$\vec{a} \times \vec{b} = (y_1 z_2 - y_2 z_1)\vec{i} + (z_1 x_2 - z_2 x_1)\vec{j} + (x_1 y_2 - x_2 y_1)\vec{k}.$$

It should be noted that the vector product is a meaningful operation for vectors in three-dimensional space. The vector product of two vectors is defined only for vectors in three-dimensional space.

EXAMPLE 1 / Find $\vec{a} \times \vec{b}$ where $\vec{a} = 3\vec{i} + 2\vec{j} - \vec{k}$ and $\vec{b} = \vec{i} + 4\vec{j} + \vec{k}$.

$$\vec{a} \times \vec{b} = [(2)(1) - (4)(-1)]\vec{i} + [(-1)(1) - (1)(3)]\vec{j} + [(3)(4) - (1)(2)]\vec{k}$$
$$= 6\vec{i} - 4\vec{j} + 10\vec{k}.$$

EXAMPLE 2 / Find a vector perpendicular to line AB for A: $(0, -1, 3)$ and B: $(2, 0, 4)$, and also perpendicular to line CD for C: $(2, -1, 4)$ and D: $(3, 3, 2)$.

The vectors $\overrightarrow{AB}$ and $\overrightarrow{CD}$ are on lines AB and CD, respectively:

$$\overrightarrow{AB} = 2\vec{i} + \vec{j} + \vec{k};$$
$$\overrightarrow{CD} = \vec{i} + 4\vec{j} - 2\vec{k}.$$

Then
$$\overrightarrow{AB} \times \overrightarrow{CD} = -6\vec{i} + 5\vec{j} + 7\vec{k}.$$

Since $\overrightarrow{AB} \times \overrightarrow{CD}$ is perpendicular to $\overrightarrow{AB}$ and $\overrightarrow{CD}$, any real nonzero multiple of $\overrightarrow{AB} \times \overrightarrow{CD}$ is perpendicular to lines AB and CD.

EXAMPLE 3 / Prove that $(\vec{a} - \vec{b}) \times (\vec{a} + \vec{b}) = (2\vec{a}) \times \vec{b}$.

$$(\vec{a} - \vec{b}) \times (\vec{a} + \vec{b}) = (\vec{a} - \vec{b}) \times \vec{a} + (\vec{a} - \vec{b}) \times \vec{b}$$
$$= \vec{a} \times (\vec{b} - \vec{a}) + \vec{b} \times (\vec{b} - \vec{a})$$
$$= \vec{a} \times \vec{b} - \vec{a} \times \vec{a} + \vec{b} \times \vec{b} - \vec{b} \times \vec{a}$$
$$= \vec{a} \times \vec{b} - \vec{b} \times \vec{a}$$
$$= \vec{a} \times \vec{b} + \vec{a} \times \vec{b}$$
$$= 2(\vec{a} \times \vec{b}) = (2\vec{a}) \times \vec{b}.$$

EXAMPLE 4 / Find the area of the parallelogram with adjacent sides associated with $\vec{a} = \vec{i} - \vec{j} + \vec{k}$ and $\vec{b} = 2\vec{j} - 3\vec{k}$.

The magnitude of $\vec{a} \times \vec{b}$ represents the area of the parallelogram with adjacent sides $\vec{a}$ and $\vec{b}$. Since

$$\vec{a} \times \vec{b} = \vec{i} + 3\vec{j} + 2\vec{k}$$

and $|\vec{a} \times \vec{b}| = |\vec{i} + 3\vec{j} + 2\vec{k}| = \sqrt{14}$, the area of the parallelogram is $\sqrt{14}$ square units.

EXAMPLE 5 / Prove that $|\vec{a} \times \vec{b}|^2 + |\vec{a} \cdot \vec{b}|^2 = |\vec{a}|^2 |\vec{b}|^2$.

$$|\vec{a} \times \vec{b}|^2 + |\vec{a} \cdot \vec{b}|^2 = |\vec{a}|^2 |\vec{b}|^2 \sin^2(\vec{a}, \vec{b}) + |\vec{a}|^2 |\vec{b}|^2 \cos^2(\vec{a}, \vec{b})$$
$$= |\vec{a}|^2 |\vec{b}|^2 [\sin^2(\vec{a}, \vec{b}) + \cos^2(\vec{a}, \vec{b})]$$
$$= |\vec{a}|^2 |\vec{b}|^2.$$

EXERCISES

1. Find $\vec{a} \times \vec{b}$ where $\vec{a} = 2\vec{i} + 3\vec{j} - \vec{k}$ and $\vec{b} = \vec{i} - \vec{j} + \vec{k}$.
2. Find $\vec{a} \times \vec{b}$ where $\vec{a} = \vec{i} + \vec{k}$ and $\vec{b} = 2\vec{i} - 5\vec{j}$.
3. Verify Theorem 2.12 for $\vec{a} = 3\vec{i}$ and $\vec{b} = \vec{i} + 2\vec{j}$.
4. Verify that the vector product is distributive with respect to the addition of vectors, using $\vec{a} = 3\vec{i} + \vec{j} - \vec{k}$, $\vec{b} = \vec{i} + 2\vec{j} + \vec{k}$, and $\vec{c} = \vec{i} - \vec{j} + 2\vec{k}$.
5. Find a vector perpendicular to lines AB and CD for $A: (0, 2, 4)$, $B: (3, -1, 2)$, $C: (2, 0, 1)$, and $D: (4, 2, 0)$.
6. Determine a unit vector perpendicular to $\vec{a} = \vec{i} + \vec{j}$ and $\vec{b} = 3\vec{i} + 2\vec{j} + \vec{k}$.
7. Prove that $(\vec{a} + \vec{b}) \times (\vec{c} + \vec{d}) = (\vec{a} \times \vec{c}) + (\vec{a} \times \vec{d}) + (\vec{b} \times \vec{c}) + (\vec{b} \times \vec{d})$.
8. Find the area of a triangle with adjacent sides $\vec{a} = 3\vec{i} + 2\vec{j}$ and $\vec{b} = 2\vec{j} - 4\vec{k}$.
9. Prove that a necessary and sufficient condition that two vectors be linearly dependent is that their vector product be the null vector.
10. If $\vec{a} + \vec{b} + \vec{c} = \vec{0}$, prove that $\vec{a} \times \vec{b} = \vec{b} \times \vec{c} = \vec{c} \times \vec{a}$.

2-6 / APPLICATIONS OF THE VECTOR PRODUCT

Many theorems in geometry and trigonometry have simple vector proofs based on the concept of the vector product. In this section we shall illustrate the proofs of several such theorems as well as the usefulness of the concept in physics. Note that a number of the proofs do not involve elements of coordinate geometry.

EXAMPLE 1 / Derive the formula $K = \frac{1}{2}bc \sin A$ for the area of a triangle.

Let A, B, and C be any three vertices of a triangle in space (Figure 2-19). Then the area K of $\triangle ABC$ is equal to one-half the area of a parallelogram with $\overrightarrow{AB}$ and $\overrightarrow{AC}$ forming adjacent edges. Since the area of the parallelogram is the magnitude of $\overrightarrow{AB} \times \overrightarrow{AC}$, it follows that

$$K = \tfrac{1}{2}|\overrightarrow{AB} \times \overrightarrow{AC}|. \tag{2-31}$$

By Definition 2.2, equation (2-31) may be written as

$$K = \tfrac{1}{2}|\overrightarrow{AB}||\overrightarrow{AC}| \sin (\overrightarrow{AB}, \overrightarrow{AC})$$
$$= \tfrac{1}{2}bc \sin A,$$

where b and c denote sides AC and AB, respectively.

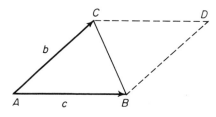

FIGURE 2-19

EXAMPLE 2 / Find the area of a triangle formed by taking A: $(0, -2, 1)$, B: $(1, -1, -2)$, and C: $(-1, 1, 0)$ as vertices.

Now, $\overrightarrow{AB} = \vec{i} + \vec{j} - 3\vec{k}$; $\overrightarrow{AC} = -\vec{i} + 3\vec{j} - \vec{k}$. Then, as in Example 1,

$$\overrightarrow{AB} \times \overrightarrow{AC} = 8\vec{i} + 4\vec{j} + 4\vec{k}$$

and

$$\tfrac{1}{2}|\overrightarrow{AB} \times \overrightarrow{AC}| = \tfrac{1}{2}\sqrt{8^2 + 4^2 + 4^2} = 2\sqrt{6}.$$

Therefore, the area of $\triangle ABC$ is $2\sqrt{6}$ square units.

EXAMPLE 3 / Derive the law of sines.

Let ABC be any triangle. Let $\vec{a}$, $\vec{b}$, and $\vec{c}$ be associated with sides BC, AC, and AB, respectively (Figure 2-20). Now, $\vec{a} = \vec{b} - \vec{c}$. Then

$$\vec{a} \times \vec{a} = \vec{a} \times (\vec{b} - \vec{c}) = \vec{a} \times \vec{b} - \vec{a} \times \vec{c}.$$

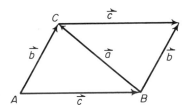

FIGURE 2-20

Since $\vec{a} \times \vec{a} = \vec{0}$, then $\vec{a} \times \vec{b} - \vec{a} \times \vec{c} = \vec{0}$ and $\vec{a} \times \vec{b} = \vec{a} \times \vec{c}$. Therefore, $|\vec{a} \times \vec{b}| = |\vec{a} \times \vec{c}|$ and $|\vec{a}||\vec{b}| \sin C = |\vec{a}||\vec{c}| \sin B$. Then, since $|\vec{a}| \neq 0$,

$$\frac{\sin C}{|\vec{c}|} = \frac{\sin B}{|\vec{b}|}.$$

In a similar manner,

$$\vec{b} = \vec{c} + \vec{a},$$
$$\vec{b} \times \vec{b} = \vec{b} \times (\vec{c} + \vec{a}) = \vec{b} \times \vec{c} + \vec{b} \times \vec{a},$$
$$\vec{0} = \vec{b} \times \vec{c} + \vec{b} \times \vec{a},$$
$$\vec{c} \times \vec{b} = \vec{b} \times \vec{a},$$
$$|\vec{c} \times \vec{b}| = |\vec{b} \times \vec{a}|,$$
$$|\vec{c}||\vec{b}| \sin A = |\vec{b}||\vec{a}| \sin C,$$
$$\frac{\sin A}{|\vec{a}|} = \frac{\sin C}{|\vec{c}|}.$$

Hence,
$$\frac{\sin A}{|\vec{a}|} = \frac{\sin B}{|\vec{b}|} = \frac{\sin C}{|\vec{c}|},$$
which is the law of sines.

EXAMPLE 4 / Prove that $\sin(\theta - \phi) = \sin\theta\cos\phi - \cos\theta\sin\phi$.

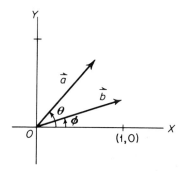

FIGURE 2-21

Let $\vec{a}$ and $\vec{b}$ be unit position vectors on the rectangular Cartesian coordinate plane making positive angles θ and ϕ with the positive half of the x-axis (Figure 2-21). Consider $\theta \geq \phi$. Then $\vec{a} = \cos\theta\vec{i} + \sin\theta\vec{j}$ and $\vec{b} = \cos\phi\vec{i} + \sin\phi\vec{j}$. By Theorem 2.15,
$$\vec{b} \times \vec{a} = (\sin\theta\cos\phi - \cos\theta\sin\phi)\vec{k}.$$
However, by Definition 2.2 with $|\vec{a}| = |\vec{b}| = 1$ and $\theta \geq \phi$,
$$\vec{b} \times \vec{a} = |\vec{b}||\vec{a}|\sin(\theta - \phi)\vec{k} = \sin(\theta - \phi)\vec{k}.$$
Hence, $\sin(\theta - \phi) = \sin\theta\cos\phi - \cos\theta\sin\phi$.

EXAMPLE 5 / Derive Hero's formula for the area of a triangle.

Let ABC be any triangle. Let $\vec{a}$, $\vec{b}$, and $\vec{c}$ be associated with sides opposite angles A, B, and C, respectively, as shown in Figure 2-22. The area K of triangle ABC may be expressed by
$$K = \tfrac{1}{2}|\vec{b} \times \vec{c}|.$$

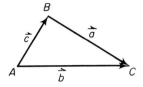

FIGURE 2-22

Then
$$2K = |\vec{b} \times \vec{c}|,$$
$$4K^2 = |\vec{b} \times \vec{c}|^2.$$

By the results of Example 5 in §2-5,

$$\begin{aligned}4K^2 &= |\vec{b}|^2|\vec{c}|^2 - (\vec{b}\cdot\vec{c})^2 \\ &= [|\vec{b}||\vec{c}| - (\vec{b}\cdot\vec{c})][|\vec{b}||\vec{c}| + (\vec{b}\cdot\vec{c})] \\ &= [|\vec{b}||\vec{c}| - |\vec{b}||\vec{c}|\cos A][|\vec{b}||\vec{c}| + |\vec{b}||\vec{c}|\cos A].\end{aligned}$$

By the law of cosines,

$$4K^2 = \left[|\vec{b}||\vec{c}| - \frac{|\vec{b}|^2 + |\vec{c}|^2 - |\vec{a}|^2}{2}\right]\left[|\vec{b}||\vec{c}| + \frac{|\vec{b}|^2 + |\vec{c}|^2 - |\vec{a}|^2}{2}\right],$$
$$16K^2 = [|\vec{a}|^2 - (|\vec{b}| - |\vec{c}|)^2][(|\vec{b}| + |\vec{c}|)^2 - |\vec{a}|^2],$$
$$16K^2 = (|\vec{a}| - |\vec{b}| + |\vec{c}|)(|\vec{a}| + |\vec{b}| - |\vec{c}|)(|\vec{b}| + |\vec{c}| - |\vec{a}|)(|\vec{b}| + |\vec{c}| + |\vec{a}|).$$

If the semiperimeter of $\triangle ABC$ is denoted by S, then

$$S = \frac{|\vec{a}| + |\vec{b}| + |\vec{c}|}{2};$$
$$K^2 = (S - |\vec{b}|)(S - |\vec{c}|)(S - |\vec{a}|)S,$$
$$K = \sqrt{(S - |\vec{b}|)(S - |\vec{c}|)(S - |\vec{a}|)S}.$$

Applications of the vector product in physics are varied and numerous. Illustrations of applications to mechanics and light are presented in the next set of examples.

EXAMPLE 6 / Find a vector expression for the velocity of a rigid body rotating with a constant angular velocity about a fixed axis (Figure 2-23).

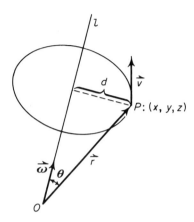

FIGURE 2-23

Consider a rigid body rotating with a constant angular velocity of ω radians per second about a fixed axis l. Let $\vec{\omega}$ be a vector with magnitude ω and direction that of the axis l. Let the origin be chosen on the axis of rotation, and let $\vec{r}$ be a position vector of the point $P: (x, y, z)$ on the path of the rotating body. Now, the velocity $\vec{v}$ of the body at P is perpendicular to the plane of $\vec{\omega}$ and $\vec{r}$. Let d be the distance between P and l. Then

$$|\vec{v}| = d\omega = |\vec{r}| \sin \theta \, |\vec{\omega}|;$$

that is,

$$\vec{v} = \vec{\omega} \times \vec{r}.$$

EXAMPLE 7 / State Snell's law for the refraction of light in vector form (Figure 2-24).

Consider a ray of light passing from one medium, where the velocity of light is v_1, to a second medium, where the velocity is v_2. Snell's law for refracted light states that the relationship between the angle of incidence θ and the angle of refraction ϕ is expressed by the equation

$$v_1 \sin \phi = v_2 \sin \theta.$$

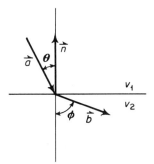

FIGURE 2-24

Let $\vec{n}$, $\vec{a}$, and $\vec{b}$ be unit vectors where $\vec{n}$ is perpendicular to the surface of separation between the media, and $\vec{a}$ and $\vec{b}$ are in the direction of the incident and refracted rays, respectively. Then Snell's law may be stated in vector form as

$$v_2 \vec{a} \times \vec{n} = v_1 \vec{b} \times \vec{n}.$$

Other applications will be considered in subsequent sections of this chapter. In Chapter 3 we shall consider the applications of the vector product to the study of the coordinate geometry of planes and lines in three-dimensional space.

EXERCISES

1. Find the area of a triangle whose vertices are $A: (1, 1, -1)$, $B: (2, 1, 0)$, and $C: (0, 1, 0)$.

2. Derive the law of sines by expressing the area of a triangle in vector product form.

3. In the xy-plane, let $\vec{a}$ and $\vec{b}$ be unit vectors making angles θ and $-\phi$ with the positive half of the x-axis. Derive the formula $\sin(\theta + \phi) = \sin\theta\cos\phi + \cos\theta\sin\phi$.

4. A rigid body is rotating with an angular velocity of 3 radians per second about an axis parallel to $2\vec{i} + \vec{j} - 2\vec{k}$ and passing through the point $O: (1, 2, -4)$. Determine the velocity of the rigid body at $P: (2, 1, 1)$.

5. State the law of reflection of light in vector form. Let $\vec{n}$, $\vec{a}$, and $\vec{b}$ be unit vectors where $\vec{n}$ is perpendicular to the surface of reflection, and $\vec{a}$ and $\vec{b}$ are in the direction of the incident and reflected rays, respectively.

6. Give a physical proof of the distributive property of the vector product by using the hydrostatic principle that a closed polyhedral surface submerged in a fluid is in equilibrium with respect to the pressures upon its faces. *Hint:* Use a triangular prism and remember that the pressures perpendicular to the faces are proportional to the areas of the faces.

7. Prove that the area of triangle *HJG* in the given figure is one-seventh the area of triangle *ABC*. The points *D*, *E*, and *F* divide the line segments *BC*, *CA*, and *AB*, respectively, in the ratio 1 to 2. The points *G*, *H*, and *J* divide the line segments *BE*, *CF*, and *AD*, respectively, in the ratio 3 to 4. The points *H*, *J*, and *G* divide the line segments *BE*, *CF*, and *AD*, respectively, in the ratio 6 to 1.

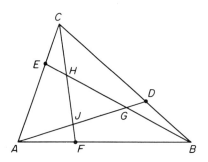

2-7 / THE SCALAR TRIPLE PRODUCT

Let $\vec{a}$, $\vec{b}$, and $\vec{c}$ be any three vectors. The expression

$$\vec{a} \cdot (\vec{b} \times \vec{c}) \tag{2-32}$$

is called the **scalar triple product** of $\vec{a}$, $\vec{b}$, and $\vec{c}$. Note that the scalar triple product is a scalar. For example, $\vec{i} \cdot (\vec{j} \times \vec{k}) = \vec{i} \cdot \vec{i} = 1$ and $\vec{j} \cdot (\vec{j} \times \vec{i}) = \vec{j} \cdot (-\vec{k}) = 0$.

If $\vec{a}$, $\vec{b}$, and $\vec{c}$ are three noncoplanar vectors, then they may be associated with the sides of a parallelepiped as shown in Figure 2-25. There exists a vector perpendicular to the plane of the parallelogram determined by $\vec{b}$ and $\vec{c}$ with magnitude $|\vec{b} \times \vec{c}|$, which is equal to B, the area of the base of the parallelepiped. The magnitude of the projection of $\vec{a}$ on $\vec{b} \times \vec{c}$ is

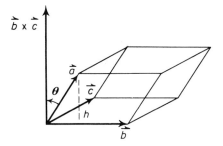

FIGURE 2-25

equal to the altitude h of the parallelepiped; that is, $h = |\vec{a}||\cos\theta|$ where θ is the angle between $\vec{a}$ and $\vec{b} \times \vec{c}$. Note that we consider $|\cos\theta|$ instead of simply $\cos\theta$ since the orientation of $\vec{b}$ and $\vec{c}$ relative to one another may be such that $\cos\theta$ would be negative. Hence, the volume V of the parallelepiped is given by

$$V = hB = |\vec{a}||\vec{b} \times \vec{c}||\cos\theta|,$$
$$V = |\vec{a}\cdot(\vec{b} \times \vec{c})|; \qquad (2\text{-}33)$$

that is, *the absolute value of the scalar triple product of any three vectors, $\vec{a}$, $\vec{b}$, and $\vec{c}$, is equal to the measure of the volume of a parallelepiped with sides $\vec{a}$, $\vec{b}$, and $\vec{c}$ having the same initial point.*

If $\vec{a}$, $\vec{b}$, and $\vec{c}$ are position vectors, where $\vec{a} = x_1\vec{i} + y_1\vec{j} + z_1\vec{k}$, $\vec{b} = x_2\vec{i} + y_2\vec{j} + z_2\vec{k}$, and $\vec{c} = x_3\vec{i} + y_3\vec{j} + z_3\vec{k}$, then, since

$$\vec{b} \times \vec{c} = (y_2z_3 - y_3z_2)\vec{i} + (z_2x_3 - z_3x_2)\vec{j} + (x_2y_3 - x_3y_2)\vec{k},$$

it follows that

$$\vec{a}\cdot(\vec{b} \times \vec{c}) = x_1(y_2z_3 - y_3z_2)$$
$$+ y_1(z_2x_3 - z_3x_2) + z_1(x_2y_3 - x_3y_2). \qquad (2\text{-}34)$$

It can be shown that the scalar triple product is not changed by a cyclic permutation of the vectors; that is,

$$\vec{a}\cdot(\vec{b} \times \vec{c}) = \vec{b}\cdot(\vec{c} \times \vec{a}) = \vec{c}\cdot(\vec{a} \times \vec{b}). \qquad (2\text{-}35)$$

Since the scalar product of two vectors is commutative,

$$\vec{c}\cdot(\vec{a} \times \vec{b}) = (\vec{a} \times \vec{b})\cdot\vec{c}.$$

Using equation (2-35), we obtain

$$\vec{a}\cdot(\vec{b} \times \vec{c}) = (\vec{a} \times \vec{b})\cdot\vec{c}. \qquad (2\text{-}36)$$

2-7 / THE SCALAR TRIPLE PRODUCT

Therefore, in the scalar triple product the operations of scalar and vector products may be interchanged without changing the value of the result. For this reason $\vec{a} \cdot (\vec{b} \times \vec{c})$ is sometimes written without the parentheses and sometimes without any multiplication signs, since only one interpretation is possible; that is,

$$\vec{a} \cdot (\vec{b} \times \vec{c}) = \vec{a} \cdot \vec{b} \times \vec{c} = (\vec{a}\vec{b}\vec{c}). \qquad (2\text{-}37)$$

It should be further noted that

$$(\vec{a}\vec{b}\vec{c}) = -(\vec{b}\vec{a}\vec{c}) = -(\vec{a}\vec{c}\vec{b}) = -(\vec{c}\vec{b}\vec{a}). \qquad (2\text{-}38)$$

The verification of these results is left to the reader as an exercise.

If the three vectors $\vec{a}$, $\vec{b}$, and $\vec{c}$ are coplanar, then the parallelepiped is a **degenerate parallelepiped** or **null parallelepiped**; that is, a parallelepiped whose volume is zero. Hence $(\vec{a}\vec{b}\vec{c}) = 0$. In particular, if two of the three vectors are parallel, the scalar triple product vanishes. The proof is left to the reader as an exercise.

EXAMPLE 1 / Find the volume of the parallelepiped whose edges are $\vec{a}$, $\vec{b}$, and $\vec{c}$ where $\vec{a} = 3\vec{i} + 2\vec{k}$, $\vec{b} = \vec{i} + 2\vec{j} + \vec{k}$, and $\vec{c} = -\vec{j} + 4\vec{k}$.

The volume of the parallelepiped is given by $|\vec{a} \cdot (\vec{b} \times \vec{c})|$:

$$\begin{aligned}
\vec{a} \cdot (\vec{b} \times \vec{c}) &= (3\vec{i} + 2\vec{k}) \cdot [(\vec{i} + 2\vec{j} + \vec{k}) \times (-\vec{j} + 4\vec{k})] \\
&= (3\vec{i} + 2\vec{k}) \cdot (9\vec{i} - 4\vec{j} - \vec{k}) \\
&= (3)(9) + (0)(-4) + (2)(-1) = 25.
\end{aligned}$$

The volume of the parallelepiped is 25 cubic units.

EXAMPLE 2 / Find a formula for the volume of a tetrahedron in terms of the vertices.

Let A, B, C, and D be the four vertices of a tetrahedron. The volume of a parallelepiped whose edges are $\overrightarrow{AB}$, $\overrightarrow{AC}$, and $\overrightarrow{AD}$ is given by the expression $|\overrightarrow{AB} \cdot (\overrightarrow{AC} \times \overrightarrow{AD})|$. The volume of a tetrahedron is one-third the product of the area of its base and its altitude. Since the area of the base of the tetrahedron is one-half the area of the base of the parallelepiped, it follows that the volume V of the tetrahedron is one-sixth the volume of the parallelepiped. Hence,

$$V = \tfrac{1}{6} |\overrightarrow{AB} \cdot (\overrightarrow{AC} \times \overrightarrow{AD})|. \qquad (2\text{-}39)$$

EXAMPLE 3 / Prove the distributive law for the vector product by letting $\vec{n} = \vec{a} \times (\vec{b} + \vec{c}) - \vec{a} \times \vec{b} - \vec{a} \times \vec{c}$, and finding $\vec{m} \cdot \vec{n}$ where $\vec{m}$ is an arbitrary nonzero vector.

Now,

$$\begin{aligned}
\vec{m} \cdot \vec{n} &= \vec{m} \cdot [\vec{a} \times (\vec{b} + \vec{c}) - \vec{a} \times \vec{b} - \vec{a} \times \vec{c}] \\
&= \vec{m} \cdot [\vec{a} \times (\vec{b} + \vec{c})] - \vec{m} \cdot [\vec{a} \times \vec{b}] - \vec{m} \cdot [\vec{a} \times \vec{c}] \\
&= [(\vec{m} \times \vec{a}) \cdot (\vec{b} + \vec{c})] - [(\vec{m} \times \vec{a}) \cdot \vec{b}] - [(\vec{m} \times \vec{a}) \cdot \vec{c}] \\
&= [(\vec{m} \times \vec{a}) \cdot \vec{b}] + [(\vec{m} \times \vec{a}) \cdot \vec{c}] - [(\vec{m} \times \vec{a}) \cdot \vec{b}] - [(\vec{m} \times \vec{a}) \cdot \vec{c}] \\
&= 0.
\end{aligned}$$

Since $\vec{m}$ is an arbitrary nonzero vector, $\vec{m}$ need not be perpendicular to $\vec{n}$. Hence, $\vec{n}$ must be the null vector, and $\vec{a} \times (\vec{b} + \vec{c}) = \vec{a} \times \vec{b} + \vec{a} \times \vec{c}$.

EXAMPLE 4 / Determine a condition under which four distinct points are coplanar (Figure 2-26).

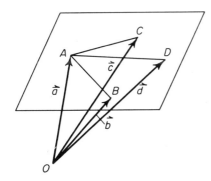

FIGURE 2-26

Let $\vec{a}$, $\vec{b}$, $\vec{c}$, and $\vec{d}$ be position vectors associated with four distinct coplanar points A, B, C, and D, respectively. Then $\overrightarrow{AC} \times \overrightarrow{AD}$ is a vector perpendicular to the plane of A, B, C, and D, and therefore is perpendicular to $\overrightarrow{AB}$. Hence, $\overrightarrow{AB} \cdot (\overrightarrow{AC} \times \overrightarrow{AD}) = 0$ and

$$(\vec{b} - \vec{a}) \cdot (\vec{c} - \vec{a}) \times (\vec{d} - \vec{a}) = 0 \qquad (2\text{-}40)$$

is a necessary condition for the four points A, B, C, and D to be coplanar. It can also be shown to be a sufficient condition for four distinct points to be coplanar.

The condition that four distinct points be coplanar may also be expressed by Theorem 1.5 in the alternate form

$$\overrightarrow{AB} = x\overrightarrow{AC} + y\overrightarrow{AD}.$$

Then

$$\vec{b} - \vec{a} = x(\vec{c} - \vec{a}) + y(\vec{d} - \vec{a}),$$
$$(x + y - 1)\vec{a} + \vec{b} - x\vec{c} - y\vec{d} = \vec{0};$$

that is, scalars $k_1, k_2, k_3,$ and k_4, not all zero, must exist such that

$$k_1\vec{a} + k_2\vec{b} + k_3\vec{c} + k_4\vec{d} = \vec{0}, \qquad (2\text{-}41)$$

where

$$k_1 + k_2 + k_3 + k_4 = 0.$$

EXERCISES

1. Find (a) $\vec{j} \cdot \vec{k} \times \vec{i}$; (b) $\vec{k} \cdot \vec{j} \times \vec{i}$; (c) $\vec{i} \cdot \vec{j} \times \vec{i}$.
2. Verify that $(\vec{a}\vec{b}\vec{c}) = -(\vec{b}\vec{a}\vec{c}) = -(\vec{a}\vec{c}\vec{b}) = -(\vec{c}\vec{b}\vec{a})$.
3. Find the volume of the parallelepiped whose edges are $\vec{a} = \vec{i} + \vec{k}$, $\vec{b} = \vec{i} + \vec{j}$, and $\vec{c} = \vec{j} + \vec{k}$.
4. Find the volume of the tetrahedron whose vertices are A: $(1, 0, 2)$, B: $(4, 3, 0)$, C: $(2, 0, 1)$, and D: $(3, 4, 0)$.
5. Find the volume of the tetrahedron whose faces lie in the coordinate planes and the plane $2x + y + 3z - 6 = 0$.
6. Describe geometrically when $\vec{a} \cdot (\vec{b} \times \vec{c})$ is (a) positive; (b) negative.
7. Verify (2-36) for $\vec{a} = 2\vec{i} + 3\vec{j} - 4\vec{k}$, $\vec{b} = \vec{i} - \vec{j} + \vec{k}$, and $\vec{c} = \vec{i} + \vec{j} + 2\vec{k}$.
8. Prove that (2-40) is a sufficient condition for the terminal points of $\vec{a}, \vec{b}, \vec{c},$ and $\vec{d}$ to be coplanar.
9. Determine if the four points A: $(1, 2, 3)$, B: $(-1, 0, 2)$, C: $(4, 1, 2)$, and D: $(4, 3, 5)$ are coplanar.
10. Determine if the position vectors $\vec{a} = \vec{i} + \vec{j} + 2\vec{k}$, $\vec{b} = 2\vec{i} - 3\vec{j} + \vec{k}$, and $\vec{c} = \vec{i} - 4\vec{j} - \vec{k}$ are coplanar.

2-8 / THE VECTOR TRIPLE PRODUCT

In this section we shall be concerned with another triple product involving vectors, which finds wide application in geometry, trigonometry, and physics. The expression

$$\vec{a} \times (\vec{b} \times \vec{c}) \tag{2-42}$$

is called a **vector triple product** of $\vec{a}$, $\vec{b}$, and $\vec{c}$. Note that the result is a vector. For example, $\vec{i} \times (\vec{j} \times \vec{k}) = \vec{i} \times \vec{i} = \vec{0}$ and $\vec{j} \times (\vec{j} \times \vec{i}) = \vec{j} \times (-\vec{k}) = -\vec{i}$. Furthermore, note that the vector product is not associative; that is,

$$\vec{a} \times (\vec{b} \times \vec{c}) \neq (\vec{a} \times \vec{b}) \times \vec{c}.$$

For example, $(\vec{i} \times \vec{i}) \times \vec{j} = \vec{0}$, whereas $\vec{i} \times (\vec{i} \times \vec{j}) = -\vec{j}$.

Since $\vec{b} \times \vec{c}$ is perpendicular to the plane determined by $\vec{b}$ and $\vec{c}$, and $\vec{a} \times (\vec{b} \times \vec{c})$ is perpendicular to $\vec{a}$ and $\vec{b} \times \vec{c}$, then $\vec{a} \times (\vec{b} \times \vec{c})$ must be a vector on the plane determined by $\vec{b}$ and $\vec{c}$. Hence, $\vec{a} \times (\vec{b} \times \vec{c})$ is a linear function of $\vec{b}$ and $\vec{c}$.

THEOREM 2.16 / $\vec{a} \times (\vec{b} \times \vec{c}) = (\vec{a} \cdot \vec{c})\vec{b} - (\vec{a} \cdot \vec{b})\vec{c}.$ (2-43)

PROOF / Let $\vec{a} \times (\vec{b} \times \vec{c}) = x\vec{b} + y\vec{c}$. Since $\vec{a} \times (\vec{b} \times \vec{c})$ is perpendicular to $\vec{a}$, then

$$\vec{a} \cdot [\vec{a} \times (\vec{b} \times \vec{c})] = 0$$

and

$$x(\vec{a} \cdot \vec{b}) + y(\vec{a} \cdot \vec{c}) = 0.$$

Now, x and y must necessarily be of the form of multiples of $\vec{a} \cdot \vec{c}$ and $\vec{a} \cdot \vec{b}$, respectively; that is, $x = m(\vec{a} \cdot \vec{c})$ and $y = -m(\vec{a} \cdot \vec{b})$, where m is an arbitrary scalar. Therefore,

$$\vec{a} \times (\vec{b} \times \vec{c}) = m[(\vec{a} \cdot \vec{c})\vec{b} - (\vec{a} \cdot \vec{b})\vec{c}]. \tag{2-44}$$

Since corresponding components of this vector equation must be equal, we need only look at the coefficients of $\vec{i}$, $\vec{j}$, or $\vec{k}$ to determine m. Let $\vec{a} = x_1\vec{i} + y_1\vec{j} + z_1\vec{k}$, $\vec{b} = x_2\vec{i} + y_2\vec{j} + z_2\vec{k}$, and $\vec{c} = x_3\vec{i} + y_3\vec{j} + z_3\vec{k}$. Then, equating the coefficients of $\vec{i}$ after substituting for $\vec{a}$, $\vec{b}$, and $\vec{c}$ in (2-44), we have

$$y_1(x_2y_3 - x_3y_2) - z_1(z_2x_3 - z_3x_2) = mx_2(x_1x_3 + y_1y_3 + z_1z_3)$$
$$-mx_3(x_1x_2 + y_1y_2 + z_1z_2)$$
$$= m[y_1(x_2y_3 - x_3y_2) - z_1(z_2x_3 - z_3x_2)].$$

Hence, $m = 1$, and $\vec{a} \times (\vec{b} \times \vec{c}) = (\vec{a} \cdot \vec{c})\vec{b} - (\vec{a} \cdot \vec{b})\vec{c}$.

In a similar manner we may show that

$$(\vec{a} \times \vec{b}) \times \vec{c} = (\vec{a} \cdot \vec{c})\vec{b} - (\vec{b} \cdot \vec{c})\vec{a}. \qquad (2\text{-}45)$$

EXERCISES

1. Find (a) $\vec{i} \times (\vec{j} \times \vec{k})$; (b) $\vec{i} \times (\vec{j} \times \vec{i})$; (c) $(\vec{i} \times \vec{j}) \times \vec{i}$.
2. Verify (2-43) for $\vec{a} = 3\vec{i} - 2\vec{j} + \vec{k}$, $\vec{b} = 2\vec{i} - 2\vec{k}$, and $\vec{c} = \vec{i} + 3\vec{j}$.
3. Prove the identity $\vec{a} \times (\vec{b} \times \vec{c}) + \vec{b} \times (\vec{c} \times \vec{a}) + \vec{c} \times (\vec{a} \times \vec{b}) = \vec{0}$.
4. Prove that if $\vec{a}, \vec{b}, \vec{c}$, and $\vec{d}$ are coplanar, then $(\vec{a} \times \vec{b}) \times (\vec{c} \times \vec{d}) = \vec{0}$.

2-9 / QUADRUPLE PRODUCTS

The concepts of the scalar triple product and the vector triple product make it possible to evaluate combinations of multiple vector and scalar products. As an illustration of the process, we shall in this section present an example of a scalar product of four vectors and an example of a vector product of four vectors. These illustrations will then be applied to the subject of spherical trigonometry.

Consider the scalar product $(\vec{a} \times \vec{b}) \cdot (\vec{c} \times \vec{d})$ of the four vectors $\vec{a}, \vec{b}, \vec{c}$, and $\vec{d}$. Now,

$$(\vec{a} \times \vec{b}) \cdot (\vec{c} \times \vec{d}) = \vec{a} \cdot [\vec{b} \times (\vec{c} \times \vec{d})]$$
$$= \vec{a} \cdot [(\vec{b} \cdot \vec{d})\vec{c} - (\vec{b} \cdot \vec{c})\vec{d}];$$

thus,

$$(\vec{a} \times \vec{b}) \cdot (\vec{c} \times \vec{d}) = (\vec{a} \cdot \vec{c})(\vec{b} \cdot \vec{d}) - (\vec{a} \cdot \vec{d})(\vec{b} \cdot \vec{c}). \qquad (2\text{-}46)$$

EXAMPLE 1 / Derive the cosine law of spherical trigonometry.

Let ABC be any spherical triangle on a unit sphere with sides $a, b,$

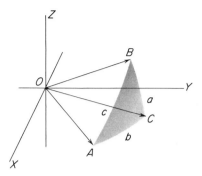

FIGURE 2-27

and c (arcs of great circles). Then, since the sphere is a unit sphere, the lengths of sides a, b, and c are equal to the radian measures of angles BOC, COA, and AOB, respectively (Figure 2-27). Now,

$$(\overrightarrow{OA} \times \overrightarrow{OB})\cdot(\overrightarrow{OA} \times \overrightarrow{OC}) = (\overrightarrow{OA}\cdot\overrightarrow{OA})(\overrightarrow{OB}\cdot\overrightarrow{OC}) - (\overrightarrow{OA}\cdot\overrightarrow{OC})(\overrightarrow{OB}\cdot\overrightarrow{OA})$$

where $|\overrightarrow{OA} \times \overrightarrow{OB}| = \sin c$, $|\overrightarrow{OA} \times \overrightarrow{OC}| = \sin b$, $\overrightarrow{OB}\cdot\overrightarrow{OC} = \cos a$, $\overrightarrow{OA}\cdot\overrightarrow{OC} = \cos b$, and $\overrightarrow{OB}\cdot\overrightarrow{OA} = \cos c$. Note that the angle between $(\overrightarrow{OA} \times \overrightarrow{OB})$ and $(\overrightarrow{OA} \times \overrightarrow{OC})$ has the same measure as the dihedral angle between the planes OAB and OAC; that is, the angle A of the spherical triangle ABC. Therefore,

$$\sin c \sin b \cos A = \cos a - \cos b \cos c.$$

By a cyclic permutation of the elements, two similar formulas may be obtained:

$$\sin a \sin c \cos B = \cos b - \cos c \cos a;$$
$$\sin b \sin a \cos C = \cos c - \cos a \cos b.$$

EXAMPLE 2 / Use the quadruple product

$$(\vec{b} \times \vec{c})\cdot(\vec{b} \times \vec{c}) = (\vec{c} \times \vec{a})\cdot(\vec{c} \times \vec{a})$$

to obtain a formula for the area of a triangle.

Let $\vec{a}$, $\vec{b}$, and $\vec{c}$ be associated with the sides of a triangle (Figure 2-28). Then the area K may be expressed as $\frac{1}{2}|\vec{b} \times \vec{c}|$, as $\frac{1}{2}|\vec{c} \times \vec{a}|$,

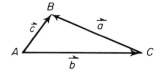

FIGURE 2-28

and as $\frac{1}{2}|\vec{b} \times \vec{a}|$. Since $|\vec{b} \times \vec{c}|^2 = (\vec{b} \times \vec{c}) \cdot (\vec{b} \times \vec{c})$ and $|\vec{c} \times \vec{a}|^2 = (\vec{c} \times \vec{a}) \cdot (\vec{c} \times \vec{a})$, then $(\vec{b} \times \vec{c}) \cdot (\vec{b} \times \vec{c}) = (\vec{c} \times \vec{a}) \cdot (\vec{c} \times \vec{a})$,

$$2K(|\vec{b}||\vec{c}|\sin A) = (|\vec{c}||\vec{a}|\sin B)(|\vec{b}||\vec{a}|\sin C),$$

$$K = \frac{|\vec{a}|^2 \sin B \sin C}{2 \sin A}.$$

Consider the vector product $(\vec{a} \times \vec{b}) \times (\vec{c} \times \vec{d})$ of the four vectors $\vec{a}, \vec{b}, \vec{c}$, and $\vec{d}$. The resulting vector is perpendicular to $\vec{a} \times \vec{b}$ and $\vec{c} \times \vec{d}$ and therefore lies on both the plane determined by $\vec{a}$ and $\vec{b}$ and the plane determined by $\vec{c}$ and $\vec{d}$. Hence, $(\vec{a} \times \vec{b}) \times (\vec{c} \times \vec{d})$ may be expressed both as a linear function of $\vec{a}$ and $\vec{b}$ and as a linear function of $\vec{c}$ and $\vec{d}$. Now, considering $(\vec{a} \times \vec{b}) \times (\vec{c} \times \vec{d})$ as a vector triple product of $\vec{a}, \vec{b}$, and $\vec{c} \times \vec{d}$, we have, by (2-45),

$$(\vec{a} \times \vec{b}) \times (\vec{c} \times \vec{d}) = (\vec{a}\vec{c}\vec{d})\vec{b} - (\vec{b}\vec{c}\vec{d})\vec{a}. \tag{2-47}$$

In a similar manner, considering $(\vec{a} \times \vec{b}) \times (\vec{c} \times \vec{d})$ as a vector triple product of $\vec{a} \times \vec{b}, \vec{c}$, and $\vec{d}$, we have

$$(\vec{a} \times \vec{b}) \times (\vec{c} \times \vec{d}) = (\vec{a}\vec{b}\vec{d})\vec{c} - (\vec{a}\vec{b}\vec{c})\vec{d}. \tag{2-48}$$

Hence, by equations (2-47) and (2-48),

$$(\vec{b}\vec{c}\vec{d})\vec{a} - (\vec{a}\vec{c}\vec{d})\vec{b} + (\vec{a}\vec{b}\vec{d})\vec{c} - (\vec{a}\vec{b}\vec{c})\vec{d} = \vec{0}. \tag{2-49}$$

EXAMPLE 3 / Show that $(\vec{e} \times \vec{f}) \times (\vec{f} \times \vec{g}) = (\vec{f}\vec{g}\vec{e})\vec{f}$.

By equation (2-48), with $\vec{e} = \vec{a}, \vec{f} = \vec{b} = \vec{c}$, and $\vec{g} = \vec{d}$,

$$(\vec{e} \times \vec{f}) \times (\vec{f} \times \vec{g}) = (\vec{e}\vec{f}\vec{g})\vec{f} - (\vec{e}\vec{f}\vec{f})\vec{g}$$
$$= (\vec{e}\vec{f}\vec{g})\vec{f}$$
$$= (\vec{f}\vec{g}\vec{e})\vec{f}.$$

EXAMPLE 4 / Derive the sine law of spherical trigonometry.

Let ABC be any spherical triangle on a unit sphere as in Example 1 and Figure 2-27. By equation (2-48),

$$(\overrightarrow{OA} \times \overrightarrow{OB}) \times (\overrightarrow{OA} \times \overrightarrow{OC}) = (\overrightarrow{OA} \cdot \overrightarrow{OB} \times \overrightarrow{OC})\overrightarrow{OA}$$

and

$$|(\overrightarrow{OA} \times \overrightarrow{OB}) \times (\overrightarrow{OA} \times \overrightarrow{OC})| = |\overrightarrow{OA} \cdot \overrightarrow{OB} \times \overrightarrow{OC}|.$$

Now, $|\overrightarrow{OA} \times \overrightarrow{OB}| = \sin c$, $|\overrightarrow{OA} \times \overrightarrow{OC}| = \sin b$, and the angle between $(\overrightarrow{OA} \times \overrightarrow{OB})$ and $(\overrightarrow{OA} \times \overrightarrow{OC})$ is angle A of the spherical triangle ABC. Therefore,

$$\sin c \sin b \sin A = |\overrightarrow{OA} \cdot \overrightarrow{OB} \times \overrightarrow{OC}|.$$

By equation (2-48),

$$(\overrightarrow{OB} \times \overrightarrow{OC}) \times (\overrightarrow{OB} \times \overrightarrow{OA}) = (\overrightarrow{OB} \cdot \overrightarrow{OC} \times \overrightarrow{OA})\overrightarrow{OB}$$

and

$$|(\overrightarrow{OB} \times \overrightarrow{OC}) \times (\overrightarrow{OB} \times \overrightarrow{OA})| = |\overrightarrow{OB} \cdot \overrightarrow{OC} \times \overrightarrow{OA}| = |\overrightarrow{OA} \cdot \overrightarrow{OB} \times \overrightarrow{OC}|.$$

Now, $|\overrightarrow{OB} \times \overrightarrow{OC}| = \sin a$, $|\overrightarrow{OB} \times \overrightarrow{OA}| = \sin c$, and the angle between $(\overrightarrow{OB} \times \overrightarrow{OC})$ and $(\overrightarrow{OB} \times \overrightarrow{OA})$ is angle B of the spherical triangle ABC. Therefore,

$$\sin a \sin c \sin B = |\overrightarrow{OA} \cdot \overrightarrow{OB} \times \overrightarrow{OC}|.$$

Hence,

$$\sin c \sin b \sin A = \sin a \sin c \sin B;$$

that is,

$$\frac{\sin A}{\sin a} = \frac{\sin B}{\sin b}.$$

By a cyclic permutation of the elements, it follows that

$$\frac{\sin B}{\sin b} = \frac{\sin C}{\sin c},$$

and thus that

$$\frac{\sin A}{\sin a} = \frac{\sin B}{\sin b} = \frac{\sin C}{\sin c}.$$

EXERCISES

In Exercises 1 through 3 prove the given identities. Use the definition $\vec{a}^2 = \vec{a} \cdot \vec{a}$.

1. $(\vec{a} \times \vec{b}) \cdot (\vec{b} \times \vec{c}) \times (\vec{c} \times \vec{a}) = (\vec{a}\vec{b}\vec{c})^2$.
2. $(\vec{a} \times \vec{b}) \times (\vec{a} \times \vec{c}) = (\vec{a}\vec{b}\vec{c})\vec{a}$.
3. $(\vec{a} \times \vec{b}) \cdot (\vec{c} \times \vec{d}) \times (\vec{e} \times \vec{f}) = (\vec{a}\vec{b}\vec{d})(\vec{c}\vec{e}\vec{f}) - (\vec{a}\vec{b}\vec{c})(\vec{d}\vec{e}\vec{f})$.

4. Prove that if $\vec{a}$ is perpendicular to $\vec{b}$ and $\vec{c}$, then $(\vec{a}\vec{b}\vec{c})^2 = \vec{a}^2(\vec{b} \times \vec{c})^2$.
5. Verify (2-46) for $\vec{a} = 3\vec{i} - 2\vec{j} + \vec{k}$, $\vec{b} = 2\vec{i} - 2\vec{k}$, $\vec{c} = \vec{i} + 3\vec{j}$, and $\vec{d} = \vec{i} + \vec{j} - \vec{k}$.

2-10 / QUATERNIONS

The forerunner of vector algebra was the subject of quaternions, the major contribution to mathematics of the Irish physicist and mathematician William Rowan Hamilton. Hamilton was mainly interested in developing an algebraic system that could describe rotations in space. For example, the unit vector $\vec{i}$ in the vector product $\vec{i} \times \vec{j}$ may be considered as an operator which rotates $\vec{j}$ through an angle of 90° in the plane perpendicular to $\vec{i}$ so as to coincide with $\vec{k}$. The story goes that after years of study on the problem, the key concept of the subject became clear to Hamilton one evening in 1843 while walking along the Royal Canal in Dublin. He immediately engraved the fundamental formula, $i^2 = j^2 = k^2 = ijk = -1$, on a stone in Brougham Bridge. Hamilton's "Lectures on Quaternions," published in 1853, and his "Elements of Quaternions," published in 1866, a year after his death, demonstrated the applications of quaternion algebra to geometry and mechanics. The major objection to his algebra was that quaternions were simply too difficult to calculate with. It was the opposition to the use of quaternion algebra by leading mathematicians of the times that led J. W. Gibbs of Yale University, among others, to modify the system of quaternions. Gibbs rejected Hamilton's concept of a single product of two vectors, and he defined an algebra of vectors essentially equivalent to the one studied today.

Hamilton's quaternion algebra depended upon four fundamental units, 1, i, j, and k, whose products are defined by Table 2-1. Note that the product

TABLE 2-1

	1	i	j	k
1	1	i	j	k
i	i	-1	k	$-j$
j	j	$-k$	-1	i
k	k	j	$-i$	-1

of two of these units is generally not commutative. A **quaternion** is an element of the form $a + bi + cj + dk$, where a, b, c, and d are real numbers; that is, the sum of a scalar a and a *pure quaternion* or *vector* $bi + cj + dk$. Consider the product of two pure quaternions $x_1 i + y_1 j + z_1 k$ and $x_2 i + y_2 j + z_2 k$. Assuming the distributive property of the product with respect to addition, and using Table 2-1, we obtain

$$(x_1 i + y_1 j + z_1 k)(x_2 i + y_2 j + z_2 k) = (y_1 z_2 - y_2 z_1)i$$
$$+ (z_1 x_2 - z_2 x_1)j + (x_1 y_2 - x_2 y_1)k - (x_1 x_2 + y_1 y_2 + z_1 z_2).$$

Therefore, the product of two pure quaternions is a quaternion. Note that the two parts of the product correspond to what is considered in vector algebra as the vector product and the scalar product of two vectors; that is, Hamilton's concept of the product of two pure quaternions or vectors is equivalent to the difference of the vector product and the scalar product of the two vectors.

The value of Hamilton's algebra of quaternions rests on two points. First, the study of quaternions led to the development of vector algebra; second, quaternion algebra destroyed the confines in which algebra had been placed, and illustrated that a consistent algebra was possible whose structure differed from that of the algebra of complex numbers.

Quaternion algebra may also be viewed as a study of *ordered quadruples* of real numbers (a, b, c, d). Then equality, scalar multiplication, addition, and multiplication may be defined:

(i) $(a, b, c, d) = (e, f, g, h)$ if, and only if, $a = e$, $b = f$, $c = g$, and $d = h$;
(ii) $k(a, b, c, d) = (ka, kb, kc, kd)$, for any real scalar k;
(iii) $(a, b, c, d) + (e, f, g, h) = (a + e, b + f, c + g, d + h)$;
(iv) $(a, b, c, d)(e, f, g, h) = (ae - bf - cg - dh, af + be + ch - dg,$
$$ag + ce + df - bh, ah + de + bg - cf).$$

Essentially, we may consider a correspondence between quaternions as ordered quadruples of real numbers and quaternions as real multiples of 1, i, j, and k:

$$a + bi + cj + dk \leftrightarrow (a, b, c, d) = a(1, 0, 0, 0) + b(0, 1, 0, 0)$$
$$+ c(0, 0, 1, 0) + d(0, 0, 0, 1),$$

where

$$1 \leftrightarrow (1, 0, 0, 0); \quad i \leftrightarrow (0, 1, 0, 0); \quad j \leftrightarrow (0, 0, 1, 0); \quad k \leftrightarrow (0, 0, 0, 1).$$

It should be evident that we may consider three-dimensional vectors as ordered triples of real numbers; that is,

$$x\vec{i} + y\vec{j} + z\vec{k} \leftrightarrow (x, y, z).$$

In general, we may define an **n-dimensional vector** as an ordered n-tuple of real numbers $(a_1, a_2, \ldots, a_n)$. Then the subject of n-dimensional vectors may be developed by considering the following definitions for equality, scalar multiplication, addition, and multiplication, and the properties of real numbers:

(i) $(a_1, a_2, \ldots, a_n) = (b_1, b_2, \ldots, b_n)$ if, and only if, $a_i = b_i$ for $i = 1, 2, \ldots, n$;
(ii) $k(a_1, a_2, \ldots, a_n) = (ka_1, ka_2, \ldots, ka_n)$, for any real scalar k;
(iii) $(a_1, a_2, \ldots, a_n) + (b_1, b_2, \ldots, b_n) = (a_1 + b_1, a_2 + b_2, \ldots, a_n + b_n)$;
(iv) $(a_1, a_2, \ldots, a_n)(b_1, b_2, \ldots, b_n) = a_1b_1 + a_2b_2 + \cdots + a_nb_n$.

A study of vectors as ordered n-tuples of real numbers leads to a generalization of the concept of a vector space for n-dimensions.

EXERCISES

1. Consider quaternions as ordered quadruples of real numbers and find
(a) $(2, 3, -1, 4) + (0, 2, 1, -2)$; (b) $(3, 1, 2, 0)(2, 1, -1, 4)$;
(c) $3(2, 1, 0, -3)$.

2. If the **conjugate** of a quaternion $x = a + bi + cj + dk$ is defined as $x^* = a - bi - cj - dk$, then find the product of a quaternion and its conjugate; that is, xx^*.

3. Find the multiplicative inverse of the nonzero quaternion $a + bi + cj + dk$ if unity is the identity element. (*Hint:* See Exercise 2.)

4. Show that the real numbers are embedded within the quaternions by considering both the sum and the product of two quaternions of the form $x + 0i + 0j + 0k$.

5. Show that the complex numbers are embedded within the quaternions by considering both the sum and the product of two quaternions of the form $x + yi + 0j + 0k$.

6. Verify that the product of quaternions is associative for $2 + i + j + k$, $1 - i - j + k$, and $1 + 2i + j - 2k$.

7. Prove that the conjugate of the product of two quaternions is equal to the product of the conjugates of the two quaternions taken in the reverse order; that is, $(xy)^* = y^*x^*$. (See Exercise 2.)

2-11 / REAL VECTOR SPACES

It has already been stated that the set of vectors of the form $m\vec{a}$ where $\vec{a}$ is a nonzero vector is a one-dimensional vector space; the set of vectors of the form $m\vec{a} + n\vec{b}$ where $\vec{a}$ and $\vec{b}$ are independent vectors is a two-dimensional vector space; the set of vectors of the form $m\vec{a} + n\vec{b} + p\vec{c}$ where $\vec{a}$, $\vec{b}$, and $\vec{c}$ are independent vectors is a three-dimensional vector space. These sets of vectors constitute vector spaces in that the operations of addition and multiplication by a real number defined on the sets have certain properties. In this section we state a formal definition of a *real vector space* and illustrate several models of this abstract mathematical system.

Let m and n be arbitrary elements of the set R of real numbers; let a, b, and c be arbitrary elements of a set V of elements called *vectors*. Consider a binary operation $+$, called the *addition of vectors*, defined on V. Consider another operation, called the *multiplication of a vector by a real number*, whereby $ma \in V$. Then V is a **real vector space** provided that

(1) $a + b \in V$;
(2) $a + b = b + a$;
(3) $(a + b) + c = a + (b + c)$;
(4) there exists a vector $0 \in V$ such that $a + 0 = 0 + a = a$ for every $a \in V$;
(5) for each $a \in V$ there exists a vector $(-a) \in V$ such that $a + (-a) = (-a) + a = 0$;
(6) $1a = a$;
(7) $(mn)a = m(na)$;
(8) $m(a + b) = ma + mb$;
(9) $(m + n)a = ma + na$.

Properties (1) through (5) are properties of the binary operation of addition of vectors; properties (6) through (9) are properties of the operation of multiplication of a vector by a real number. It is important to note that in property (9), for example, the expression $m + n$ denotes the addition of two real numbers, whereas $ma + na$ denotes the addition of two vectors. The context of a discussion will dictate whether we are adding real numbers or vectors; hence, the same symbol is used to designate the two different operations.

Many models of a real vector space exist in mathematics. It is for this reason that the study of abstract real vector spaces is important. The following examples contain models of real vector spaces and are presented to illustrate the variety of interpretations of a real vector space that are possible. In Chapters 4 and 5 we will discuss additional models of real vector spaces.

EXAMPLE 1 / THE SET OF ORDERED n-TUPLES OF REAL NUMBERS

Let V be the set of ordered n-tuples of real numbers $(a_1, a_2, \ldots, a_n)$. Using the definitions of equality of vectors, addition of vectors, and multiplication of a vector by a real number (scalar multiplication) stated in §2-10, the set V of ordered n-tuples of real numbers is a real vector space.

The real vector space V is said to be of **dimension** n since each vector can be expressed as a linear function of n independent vectors. The n independent vectors constitute a *basis* for the vector space. For example, the simplest basis for this vector space V is

$$(1, 0, \ldots, 0), (0, 1, \ldots, 0), \ldots, (0, 0, \ldots, 1);$$

that is, for each vector $(a_1, a_2, \ldots, a_n)$,

$$(a_1, a_2, \ldots, a_n) = a_1(1, 0, \ldots, 0) + a_2(0, 1, \ldots, 0)$$
$$+ \cdots + a_n(0, 0, \ldots, 1).$$

EXAMPLE 2 / THE SET OF COMPLEX NUMBERS

Let C be the set of complex numbers $a + bi$, where $a, b \in R$ and i is such that $i^2 = -1$. If $a + bi, c + di \in C$, then $a + bi = c + di$ if, and only if, $a = c$ and $b = d$. If $a + bi, c + di \in C$, then the sum of the two complex numbers is defined as a complex number:

$$(a + bi) + (c + di) = (a + c) + (b + d)i.$$

If $a + bi \in C$ and $m \in R$, then the product of the complex number and the real number is defined as a complex number:

$$m(a + bi) = ma + mbi.$$

Under these definitions of addition and multiplication by a real number, the set C of complex numbers is a real vector space.

The real vector space C is a two-dimensional vector space since two independent vectors constitute a basis for the vector space. For example, the set of elements 1 and i, that is, $1 + 0i$ and $0 + 1i$, is a basis for the vector space C. If $a + bi \in C$, then

$$a + bi = a(1 + 0i) + b(0 + 1i).$$

Other bases exist.

EXAMPLE 3 / THE SET OF QUATERNIONS

Let Q be the set of quaternions $a + bi + cj + dk$. Using the definitions of equality, addition, and multiplication by a real number stated in §2-10, the set Q of quaternions is a real vector space. The vector space Q is a four-dimensional vector space.

A **subspace** of a real vector space V is a subset W of V for which W is itself a real vector space under the operations of addition and multiplication by a real number defined on V. It should be evident that the real vector space C in Example 2 is a subspace of the real vector space Q in Example 3 since each complex number $a + bi$ may be considered a quaternion of the form $a + bi + 0j + 0k$.

The next example of a real vector space differs from the examples considered so far in that it is not a *finite-dimensional vector space*; that is, no finite basis exists for the vectors in this vector space.

EXAMPLE 4 / THE SET OF REAL-VALUED FUNCTIONS DEFINED ON AN INTERVAL

Let F be the set of real-valued functions defined on an interval, for example, $0 \leq x \leq 1$. If $g, h \in F$, then the sum $g + h$ of the functions is defined by

$$(g + h)(x) = g(x) + h(x).$$

If $g \in F$ and $m \in R$, then the product of the function and the real number is defined as the function mg whereby

$$(mg)(x) = m[g(x)].$$

The set F is a real vector space under these definitions of the addition of functions and the multiplication of a function by a real number.

EXAMPLE 5 / **THE SET OF REAL POLYNOMIAL FUNCTIONS OVER R**

Let P be the set of real polynomial functions over R; that is, let P be the set of functions p of the form

$$p(x) = a_0 + a_1 x + a_2 x^2 + \cdots + a_n x^n,$$

where $a_0, a_1, a_2, \ldots, a_n$ are real numbers, with domain R. If the addition of two polynomial functions and the multiplication of a polynomial function by a real number is defined as in Example 4, then P is a real vector space.

EXERCISES

In Exercises 1 through 5 prove that the set of elements in the specified example of this section is a real vector space.

1. Example 1.
2. Example 2.
3. Example 3.
4. Example 4.
5. Example 5.

6. Prove that the set R of real numbers is a real vector space.

7. Let S be the set of ordered pairs of real numbers. Consider the following definitions of addition and multiplication by a real number:

If $(a, b), (c, d) \in S$, then $(a, b) + (c, d) = (a + c, b + d)$.
If $(a, b) \in S$ and $m \in R$, then $m(a, b) = (ma, b)$.

Determine whether or not S is a real vector space.

8. Let S be the set of ordered pairs of integers. Consider the following definitions of addition and multiplication by a real number:

If $(a, b), (c, d) \in S$, then $(a, b) + (c, d) = (a + c, b + d)$.

If $(a, b) \in S$ and $m \in R$, then $m(a, b) = (ma, mb)$.

Determine whether or not S is a real vector space.

In Exercises 9 through 12 prove each statement about a real vector space V, where $m \in R$ and $a \in V$.

9. $m0 = 0$.

10. $0a = 0$.

11. $-(ma) = (-m)a = m(-a)$.

12. If $ma = 0$, then $m = 0$ or $a = 0$.

13. Determine whether or not the real vector space of Example 5 is a finite-dimensional vector space.

14. Consider a quaternion as an ordered quadruple of real numbers. Show that the set Q of quaternions is a subspace of the real vector space V in Example 1.

15. Show that the complex numbers $1 + i$ and $3 - 5i$ form a basis for the real vector space C in Example 2.

16. Prove that a nonempty subset W of a real vector space V is a subspace of V if $a + b, ma \in W$ for every $a, b \in W$ and $m \in R$.

17. Consider the real vector space V in Example 1. Determine whether or not each of the following subsets of V is a subspace of V:

(a) the set of elements of V for which $a_i = 0$, where $i \geq 3$;

(b) the set of elements of V for which $a_i = 0$, where $i \geq 4$;

(c) the set of elements of V for which $a_1 > 0$;

(d) the set of elements of V for which $a_1 a_2 = 0$;

(e) the set of elements of V for which $a_1 + a_2 = a_3$;

(f) the set of elements of V for which $a_1 = a_2 = \cdots = a_n$.

THREE

PLANES AND LINES IN SPACE

3-1 / DIRECTION COSINES AND NUMBERS

In this chapter we shall study many of the properties of planes and lines in space through the use of position vectors.

Given a nonzero position vector $\vec{r} = x\vec{i} + y\vec{j} + z\vec{k}$, then, since $\vec{i} \cdot \vec{i} = 1$, $\vec{j} \cdot \vec{i} = 0$, and $\vec{k} \cdot \vec{i} = 0$,

$$\vec{r} \cdot \vec{i} = (x\vec{i} + y\vec{j} + z\vec{k}) \cdot \vec{i} = x.$$

However, by definition, we may express

$$\vec{r} \cdot \vec{i} = |\vec{r}||\vec{i}|\cos(\vec{r}, \vec{i});$$

$$\cos(\vec{r}, \vec{i}) = \frac{\vec{r} \cdot \vec{i}}{|\vec{r}||\vec{i}|} = \frac{x}{|\vec{r}|}. \tag{3-1}$$

In a similar manner it can be shown that

$$\cos(\vec{r}, \vec{j}) = \frac{y}{|\vec{r}|}; \qquad (3\text{-}2)$$

$$\cos(\vec{r}, \vec{k}) = \frac{z}{|\vec{r}|}. \qquad (3\text{-}3)$$

Note that in Figure 3-1 angle $(\vec{r}, \vec{i})$ is the angle formed with the position vector $\vec{r}$ on its initial side and the unit vector $\vec{i}$ along the x-axis on its terminal side. Similar statements hold for the angles $(\vec{r}, \vec{j})$ and $(\vec{r}, \vec{k})$, where $\vec{j}$ and $\vec{k}$ are unit vectors along the y- and z-axes, respectively. The angles $(\vec{r}, \vec{i})$, $(\vec{r}, \vec{j})$, and $(\vec{r}, \vec{k})$ are called **direction angles** of $\vec{r}$ and of the line l through the origin and the point $P: (x, y, z)$. The numbers $\cos(\vec{r}, \vec{i})$, $\cos(\vec{r}, \vec{j})$, and $\cos(\vec{r}, \vec{k})$ are called **direction cosines** of the vector $\vec{r}$ and of the line l. Notice that

$$\cos^2(\vec{r}, \vec{i}) + \cos^2(\vec{r}, \vec{j}) + \cos^2(\vec{r}, \vec{k}) = \frac{x^2 + y^2 + z^2}{|\vec{r}|^2} = 1; \qquad (3\text{-}4)$$

that is, *the sum of the squares of the direction cosines of any nonzero position vector $\vec{r}$ is equal to* 1. Hence, the three direction cosines are not independent. If two of them are given, then the numerical value of the third may be determined from equation (3-4). Furthermore, if three angles are to be considered as direction angles of a vector or of a line, the sum of the squares of their cosines must be equal to 1.

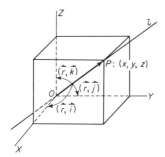

FIGURE 3-1

In rectangular Cartesian coordinate form the direction cosines of vector $\vec{r}$ are given by the ordered triple

$$\left(\frac{x}{\sqrt{x^2 + y^2 + z^2}} : \frac{y}{\sqrt{x^2 + y^2 + z^2}} : \frac{z}{\sqrt{x^2 + y^2 + z^2}} \right), \qquad (3\text{-}5)$$

whose members are proportional to the components x, y, and z of vector $\vec{r}$.

Now, let $\vec{r}_1 = x_1\vec{i} + y_1\vec{j} + z_1\vec{k}$ and $\vec{r}_2 = x_2\vec{i} + y_2\vec{j} + z_2\vec{k}$ be any two nonzero position vectors. Let $(l_1:m_1:n_1)$ and $(l_2:m_2:n_2)$ be their respective direction cosines. Then the cosine of the angle between the two position vectors may be derived as follows:

$$\cos(\vec{r}_1, \vec{r}_2) = \frac{\vec{r}_1 \cdot \vec{r}_2}{|\vec{r}_1||\vec{r}_2|} = \frac{x_1 x_2 + y_1 y_2 + z_1 z_2}{|\vec{r}_1||\vec{r}_2|},$$

$$\cos(\vec{r}_1, \vec{r}_2) = \frac{x_1}{|\vec{r}_1|}\frac{x_2}{|\vec{r}_2|} + \frac{y_1}{|\vec{r}_1|}\frac{y_2}{|\vec{r}_2|} + \frac{z_1}{|\vec{r}_1|}\frac{z_2}{|\vec{r}_2|},$$

$$\cos(\vec{r}_1, \vec{r}_2) = l_1 l_2 + m_1 m_2 + n_1 n_2. \tag{3-6}$$

From the first of the last three equations, we obtain in rectangular Cartesian coordinate form

$$\cos(\vec{r}_1, \vec{r}_2) = \frac{x_1 x_2 + y_1 y_2 + z_1 z_2}{\sqrt{x_1^2 + y_1^2 + z_1^2}\sqrt{x_2^2 + y_2^2 + z_2^2}}. \tag{3-7}$$

Two nonzero position vectors are perpendicular if, and only if, the angle $(\vec{r}_1, \vec{r}_2)$ has measure 90°. Since $\cos 90° = 0$, the condition for perpendicularity may be expressed from (3-6) and (3-7) as either

$$l_1 l_2 + m_1 m_2 + n_1 n_2 = 0 \tag{3-8}$$

or

$$x_1 x_2 + y_1 y_2 + z_1 z_2 = 0. \tag{3-9}$$

The components x, y, and z of $\vec{r}$ are sometimes called **direction numbers** of the position vector and may be written as $(x:y:z)$. Any ordered set of three numbers that can be obtained from these by multiplying all of them by the same positive constant k is also a set of direction numbers for the vector $\vec{r}$, in that they define the direction of the vector. Any ordered set of three numbers that can be obtained from $(x:y:z)$ by multiplying each of them by the same nonzero constant k is a set of direction numbers for any line parallel to $\vec{r}$. Note that a line is generally not oriented.

Consider a line passing through $P_1: (x_1, y_1, z_1)$ and $P_2: (x_2, y_2, z_2)$, with the length of the line segment $P_1 P_2$ equal to d. Then the position vector $(x_2 - x_1)\vec{i} + (y_2 - y_1)\vec{j} + (z_2 - z_1)\vec{k}$ is equivalent to $\vec{P_1 P_2}$. The direction cosines of the position vector and $\vec{P_1 P_2}$ are then given by the equations

$$\cos\alpha = \frac{x_2 - x_1}{d}, \quad \cos\beta = \frac{y_2 - y_1}{d}, \quad \cos\gamma = \frac{z_2 - z_1}{d}, \tag{3-10}$$

where α, β, and γ are the angles the position vector makes with the positive x-, y-, and z-axes, respectively. The direction cosines of the line P_1P_2 may be taken as those of any position vector parallel to it. Note that since a position vector equivalent to $\overrightarrow{P_2P_1}$ could have been determined, then there exist two sets of direction cosines for each line but only one set of direction cosines for any position vector parallel to the line.

EXAMPLE / Find a set of direction cosines for the line through $A: (3, -2, 4)$ and $B: (5, -3, 2)$.

The position vector equivalent to $\overrightarrow{AB}$ is $(5 - 3)\vec{i} + (-3 + 2)\vec{j} + (2 - 4)\vec{k}$; that is, $2\vec{i} - \vec{j} - 2\vec{k}$. Its direction cosines are equal to each of its components divided by the square root of the sum of the squares of its components. Since its components are 2, -1, and -2, then $\sqrt{2^2 + (-1)^2 + (-2)^2} = 3$; $\cos \alpha = \frac{2}{3}$; $\cos \beta = -\frac{1}{3}$; $\cos \gamma = -\frac{2}{3}$. The direction cosines of the line AB may be taken as $(\frac{2}{3}: -\frac{1}{3}: -\frac{2}{3})$. Since line AB is also parallel to $\overrightarrow{BA}$ with direction cosines $(-\frac{2}{3}: \frac{1}{3}: \frac{2}{3})$, this ordered triple represents another set of direction cosines of the line. One set of direction numbers for the line is $(2: -1: -2)$.

In general, any ordered set of scalars $(2k: -k: -2k)$, where $k \neq 0$, represents a set of direction numbers for line AB.

EXERCISES

In Exercises 1 through 4 use points $A: (1, -5, 3)$ and $B: (4, 7, -1)$.

1. Find the direction cosines of $\overrightarrow{AB}$.
2. Find the two sets of direction cosines for line AB.
3. Find the general form of the direction numbers of $\overrightarrow{AB}$.
4. Find the general form of the direction numbers of line AB.
5. Given the following sets of angles, choose those which may be a set of direction angles:
   (a) $(30°: 45°: 60°)$;
   (b) $(120°: 135°: 60°)$;
   (c) $(30°: 150°: 0°)$;
   (d) $(0°: 0°: 90°)$.
6. Determine which of the following sets of values may be a set of direction cosines:

(a) $(\frac{2}{3}: \frac{1}{3}: -\frac{2}{3})$; (b) $(1: -\frac{1}{2}: \frac{1}{2})$;
(c) $(\frac{5}{8}: \frac{1}{3}: \frac{1}{2})$; (d) $(0: 1: 0)$;
(e) $(\cos \theta: \sin \theta: 0)$.

7. If two direction angles of a vector are 60° and 60°, find (a) a third direction angle of the vector; (b) a third direction angle of any line parallel to the vector.

8. Find the direction cosines of the unit vectors on the coordinate axes: (a) $\vec{i}$; (b) $\vec{j}$; (c) $\vec{k}$.

9. If a line has a set of direction numbers $(1: 0: -3)$, find a set of direction cosines of the line.

10. Show that the direction cosines of a line in the xy-plane are a special case of the direction cosines of a line in space.

11. Given the point $(2, 1, -4)$ on a line with a set of direction numbers $(3: -1: 1)$, find three other points on the line.

12. If the direction cosines of two intersecting lines are $(1/2: \sqrt{2}/2: 1/2)$ and $(-1/2: 1/2: \sqrt{2}/2)$, find the angles between the two lines.

13. If a set of direction numbers of two intersecting lines are $(1: 1: 4)$ and $(0: 1: -1)$, find the acute angle between the two lines.

14. Show that the lines AB and CD are parallel where $A: (2, 3, 6)$, $B: (4, 1, 2)$, $C: (3, 0, 1)$, and $D: (2, 1, 3)$.

15. Determine the cosine of the angle between the diagonal of a cube and one of its intersecting edges.

3-2 / EQUATION OF A PLANE

There are several ways of determining a plane in space. For example, a plane may be described or specified by three given points on the plane which do not lie on a single straight line, or by a line and a point on the plane, providing the point does not lie on the line, or by other means. A convenient way of describing a plane is by designating a point through which it passes and a vector that is perpendicular to the plane. To determine the vector equation of such a plane, let $\vec{n}$ be a vector perpendicular to the plane (Figure 3-2). Let $P: (x, y, z)$ be a general point on the plane. Since $\overrightarrow{P_0P}$ is perpendicular to $\vec{n}$,

$$\overrightarrow{P_0P} \cdot \vec{n} = 0; \qquad (3\text{-}11)$$

$$(\overrightarrow{OP} - \overrightarrow{OP_0}) \cdot \vec{n} = 0. \qquad (3\text{-}12)$$

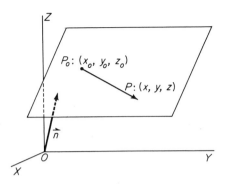

FIGURE 3-2

Equations (3-11) and (3-12) represent vector forms of the equation of the plane through $P_0: (x_0, y_0, z_0)$ with vector $\vec{n}$ perpendicular to the plane.

The general rectangular Cartesian coordinate form of the equation of a plane can be obtained from the vector form (3-12). Let $\vec{n} = a\vec{i} + b\vec{j} + c\vec{k}$. Since $\overrightarrow{OP} = x\vec{i} + y\vec{j} + z\vec{k}$ and $\overrightarrow{OP_0} = x_0\vec{i} + y_0\vec{j} + z_0\vec{k}$, then

$$\overrightarrow{OP} - \overrightarrow{OP_0} = (x - x_0)\vec{i} + (y - y_0)\vec{j} + (z - z_0)\vec{k},$$
$$a(x - x_0) + b(y - y_0) + c(z - z_0) = 0, \qquad (3\text{-}13)$$

and

$$ax + by + cz + d = 0, \qquad (3\text{-}14)$$

where $d = -ax_0 - by_0 - cz_0$. This is the general rectangular Cartesian coordinate form of the equation of a plane with arbitrary constants a, b, c, and d. Hereafter, the expression **coordinate form** will be used to mean rectangular Cartesian coordinate form.

Equation (3-13) is sometimes called the **point-direction number form of the equation of a plane** since it involves the coordinates of a point on the plane and the direction numbers of a vector perpendicular to the plane.

It is important to remember that the coefficients a, b, and c of x, y, and z, respectively, in any equation of a plane, are the components of a vector perpendicular to the plane.

If $abcd \neq 0$, the general equation may be written in the form

$$\frac{x}{e} + \frac{y}{f} + \frac{z}{g} = 1, \qquad (3\text{-}15)$$

where $e = -d/a$, $f = -d/b$, and $g = -d/c$. This equation is called the

intercept form of the equation of a plane since e, f, and g are the x, y, and z intercepts, respectively, of the given plane. That is to say, the given plane intersects the coordinate axes at points whose coordinates are $(e, 0, 0)$, $(0, f, 0)$, and $(0, 0, g)$.

EXAMPLE 1 / Find the equation of the plane through $N: (-2, 1, 2)$ and perpendicular to $\overrightarrow{ON}$.

Let $P: (x, y, z)$ be a general point on the plane through N perpendicular to $\overrightarrow{ON}$. Then $\overrightarrow{ON}$ and $\overrightarrow{NP}$ are perpendicular to each other and $\overrightarrow{ON} \cdot \overrightarrow{NP} = 0$. Since

$$\overrightarrow{ON} = -2\vec{i} + \vec{j} + 2\vec{k}$$

and

$$\begin{aligned}\overrightarrow{NP} = \overrightarrow{OP} - \overrightarrow{ON} &= (x\vec{i} + y\vec{j} + z\vec{k}) - (-2\vec{i} + \vec{j} + 2\vec{k}) \\ &= (x+2)\vec{i} + (y-1)\vec{j} + (z-2)\vec{k}, \\ \overrightarrow{ON} \cdot \overrightarrow{NP} &= -2(x+2) + (y-1) + 2(z-2) = 0;\end{aligned}$$

that is,

$$2x - y - 2z + 9 = 0.$$

EXAMPLE 2 / Find the equation of the plane that passes through the point $P: (4, 2, 1)$ and is parallel to the plane $2x + 3y - z + 5 = 0$.

A vector perpendicular to one of two parallel planes is a vector perpendicular to the other plane. Therefore, the components 2, 3, and -1 of a vector perpendicular to the given plane are also the components of a vector perpendicular to the desired plane, and thus are the coefficients of x, y, and z, respectively, of the plane to be determined. It remains to find d in the equation

$$2x + 3y - z + d = 0.$$

Since P lies on the plane, the coordinates of P satisfy the equation of the plane; that is,

$$2(4) + 3(2) - (1) + d = 0 \quad \text{and} \quad d = -13.$$

Therefore, the equation of the desired plane is

$$2x + 3y - z - 13 = 0.$$

A plane may also be determined if three distinct noncollinear points of the plane are known. Several vector forms of the equation of the plane may be found. One vector form depends upon the concept of a set of linearly dependent vectors (§1-5).

Let $A: (x_1, y_1, z_1)$, $B: (x_2, y_2, z_2)$, and $C: (x_3, y_3, z_3)$ be three distinct noncollinear points. Consider $P: (x, y, z)$ a general point on the plane ABC (Figure 3-3). Since $\overrightarrow{AP}$ lies on the plane determined by $\overrightarrow{AB}$ and $\overrightarrow{AC}$,

$$\overrightarrow{AP} = m\overrightarrow{AB} + n\overrightarrow{AC} \tag{3-16}$$

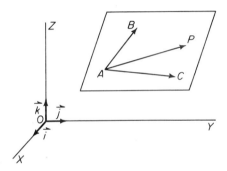

FIGURE 3-3

represents a vector form of the equation of the plane. Equation (3-16) may also be written as

$$\overrightarrow{OP} - \overrightarrow{OA} = m(\overrightarrow{OB} - \overrightarrow{OA}) + n(\overrightarrow{OC} - \overrightarrow{OA}) \tag{3-17}$$

or

$$\overrightarrow{OP} = (1 - m - n)\overrightarrow{OA} + m\overrightarrow{OB} + n\overrightarrow{OC}. \tag{3-18}$$

Now,

$$\overrightarrow{OP} = x\vec{i} + y\vec{j} + z\vec{k}, \qquad \overrightarrow{OA} = x_1\vec{i} + y_1\vec{j} + z_1\vec{k},$$
$$\overrightarrow{OB} = x_2\vec{i} + y_2\vec{j} + z_2\vec{k}, \quad \text{and} \quad \overrightarrow{OC} = x_3\vec{i} + y_3\vec{j} + z_3\vec{k}.$$

The corresponding components, after substituting in equation (3-18), must be equal:

$$\begin{cases} x = (1 - m - n)x_1 + mx_2 + nx_3 \\ y = (1 - m - n)y_1 + my_2 + ny_3 \\ z = (1 - m - n)z_1 + mz_2 + nz_3. \end{cases} \quad (3\text{-}19)$$

The equations of (3-19) are one set of **parametric equations of a plane** through A, B, and C with parameters m and n. Each point of the plane corresponds to an ordered pair of values (m, n) of the parameters.

A second vector form of the equation of a plane through three distinct noncollinear points depends upon the concept of the scalar triple product. In Figure 3-3, $\overrightarrow{AB} \times \overrightarrow{AC}$ is a vector perpendicular to the plane determined by A, B, and C; $\overrightarrow{AP}$ lies on the plane. Hence,

$$\overrightarrow{AP} \cdot (\overrightarrow{AB} \times \overrightarrow{AC}) = 0. \quad (3\text{-}20)$$

Equation (3-20) is another vector form of the equation of the plane ABC.

EXAMPLE 3 / Find a set of parametric equations of the plane through $A: (3, -1, 2)$, $B: (1, 4, 0)$, and $C: (0, -2, 1)$.

By equations (3-19),

$$\begin{cases} x = (1 - m - n)(3) + m(1) + n(0) \\ y = (1 - m - n)(-1) + m(4) + n(-2) \\ z = (1 - m - n)(2) + m(0) + n(1); \end{cases}$$

that is,

$$\begin{cases} x = 3 - 2m - 3n \\ y = -1 + 5m - n \\ z = 2 - 2m - n. \end{cases}$$

Note that equations (3-19) were derived from equation (3-18). Therefore, if we let $m = n = 0$, the coordinates of A should be obtained; if we let $m = 1$ and $n = 0$, the coordinates of B should be obtained; if we let $m = 0$ and $n = 1$, the coordinates of C should be obtained. This procedure serves as a check for the validity of the set of parametric equations derived.

Furthermore, note that a set of parametric equations of a plane is not unique. For example, a simple exchange of the roles of the points A and B or A and C in equation (3-16) yields a different set of parametric equations.

EXAMPLE 4 / Find the coordinate form of the equation of the plane through the points $A: (1, 2, 0)$, $B: (3, -1, 2)$, and $C: (2, 4, 3)$.

Let P be any point of the plane determined by A, B, and C. Then $\overrightarrow{AB} = 2\vec{i} - 3\vec{j} + 2\vec{k}$, $\overrightarrow{AC} = \vec{i} + 2\vec{j} + 3\vec{k}$, $\overrightarrow{AP} = (x-1)\vec{i} + (y-2)\vec{j} + z\vec{k}$, and

$$\begin{aligned}\overrightarrow{AP} \cdot (\overrightarrow{AB} \times \overrightarrow{AC}) &= [(x-1)\vec{i} + (y-2)\vec{j} + z\vec{k}] \\ &\quad \cdot [(2\vec{i} - 3\vec{j} + 2\vec{k}) \times (\vec{i} + 2\vec{j} + 3\vec{k})] \\ &= [(x-1)\vec{i} + (y-2)\vec{j} + z\vec{k}] \\ &\quad \cdot (-13\vec{i} - 4\vec{j} + 7\vec{k}) \\ &= -13(x-1) - 4(y-2) + 7z.\end{aligned}$$

Then, by equation (3-20),

$$-13(x-1) - 4(y-2) + 7z = 0;$$

that is,

$$13x + 4y - 7z - 21 = 0.$$

EXERCISES

1. Find the equation of the plane through $N: (3, -2, 1)$ and perpendicular to $\overrightarrow{ON}$.

2. Find the equation of the plane through $N: (2, 1, -1)$ and perpendicular to $\vec{r} = \vec{i} + \vec{j} - 3\vec{k}$.

3. Find the equation of the plane through $N: (7, -4, 3)$ and perpendicular to line AB where $A: (2, 0, -1)$ and $B: (5, 6, 0)$.

4. Determine the equation of the plane that passes through $P: (1, 2, -2)$ and is parallel to the plane $x + 4y - 2z = 0$.

5. Determine a set of parametric equations of the plane through $A: (1, 2, 0)$, $B: (-3, 2, 4)$, and $C: (5, 1, -1)$.

6. Find the general coordinate form of the equation of the plane in Exercise 5.

7. State the condition that must exist in each case among the coefficients of the equation of the plane $ax + by + cz + d = 0$ in order that

(a) the plane may have intercept 2 on the y-axis;
(b) the plane may have equal intercepts on the y- and z-axes;
(c) the plane may have equal intercepts on all three coordinate axes;
(d) the plane may be parallel to the plane $2x - y + z - 7 = 0$;
(e) the plane may be perpendicular to the vector $\vec{i} + 3\vec{j} + 5\vec{k}$;
(f) the plane may pass through the origin;
(g) the plane may pass through the point $(2, -1, 4)$;
(h) the plane may be parallel to the y-axis;
(i) the plane may be parallel to the yz-plane;
(j) the plane may contain the z-axis.

8. Show that the points $A: (1, -1, 1)$, $B: (2, 0, 0)$, $C: (-1, 1, 5)$ and $D: (0, 0, 3)$ are coplanar.

9. Determine the equation of the plane through $A: (2, 1, 5)$, $B: (3, -2, 4)$, and $C: (1, -3, 3)$.

10. Find a vector form of the equation of the plane through the origin and parallel to the position vectors $\vec{OA}$ and $\vec{OB}$.

11. Find a vector form of the equation of the plane through the point C and parallel to the position vectors $\vec{OA}$ and $\vec{OB}$.

3-3 / EQUATION OF A SPHERE

A sphere is the locus of points in space that are equidistant from one fixed point, the center. Let S be a sphere of radius a with center at a point $C: (x_0, y_0, z_0)$ as shown in Figure 3-4. If $P: (x, y, z)$ is a general point on the sphere S, $\vec{OP}$ is the position vector of this point, and $\vec{OC}$ is the position vector of the center C, then

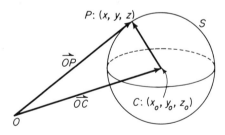

FIGURE 3-4

$$|\overrightarrow{OP} - \overrightarrow{OC}| = a \qquad (3\text{-}21)$$

and

$$(\overrightarrow{OP} - \overrightarrow{OC}) \cdot (\overrightarrow{OP} - \overrightarrow{OC}) = a^2. \qquad (3\text{-}22)$$

Equation (3-22) represents a vector form of the equation of sphere S.

Since $\overrightarrow{OP} = x\vec{i} + y\vec{j} + z\vec{k}$ and $\overrightarrow{OC} = x_0\vec{i} + y_0\vec{j} + z_0\vec{k}$, then $\overrightarrow{OP} - \overrightarrow{OC} = (x - x_0)\vec{i} + (y - y_0)\vec{j} + (z - z_0)\vec{k}$. Therefore, by Theorem 2.7, equation (3-22) may be expressed in terms of rectangular Cartesian coordinates as

$$(x - x_0)^2 + (y - y_0)^2 + (z - z_0)^2 = a^2. \qquad (3\text{-}23)$$

Equation (3-23) is the coordinate form of the equation of a sphere with center at $C: (x_0, y_0, z_0)$ and radius equal to a.

EXAMPLE 1 / Find the equation of the sphere with center at $C: (3, -1, 2)$ and radius equal to 5.

Let $P: (x, y, z)$ be a general point on the sphere. Then the equation of the sphere in vector form is given by equation (3-22),

$$(\overrightarrow{OP} - \overrightarrow{OC}) \cdot (\overrightarrow{OP} - \overrightarrow{OC}) = a^2,$$

where

$$\overrightarrow{OP} = x\vec{i} + y\vec{j} + z\vec{k}, \qquad \overrightarrow{OC} = 3\vec{i} - \vec{j} + 2\vec{k}, \quad \text{and} \quad a = 5.$$

Therefore,

$$(x - 3)^2 + (y + 1)^2 + (z - 2)^2 = 25;$$

that is,

$$x^2 + y^2 + z^2 - 6x + 2y - 4z - 11 = 0$$

is the coordinate form of the equation of the sphere.

EXAMPLE 2 / Find the equation of the sphere with center at the origin and radius equal to a.

If $P: (x, y, z)$ is a general point on the sphere, then a vector form of the equation of the sphere with center at the origin and radius equal to a is

$$\overrightarrow{OP} \cdot \overrightarrow{OP} = a^2.$$

In coordinate form this equation may be expressed as

$$x^2 + y^2 + z^2 = a^2.$$

The equation of the plane π which intersects a given sphere S at one, and only one, point P is said to be tangent to the sphere at that point. To determine the equation of such a plane, let $P: (x, y, z)$ be a general point on the plane π, let $P_1: (x_1, y_1, z_1)$ be the point of tangency, and let $C: (x_0, y_0, z_0)$ be the center of the sphere S as shown in Figure 3-5. Then

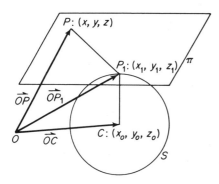

FIGURE 3-5

$$(\overrightarrow{OP} - \overrightarrow{OP_1}) \cdot (\overrightarrow{OC} - \overrightarrow{OP_1}) = 0, \quad (3\text{-}24)$$

since the radius from C to P_1 will be perpendicular to every line in π through P_1. Equation (3-24) is a vector form of the equation of the tangent plane π. Since $\overrightarrow{OP} = x\vec{i} + y\vec{j} + z\vec{k}$, $\overrightarrow{OP_1} = x_1\vec{i} + y_1\vec{j} + z_1\vec{k}$, and $\overrightarrow{OC} = x_0\vec{i} + y_0\vec{j} + z_0\vec{k}$, then $\overrightarrow{OP} - \overrightarrow{OP_1} = (x - x_1)\vec{i} + (y - y_1)\vec{j} + (z - z_1)\vec{k}$ and $\overrightarrow{OC} - \overrightarrow{OP_1} = (x_0 - x_1)\vec{i} + (y_0 - y_1)\vec{j} + (z_0 - z_1)\vec{k}$. Therefore, the equation

$$(x - x_1)(x_0 - x_1) + (y - y_1)(y_0 - y_1) + (z - z_1)(z_0 - z_1) = 0 \quad (3\text{-}25)$$

represents the coordinate form of the equation of the plane π tangent at (x_1, y_1, z_1) to the sphere with center at the point (x_0, y_0, z_0).

EXAMPLE 3 / Find the equation of the plane tangent at $P_1: (2, 4, 1)$ to the sphere with center at $C: (3, -1, 2)$.

The tangent plane is expressed by equation (3-24) as

$$(\overrightarrow{OP} - \overrightarrow{OP_1}) \cdot (\overrightarrow{OC} - \overrightarrow{OP_1}) = 0,$$

where

$$\overrightarrow{OP} = x\vec{i} + y\vec{j} + z\vec{k}, \qquad \overrightarrow{OC} = 3\vec{i} - \vec{j} + 2\vec{k},$$

and

$$\overrightarrow{OP_1} = 2\vec{i} + 4\vec{j} + \vec{k}.$$

Since

$$\overrightarrow{OP} - \overrightarrow{OP_1} = (x-2)\vec{i} + (y-4)\vec{j} + (z-1)\vec{k}$$

and

$$\overrightarrow{OC} - \overrightarrow{OP_1} = \vec{i} - 5\vec{j} + \vec{k},$$

then

$$(\overrightarrow{OP} - \overrightarrow{OP_1}) \cdot (\overrightarrow{OC} - \overrightarrow{OP_1}) = (x-2) - 5(y-4) + (z-1).$$

Therefore,

$$x - 5y + z + 17 = 0$$

is the equation of the plane tangent at P_1: (2, 4, 1) to the sphere with center at C: (3, −1, 2).

EXAMPLE 4 / Find the equation of the tangent plane to the sphere with center at the origin and radius equal to a.

Let $P_1: (x_1, y_1, z_1)$ be any point on the sphere with center at the origin and radius equal to a. Using equation (3-25), where the origin is $C: (x_0, y_0, z_0)$, we obtain

$$(x - x_1)(-x_1) + (y - y_1)(-y_1) + (z - z_1)(-z_1) = 0,$$
$$x_1 x + y_1 y + z_1 z = x_1^2 + y_1^2 + z_1^2,$$

and

$$x_1 x + y_1 y + z_1 z = a^2. \qquad (3\text{-}26)$$

Equation (3-26) represents the coordinate form of the equation of the tangent plane to the sphere with center at the origin, radius equal to a, and tangent point at $P_1: (x_1, y_1, z_1)$.

EXERCISES

1. Find the equation of the sphere with center at $C: (1, -2, 4)$ and radius equal to 3.

2. Find the equation of the sphere with center at $C: (0, 0, a)$ and radius equal to a.

3. Find the equation of the plane tangent at $P_1: (2, 6, 5)$ to the sphere with center at $C: (1, 5, 3)$.

4. Find the center and radius of the sphere whose equation is $x^2 + y^2 + z^2 - 6x + 6y - 2z + 3 = 0$.

5. Determine the equation of the plane tangent at $P_1: (2, 3, 0)$ to the sphere $x^2 + y^2 + z^2 + 4x - 6y = 3$.

6. Determine the equation of the sphere whose diameter is AB, where $A: (-4, 5, 0)$ and $B: (4, 1, 8)$.

7. If the line segment joining $A: (-2, -1, 2)$ and $B: (2, 1, 2)$ subtends a right angle at $P: (x, y, z)$ when P does not coincide with A or B, find a relationship among the coordinates of P.

8. If the coordinates of the midpoint of line segment OP satisfy the equation $x^2 + y^2 + z^2 = 8$, determine the value of k such that the coordinates of P satisfy the equation $x^2 + y^2 + z^2 = k$.

9. Determine the equation of the plane tangent at $(1, 0, 0)$ to the sphere $x^2 + y^2 + z^2 = 1$.

3-4 / ANGLE BETWEEN TWO PLANES

When two planes intersect, two pairs of supplementary dihedral angles are formed. The measures of these dihedral angles are the same as the measures of the corresponding angles formed by the intersection of vectors perpendicular to the planes (Figure 3-6).

To determine a formula for the angle between two intersecting planes, let the equations of the planes be $a_1 x + b_1 y + c_1 z = d_1$ and $a_2 x + b_2 y + c_2 z = d_2$. The vectors $\vec{n_1} = a_1 \vec{i} + b_1 \vec{j} + c_1 \vec{k}$ and $\vec{n_2} = a_2 \vec{i} + b_2 \vec{j} + c_2 \vec{k}$ are perpendicular to these two planes, respectively. The angle between these vectors may be found by using the relationship

$$\vec{n_1} \cdot \vec{n_2} = |\vec{n_1}||\vec{n_2}| \cos(\vec{n_1}, \vec{n_2}).$$

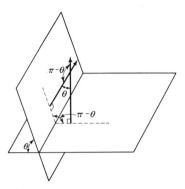

FIGURE 3-6

It follows that

$$\cos(\vec{n_1}, \vec{n_2}) = \pm \frac{\vec{n_1} \cdot \vec{n_2}}{|\vec{n_1}||\vec{n_2}|}, \tag{3-27}$$

where the two algebraic signs are considered in order to obtain the two supplementary angles. In terms of the components,

$$\cos(\vec{n_1}, \vec{n_2}) = \pm \frac{a_1 a_2 + b_1 b_2 + c_1 c_2}{\sqrt{a_1^2 + b_1^2 + c_1^2}\sqrt{a_2^2 + b_2^2 + c_2^2}}. \tag{3-28}$$

Note that the planes are perpendicular if $\cos(\vec{n_1}, \vec{n_2}) = 0$; that is, if

$$\vec{n_1} \cdot \vec{n_2} = 0. \tag{3-29}$$

In coordinate form, the planes are perpendicular if

$$a_1 a_2 + b_1 b_2 + c_1 c_2 = 0. \tag{3-30}$$

The planes are parallel if the vectors perpendicular to the planes are parallel. This condition may be expressed by the relationship

$$\frac{a_1}{a_2} = \frac{b_1}{b_2} = \frac{c_1}{c_2} = t, \tag{3-31}$$

for some scalar t.

EXAMPLE 1 / Find the measures of the dihedral angles formed by the planes $-x + 2y + 2z = 10$ and $x + y + 4z = 7$.

3-4 / ANGLE BETWEEN TWO PLANES

The vectors perpendicular to the planes are $\vec{n_1} = -\vec{i} + 2\vec{j} + 2\vec{k}$ and $\vec{n_2} = \vec{i} + \vec{j} + 4\vec{k}$. Then

$$\cos(\vec{n_1}, \vec{n_2}) = \pm \frac{\vec{n_1} \cdot \vec{n_2}}{|\vec{n_1}||\vec{n_2}|}$$

$$= \pm \frac{(-1)(1) + (2)(1) + (2)(4)}{\sqrt{(-1)^2 + (2)^2 + (2)^2} \sqrt{(1)^2 + (1)^2 + (4)^2}}$$

$$= \pm \frac{9}{\sqrt{9}\sqrt{18}} = \pm \frac{1}{\sqrt{2}}.$$

Hence, the measures of the dihedral angles formed by the planes are 45° and 135°.

EXAMPLE 2 / Show that the planes $2x - y + 2z = 3$ and $2x + 2y - z = 7$ are perpendicular.

The vectors $\vec{n_1} = 2\vec{i} - \vec{j} + 2\vec{k}$ and $\vec{n_2} = 2\vec{i} + 2\vec{j} - \vec{k}$ are perpendicular to the planes. Since $\vec{n_1} \cdot \vec{n_2} = (2)(2) + (-1)(2) + (2)(-1) = 0$, the planes are perpendicular.

EXERCISES

1. Find the measures of the dihedral angles formed by the planes $2x - 4y - 4z - 5 = 0$ and $x + 4y + z - 3 = 0$.

2. Show that the planes $x + 3y + z + 4 = 0$ and $2x - y + z + 2 = 0$ are perpendicular.

3. Show that the planes $3x - 2y + z + 7 = 0$ and $9x - 6y + 3z + 10 = 0$ are parallel.

4. Determine the value of c for which the planes $3x - 4y + cz = 0$ and $2x - 3y + 6z - 1 = 0$ are perpendicular.

5. Find the measure of the acute angle between the planes $x + 2y + 2z - 1 = 0$ and $3x - 4y - 5z = 0$.

6. Find the equation of the plane which contains the points $A: (1, -1, -2)$ and $B: (4, 2, 2)$, and which is perpendicular to the plane $3x + y - 2z = 12$.

3-5 / DISTANCE BETWEEN A POINT AND A PLANE

A formula for the shortest distance between a point and a line on a plane was determined in §2-3. In a similar manner, the shortest distance between a point and a plane in space may be determined. Let $P_1: (x_1, y_1, z_1)$ be any point in space, and let $P_0: (x_0, y_0, z_0)$ be any point on the plane $ax + by + cz + d = 0$. The shortest distance r between the point and the plane is equal to the magnitude of the projection of $\vec{P_0P_1}$ on $\vec{n}$, a unit vector perpendicular to the plane as shown in Figure 3-7. Therefore,

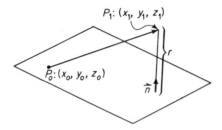

FIGURE 3-7

$$r = |\vec{P_0P_1} \cdot \vec{n}|$$
$$= \left|[(x_1 - x_0)\vec{i} + (y_1 - y_0)\vec{j} + (z_1 - z_0)\vec{k}] \cdot \frac{a\vec{i} + b\vec{j} + c\vec{k}}{\sqrt{a^2 + b^2 + c^2}}\right|$$
$$= \frac{|a(x_1 - x_0) + b(y_1 - y_0) + c(z_1 - z_0)|}{\sqrt{a^2 + b^2 + c^2}}$$
$$= \frac{|ax_1 + by_1 + cz_1 - ax_0 - by_0 - cz_0|}{\sqrt{a^2 + b^2 + c^2}}.$$

But $d = -ax_0 - by_0 - cz_0$ since $P_0: (x_0, y_0, z_0)$ lies on the plane. Therefore,

$$r = \frac{|ax_1 + by_1 + cz_1 + d|}{\sqrt{a^2 + b^2 + c^2}}. \tag{3-32}$$

EXAMPLE 1 / Find the shortest distance between the point $P_1: (1, -2, -4)$ and the plane $2x + 2y - z = 11$.

By equation (3-32), the distance r is given as

$$r = \frac{|(2)(1) + (2)(-2) + (-1)(-4) + (-11)|}{\sqrt{2^2 + 2^2 + (-1)^2}}$$
$$= \frac{|-9|}{3} = 3.$$

An alternate approach is to consider the problem in vector form; that is, essentially, to derive equation (3-32) for this particular case. Choose any point on the plane, say $P_0: (3, 2, -1)$. Then $\overrightarrow{P_0P_1} = -2\vec{i} - 4\vec{j} - 3\vec{k}$. A unit vector perpendicular to the plane is $\vec{n} = \frac{2}{3}\vec{i} + \frac{2}{3}\vec{j} - \frac{1}{3}\vec{k}$. Since r is equal to the magnitude of the projection of $\overrightarrow{P_0P_1}$ on $\vec{n}$, then

$$r = |\overrightarrow{P_0P_1} \cdot \vec{n}| = |(-2)(\tfrac{2}{3}) + (-4)(\tfrac{2}{3}) + (-3)(-\tfrac{1}{3})| = 3;$$

that is, 3 units.

EXAMPLE 2 / Find a formula for the shortest distance between the origin and any plane $ax + by + cz + d = 0$.

By equation (3-32), where $P_1: (x_1, y_1, z_1)$ is the origin,

$$r = \frac{|d|}{\sqrt{a^2 + b^2 + c^2}}. \qquad (3\text{-}33)$$

If two planes

$$a_1x + b_1y + c_1z + d_1 = 0 \quad \text{and} \quad a_2x + b_2y + c_2z + d_2 = 0$$

are parallel or coincide, then by §3-4

$$\frac{a_1}{a_2} = \frac{b_1}{b_2} = \frac{c_1}{c_2} = t.$$

Therefore, by multiplying the equation of the second plane by t, the variables x, y, and z in the equations of the two planes will contain identical coefficients. That is, the equations become

$$a_1x + b_1y + c_1z + d_1 = 0 \quad \text{and} \quad a_1x + b_1y + c_1z + d_3 = 0, \quad (3\text{-}34)$$

where $d_3 = td_2$, but is not necessarily equal to d_1 unless the planes coincide. To find the distance r between these planes, take any point $P_1: (x_1, y_1, z_1)$ on the plane $a_1x + b_1y + c_1z + d_1 = 0$. The distance r between the point P_1 and the plane $a_1x + b_1y + c_1z + d_3 = 0$ is given by the expression

$$r = \frac{|a_1x_1 + b_1y_1 + c_1z_1 + d_3|}{\sqrt{a_1^2 + b_1^2 + c_1^2}}.$$

But $a_1x_1 + b_1y_1 + c_1z_1 = -d_1$ since the point P_1 lies on the plane $a_1x + b_1y + c_1z + d_1 = 0$. Therefore,

$$r = \frac{|d_3 - d_1|}{\sqrt{a_1^2 + b_1^2 + c_1^2}}. \tag{3-35}$$

Equation (3-35) represents the formula for the distance between two parallel planes of the form in (3-34).

EXAMPLE 3 / Find the distance between the planes

$6x - 3y + 6z + 2 = 0$ and $2x - y + 2z + 4 = 0$.

The planes are parallel since $6/2 = -3/-1 = 6/2 = 3$. Multiplying the terms on each side of the second equation by 3, we find that the set of equations becomes

$6x - 3y + 6z + 2 = 0$ and $6x - 3y + 6z + 12 = 0$.

By equation (3-35), the distance r between the two parallel planes is given as

$$r = \frac{|12 - 2|}{\sqrt{(6)^2 + (-3)^2 + (6)^2}} = \frac{10}{9};$$

that is, $\frac{10}{9}$ units.

EXERCISES

1. Find the distance between $P_1: (0, -5, 2)$ and the plane $x + 2y + z - 4 = 0$.
2. Find the distance between the origin and the plane $x - 2y - 2z - 3 = 0$.
3. Determine d such that the distance between $P_1: (4, 0, 1)$ and the plane $2x + y - 2z + d = 0$ is 3.
4. Find the distance between $P_1: (0, 0, 4)$ and the plane $2x + 2y - 3 = 0$.
5. Find the distance between the parallel planes $3x - 4y + 12z + 4 = 0$ and $3x - 4y + 12z - 22 = 0$.
6. Find the distance between the planes $x + y - 2z = 2$ and $3x + 3y - 6z = 2$.

7. If the coordinates of P satisfy the equation $x + 2y + 3z - 4 = 0$, then determine the value of d such that the coordinates of the midpoint of line segment OP satisfy the equation $x + 2y + 3z + d = 0$.

8. Determine the value of d such that the plane $2x - 2y + z + d = 0$ is tangent to the sphere $x^2 + y^2 + z^2 - 2x - 4y + 2z - 10 = 0$.

9. Determine the equation of the sphere with center at $C: (2, 2, 1)$ and tangent to the plane $3x + 4y + 12z = 0$.

3-6 / EQUATION OF A LINE

As in the case of the plane in space, there are several ways of specifying a line in space, such as giving two points on the line, or two planes through the line, or a point on the line and a set of direction numbers for the line, or by other means. It will be convenient to determine the form of an equation of a straight line by designating a point through which it passes and a vector to which it is parallel.

Let $P_1: (x_1, y_1, z_1)$ be the point through which the line l passes and $\vec{r} = a\vec{i} + b\vec{j} + c\vec{k}$ be the vector parallel to the line (Figure 3-8). Let $P: (x, y, z)$ be a general point on line l. Since $\overrightarrow{OP} - \overrightarrow{OP_1}$ is parallel to $\vec{r}$, it follows that

$$(\overrightarrow{OP} - \overrightarrow{OP_1}) \times \vec{r} = \vec{0}, \qquad (3\text{-}36)$$

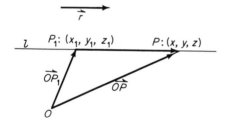

FIGURE 3-8

which is a vector form of the equation of the straight line l. Another vector form of the equation of the straight line is expressed by

$$\overrightarrow{OP} - \overrightarrow{OP_1} = m\vec{r}, \qquad (3\text{-}37)$$

where m represents any real number. Since

$$\overrightarrow{OP} = x\vec{i} + y\vec{j} + z\vec{k}, \qquad \overrightarrow{OP_1} = x_1\vec{i} + y_1\vec{j} + z_1\vec{k},$$

and
$$\vec{OP} - \vec{OP_1} = (x - x_1)\vec{i} + (y - y_1)\vec{j} + (z - z_1)\vec{k},$$
then
$$(x - x_1)\vec{i} + (y - y_1)\vec{j} + (z - z_1)\vec{k} = m(a\vec{i} + b\vec{j} + c\vec{k}).$$

By equating components, since $\vec{i}$, $\vec{j}$, and $\vec{k}$ are linearly independent vectors, we obtain

$$(x - x_1) = ma, \quad (y - y_1) = mb, \quad (z - z_1) = mc;$$

$$\begin{cases} x = x_1 + ma \\ y = y_1 + mb \\ z = z_1 + mc. \end{cases} \quad (3\text{-}38)$$

The equations of (3-38) represent a set of **parametric equations of a line** l with parameter m. Each point of the line corresponds to a value of the parameter. Note that $(a:b:c)$ represents a set of direction numbers for the line since a, b, and c are the components of a vector parallel to the line. After solving for m in each equation, it follows that

$$\frac{x - x_1}{a} = \frac{y - y_1}{b} = \frac{z - z_1}{c}, \quad (3\text{-}39)$$

which represents the **point-direction number form of the equation of a line** through $P_1: (x_1, y_1, z_1)$ with direction numbers $(a:b:c)$.

The equations of (3-39) are also called the **symmetric form of the equation of a line** or a set of **symmetric equations of a line**. Note that the symmetric form of the equation of a line is not unique, since the coordinates of infinitely many points on the line may be used as well as infinitely many sets of direction numbers.

If a, b, or c is zero, the symmetric form of the equation of a straight line can still be used, provided it is designated as a special convention that whenever a denominator in the formula is zero the formal ratio is deleted and the numerator is set equal to zero. For example, the equation

$$\frac{x - x_1}{0} = \frac{y - y_1}{b} = \frac{z - z_1}{c}, \quad (3\text{-}40)$$

where $b \neq 0$ and $c \neq 0$, shall mean

$$x - x_1 = 0; \quad \frac{y - y_1}{b} = \frac{z - z_1}{c}. \quad (3\text{-}41)$$

There exist several vector forms of the equation of a straight line. Another vector form depends upon the concept of a set of linearly dependent vectors. Let $A: (x_1, y_1, z_1)$ and $B: (x_2, y_2, z_2)$ be two distinct points. Consider $P: (x, y, z)$ a general point on the line l (Figure 3-9). Since P lies on the line determined by A and B, then by Theorem 1.9

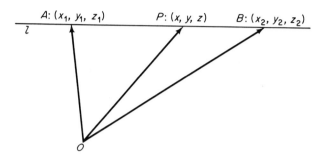

FIGURE 3-9

$$\overrightarrow{OP} = m\overrightarrow{OA} + (1 - m)\overrightarrow{OB} \qquad (3\text{-}42)$$

represents a vector form of the equation of the line. Now, $\overrightarrow{OP} = x\vec{i} + y\vec{j} + z\vec{k}$, $\overrightarrow{OA} = x_1\vec{i} + y_1\vec{j} + z_1\vec{k}$, and $\overrightarrow{OB} = x_2\vec{i} + y_2\vec{j} + z_2\vec{k}$. Equating corresponding components after substituting in equation (3-42), we obtain

$$\begin{cases} x = mx_1 + (1 - m)x_2 \\ y = my_1 + (1 - m)y_2 \\ z = mz_1 + (1 - m)z_2. \end{cases} \qquad (3\text{-}43)$$

The equations of (3-43) are another form of a set of **parametric equations of a line** through A and B with parameter m. Each point of the line corresponds to a value of the parameter. Solving for m in each equation of (3-43), it follows that

$$\frac{x - x_2}{x_1 - x_2} = \frac{y - y_2}{y_1 - y_2} = \frac{z - z_2}{z_1 - z_2}, \qquad (3\text{-}44)$$

which represents the **two-point form of the equation of a line** through $A: (x_1, y_1, z_1)$ and $B: (x_2, y_2, z_2)$. Note that the two-point form of the equation of a straight line on a plane is a special case of (3-44).

EXAMPLE 1 / Determine a set of symmetric equations of the line through $P_1: (4, 1, 5)$ and parallel to $\vec{r} = 2\vec{i} - 2\vec{j} + 3\vec{k}$.

The components of $\vec{r}$ provide a set of direction numbers for the line: (2: −2: 3). The equations of (3-39) where $x_1 = 4$, $y_1 = 1$, and $z_1 = 5$ provide the symmetric equations of the line:

$$\frac{x-4}{2} = \frac{y-1}{-2} = \frac{z-5}{3}.$$

EXAMPLE 2 / Determine a set of parametric equations of the line in Example 1.

By the equations of (3-38), a set of parametric equations of the line is

$$\begin{cases} x = 4 + 2m \\ y = 1 - 2m \\ z = 5 + 3m. \end{cases}$$

EXAMPLE 3 / Determine a vector form of the equation of the line through A: (2, 3, 4) and B: (−1, 2, 1). Use the vector form to determine a set of symmetric equations of the line. Compare the results with those obtained by using the equations of (3-44).

Let P: (x, y, z) be any point on the line through A and B. Then

$$\vec{OP} = m\vec{OA} + (1 - m)\vec{OB}.$$

Since $\vec{OP} = x\vec{i} + y\vec{j} + z\vec{k}$, $\vec{OA} = 2\vec{i} + 3\vec{j} + 4\vec{k}$, and $\vec{OB} = -\vec{i} + 2\vec{j} + \vec{k}$, then

$$x\vec{i} + y\vec{j} + z\vec{k} = 2m\vec{i} + 3m\vec{j} + 4m\vec{k} - (1 - m)\vec{i} + 2(1 - m)\vec{j} + (1 - m)\vec{k}$$
$$= (3m - 1)\vec{i} + (m + 2)\vec{j} + (3m + 1)\vec{k};$$

that is,

$$(x - 3m + 1)\vec{i} + (y - m - 2)\vec{j} + (z - 3m - 1)\vec{k} = \vec{0},$$

which represents a vector form of the equation of the line. Since $\vec{i}$, $\vec{j}$, and $\vec{k}$ are linearly independent vectors,

$$\begin{cases} x - 3m + 1 = 0 \\ y - m - 2 = 0 \\ z - 3m - 1 = 0; \end{cases}$$

3-6 / EQUATION OF A LINE

that is,

$$\begin{cases} \dfrac{x+1}{3} = m \\ \dfrac{y-2}{1} = m \\ \dfrac{z-1}{3} = m. \end{cases}$$

Hence,

$$\frac{x+1}{3} = \frac{y-2}{1} = \frac{z-1}{3}$$

is a set of symmetric equations of the line through A and B.

By the equations of (3-44) with $(x_1, y_1, z_1) = (2, 3, 4)$ and $(x_2, y_2, z_2) = (-1, 2, 1)$,

$$\frac{x - (-1)}{2 - (-1)} = \frac{y - 2}{3 - 2} = \frac{z - 1}{4 - 1};$$

that is,

$$\frac{x+1}{3} = \frac{y-2}{1} = \frac{z-1}{3},$$

which agrees with the results obtained when the vector form of the equation of the line is used.

EXAMPLE 4 / Determine a set of symmetric equations of the line through the origin and $P_1: (x_1, y_1, z_1)$.

Let $(x_2, y_2, z_2) = (0, 0, 0)$ in the equations of (3-44). Then

$$\frac{x}{x_1} = \frac{y}{y_1} = \frac{z}{z_1}$$

represents a set of symmetric equations of the line through the origin and $P_1: (x_1, y_1, z_1)$.

Since an equation such as $ax + by + cz + d = 0$ represents a plane, two such equations considered simultaneously represent two planes that will intersect in a straight line, if they are not parallel. The condition that the two planes be parallel is that the coefficients of x, y, and z in the equations be proportional. Therefore, except when $a_1/a_2 = b_1/b_2 = c_1/c_2$, the set of two equations

$$\begin{cases} a_1x + b_1y + c_1z + d_1 = 0 \\ a_2x + b_2y + c_2z + d_2 = 0 \end{cases} \qquad (3\text{-}45)$$

taken simultaneously represents a line, and is called the **general form of the equation of a line.**

Many pairs of planes may be passed through the same line. Therefore, a line may be represented by any one of infinitely many pairs of planes through the line.

To determine a set of symmetric equations of a line by means of a pair of equations for the planes, first locate the coordinates of two of the **piercing points**; that is, the points of intersection of the line with the coordinate planes. This can be done by letting x, y, or z be zero and solving the resulting two equations simultaneously. A set of symmetric equations of the line may then be determined by using (3-44).

EXAMPLE 5 / Find a set of symmetric equations of the line determined by the set of equations:

$$\begin{cases} 3x - y + z = 8 \\ 2x + y + 4z = 2. \end{cases}$$

If $z = 0$, then the two equations

$$\begin{cases} 3x - y = 8 \\ 2x + y = 2, \end{cases}$$

solved simultaneously, determine the piercing point $A: (2, -2, 0)$ of the xy-plane. If $x = 0$, then the two equations

$$\begin{cases} -y + z = 8 \\ y + 4z = 2, \end{cases}$$

solved simultaneously, determine the piercing point $B: (0, -6, 2)$ of the yz-plane. Since $\vec{BA} = 2\vec{i} + 4\vec{j} - 2\vec{k}$, then $(2:4:-2)$ is a set of direction numbers for the line BA. Hence, a set of symmetric equations of line BA is

$$\frac{x-2}{2} = \frac{y+2}{4} = \frac{z}{-2}.$$

EXAMPLE 6 / Determine a set of direction numbers for the line determined by the set of planes:

$$\begin{cases} a_1x + b_1y + c_1z + d_1 = 0 \\ a_2x + b_2y + c_2z + d_2 = 0. \end{cases}$$

The line determined by the two planes is perpendicular to the vectors $a_1\vec{i} + b_1\vec{j} + c_1\vec{k}$ and $a_2\vec{i} + b_2\vec{j} + c_2\vec{k}$ perpendicular to the planes. A position vector parallel to the line is given by Theorem 2.15 in the expression

$$(a_1\vec{i} + b_1\vec{j} + c_1\vec{k}) \times (a_2\vec{i} + b_2\vec{j} + c_2\vec{k});$$

that is,

$$(b_1c_2 - b_2c_1)\vec{i} + (c_1a_2 - c_2a_1)\vec{j} + (a_1b_2 - a_2b_1)\vec{k}.$$

The components of this vector comprise a set of direction numbers for the line. Hence,

$$(b_1c_2 - b_2c_1 : c_1a_2 - c_2a_1 : a_1b_2 - a_2b_1)$$

is a set of direction numbers for the line

$$\begin{cases} a_1x + b_1y + c_1z + d_1 = 0 \\ a_2x + b_2y + c_2z + d_2 = 0. \end{cases}$$

EXERCISES

1. Determine a set of parametric equations of the line through $P_1: (4, 5, 2)$ and parallel to $\vec{r} = 2\vec{i} - 3\vec{j} + \vec{k}$.

2. Determine a set of symmetric equations of the line in Exercise 1.

3. Use the concept of linear dependence of vectors to determine a set of symmetric equations of the line through $A: (1, 2, 0)$ and $B: (4, 3, -2)$.

4. Determine a set of parametric equations of the line through the origin and $P_1: (3, 5, -1)$.

5. Determine a set of symmetric equations of the line in Exercise 4.

6. Given the line $x = 2 + 4m$, $y = 2$, and $z = 3 - 3m$, find (a) a set of direction numbers; (b) a set of direction cosines; (c) three points on the line.

7. Find a set of parametric equations of the line through $A: (4, 1, 3)$ and $B: (1, 1, 2)$.

8. Find a set of symmetric equations of the line through $A: (1, 1, -5)$ and having a set of direction numbers $(4: 2: 3)$.

9. Find a set of symmetric equations of the line through $A: (3, -1, 4)$ and parallel to the line

$$\frac{x+1}{3} = \frac{y}{5} = \frac{z-7}{2}.$$

10. Find a set of parametric equations of the line through $A: (5, 1, 2)$ and perpendicular to the plane $4x + 2y + 5z - 18 = 0$.

11. Determine a set of direction cosines of the line

$$\frac{x-1}{2} = \frac{y+2}{-2} = z - 2.$$

12. Determine t if the line

$$\frac{x+1}{3} = \frac{y-2}{6} = \frac{z-3}{4}$$

is parallel to the plane $2x + 3y - tz + 7 = 0$.

13. Determine a set of direction cosines of the line common to the planes $2x - y + z + 3 = 0$ and $5x + y - z + 4 = 0$.

14. Find a set of symmetric equations of the line determined by the planes $x - y + z = 8$ and $2x + y - z = 1$.

15. Determine the coordinates of the point of intersection of the line through $A: (-2, 3, 7)$ and $B: (6, -1, 2)$ with the xy-plane.

16. Determine the coordinates of the point of intersection of the line through $A: (1, 1, 1)$ and $B: (3, 2, 1)$ with the plane $x - 3y = 0$.

17. Show that a vector parallel to the line determined by $3x - y - 5 = 0$ and $4x - z - 9 = 0$ is perpendicular to a vector parallel to the line determined by $y + z = 0$ and $x - y - 1 = 0$.

18. Show that the line determined by $5x + y - 3z + 1 = 0$ and $3x - 6y - 4z + 15 = 0$ is parallel to the line determined by $x - y - z = 0$ and $7x + 2y - 4z - 3 = 0$.

19. Find the equation of the plane which passes through the points $A: (3, 2, -1)$ and $B: (2, 5, 0)$, and is parallel to the line

$$\frac{x-2}{3} = \frac{y-1}{2} = \frac{z}{-1}.$$

20. Find the equation of the plane passing through the two parallel lines

$$\frac{x-x_0}{\cos \alpha} = \frac{y-y_0}{\cos \beta} = \frac{z-z_0}{\cos \gamma} \quad \text{and} \quad \frac{x-x_1}{\cos \alpha} = \frac{y-y_1}{\cos \beta} = \frac{z-z_1}{\cos \gamma}.$$

FOUR

MATRICES

4-1 / DEFINITIONS AND ELEMENTARY PROPERTIES

In many branches of the physical, biological, and social sciences it is necessary for scientists to express and use a set of numbers in a rectangular array. Indeed, in many everyday activities it is convenient, if not necessary, to use sets of numbers arranged in rows and columns for keeping records, for purposes of comparison, and for a variety of other reasons.

Consider a company that manufactures three models of typewriters: an electric model, a standard model, and a portable model. If the company wishes to compare the units of raw material and labor involved in one month's production of each of these models, an *array* may be used to present the data:

	Electric model	Standard model	Portable model
Units of material	20	17	12
Units of labor	6	8	5

The units used are not intended to be realistic but merely to illustrate an oversimplified application of an array of real numbers. Units of material for the three models comprise the first row of the array, units of labor the second row, and units of production for each model (material and labor) the columns of the array. If the pattern in which the units are to be recorded is clearly defined in advance, this rectangular array may be presented simply as

$$\begin{pmatrix} 20 & 17 & 12 \\ 6 & 8 & 5 \end{pmatrix}. \tag{4-1}$$

A second example of the use of rectangular arrays of real numbers is one that might be used by a basketball coach who wishes to keep a record of the scoring performances of three of his players. Consider the following array:

	Games	Field goals	Free throws
Player A	16	110	62
Player B	14	85	42
Player C	16	73	55

or simply

$$\begin{pmatrix} 16 & 110 & 62 \\ 14 & 85 & 42 \\ 16 & 73 & 55 \end{pmatrix}. \tag{4-2}$$

Rectangular arrays of elements a_{ij} such as

$$\begin{pmatrix} a_{11} & a_{12} & \cdots & a_{1n} \\ a_{21} & a_{22} & \cdots & a_{2n} \\ \cdots & \cdots & \cdots & \cdots \\ a_{m1} & a_{m2} & \cdots & a_{mn} \end{pmatrix}$$

are called **matrices** (singular: **matrix**). Each element a_{ij} has two indices: the **row index**, i, and **column index**, j. The elements $a_{i1}, a_{i2}, \ldots, a_{in}$ are the elements of the ith row, and the elements $a_{1j}, a_{2j}, \ldots, a_{mj}$ are the elements of the jth column. The element a_{ij} is the element contained simultaneously in the ith row and jth column. For example, the element a_{21} of matrix (4-1) is equal to 6; that is, a_{21} is the element in the second row and first column.

A matrix of m rows and n columns is called a matrix of **order** m by n.

Thus, matrix (4-1) is of order 2 by 3, while matrix (4-2) is of order 3 by 3. In general, when the number of rows equals the number of columns, the matrix is called a **square matrix**. A square matrix of order n by n is said, simply, to be of order n. Matrix (4-2) is an example of a square matrix of order three.

If each of the elements of a matrix is a real number, the matrix is called a **real matrix**. Unless otherwise stated, we shall be concerned only with real matrices.

Whenever it is convenient, matrices will be denoted symbolically by capital letters $A, B, C, \ldots$, or by $((a_{ij})), ((b_{ij})), ((c_{ij})), \ldots$ where $a_{ij}, b_{ij}, c_{ij}, \ldots$, respectively, represent the general elements of the matrices.

EXAMPLE 1 / Construct a square matrix $((a_{ij}))$ of order three where $a_{ij} = 3i - j^2$.

If $a_{ij} = 3i - j^2$, then

$$a_{11} = 3(1) - (1)^2 = 2, \; a_{12} = 3(1) - (2)^2 = -1,$$
$$a_{13} = 3(1) - (3)^2 = -6,$$
$$a_{21} = 3(2) - (1)^2 = 5, \; a_{22} = 3(2) - (2)^2 = 2,$$
$$a_{23} = 3(2) - (3)^2 = -3,$$
$$a_{31} = 3(3) - (1)^2 = 8, \; a_{32} = 3(3) - (2)^2 = 5,$$
$$a_{33} = 3(3) - (3)^2 = 0.$$

Hence, the desired matrix is

$$\begin{pmatrix} 2 & -1 & -6 \\ 5 & 2 & -3 \\ 8 & 5 & 0 \end{pmatrix}.$$

Two matrices $((a_{ij}))$ and $((b_{ij}))$ are said to be **equal** if, and only if, they are of the same order and $a_{ij} = b_{ij}$ for all pairs (i, j).

EXAMPLE 2 / Determine whether or not the matrices of each pair are equal:

(a) $\begin{pmatrix} 4 & 1 & 2 \\ -2 & 3 & 5 \end{pmatrix}$ and $\begin{pmatrix} 1 & 2 \\ 3 & 5 \end{pmatrix}$; (b) $\begin{pmatrix} 5 & 2 \\ 1 & 4 \end{pmatrix}$ and $\begin{pmatrix} 5 & 2 \\ 1 & 4 \end{pmatrix}$;

(c) $\begin{pmatrix} a \\ b \\ c \end{pmatrix}$ and $\begin{pmatrix} 2a \\ 2b \\ 2c \end{pmatrix}$;   (d) $\begin{pmatrix} 2 & -2 \\ 1 & 1 \\ 3 & 0 \end{pmatrix}$ and $\begin{pmatrix} 0 & -2 \\ 0 & 1 \\ 0 & 0 \end{pmatrix}$.

The matrices in (a) cannot be equal since they are not of the same order. The matrices in (b) are equal. The matrices in (c) are equal if, and only if, $a = b = c = 0$. Although the matrices in (d) are of the same order, they are not equal since not all corresponding elements are equal.

Consider again matrix (4-2) which represents the scoring performances of three basketball players for one season. Suppose that the matrix representing the scoring performances of these players in the next season of play is

$$\begin{pmatrix} 18 & 142 & 98 \\ 18 & 83 & 30 \\ 15 & 103 & 60 \end{pmatrix}.$$

A matrix representing the combined scoring performance of each of the three players during two seasons may be obtained by adding the corresponding entries of the two matrices:

$$\begin{pmatrix} 16 & 110 & 62 \\ 14 & 85 & 42 \\ 16 & 73 & 55 \end{pmatrix} + \begin{pmatrix} 18 & 142 & 98 \\ 18 & 83 & 30 \\ 15 & 103 & 60 \end{pmatrix} = \begin{pmatrix} 34 & 252 & 160 \\ 32 & 168 & 72 \\ 31 & 176 & 115 \end{pmatrix}.$$

That is, in two seasons players A, B, and C participated in 34, 32, and 31 games, scored 252, 168, and 176 field goals, and 160, 72, and 115 free throws, respectively.

In general, the addition of two matrices $((a_{ij}))$ and $((b_{ij}))$ is defined if, and only if, the matrices are of the same order. If $((a_{ij}))$ and $((b_{ij}))$ are matrices of the same order, then the **sum** $((a_{ij})) + ((b_{ij}))$ is defined as a third matrix $((c_{ij}))$ of that same order where each element c_{ij} satisfies the condition $c_{ij} = a_{ij} + b_{ij}$.

Consider any three real matrices of order m by n: $A = ((a_{ij}))$, $B = ((b_{ij}))$, and $C = ((c_{ij}))$. Since the addition of real numbers is commutative, $a_{ij} + b_{ij} = b_{ij} + a_{ij}$ for all pairs (i, j) and

$$A + B = B + A. \tag{4-4}$$

Since the addition of real numbers is associative, $(a_{ij} + b_{ij}) + c_{ij} = a_{ij} + (b_{ij} + c_{ij})$ for all pairs (i, j) and

$$(A + B) + C = A + (B + C). \tag{4-5}$$

Thus, *the addition of real matrices is commutative and associative.*

A **null matrix** or **zero matrix**, denoted by 0, is a matrix wherein all of the elements are zero. For every matrix A of order m by n there exists a null matrix of order m by n such that $A + 0 = 0 + A = A$. This null matrix of order m by n is the *additive identity element* for the set of all matrices of order m by n.

EXAMPLE 3 / Find the sum of matrices A and B where

$$A = \begin{pmatrix} 2 & 1 & 1 \\ 0 & -3 & 4 \end{pmatrix} \quad \text{and} \quad B = \begin{pmatrix} 3 & -1 & 3 \\ 2 & 0 & 5 \end{pmatrix}.$$

$$A + B = \begin{pmatrix} 2+3 & 1+(-1) & 1+3 \\ 0+2 & -3+0 & 4+5 \end{pmatrix} = \begin{pmatrix} 5 & 0 & 4 \\ 2 & -3 & 9 \end{pmatrix}.$$

Consider again matrix (4-1), which represents the units of material and labor involved in one month's production of three models of typewriters. Suppose the manufacturing company wishes to double its production of each model. Then the matrix

$$\begin{pmatrix} 40 & 34 & 24 \\ 12 & 16 & 10 \end{pmatrix}$$

would represent the units of material and labor involved in a single month's production of the three models of typewriters. It is convenient to represent the doubling of the entries in matrix (4-1) as a product of the matrix and the real number 2; that is,

$$2\begin{pmatrix} 20 & 17 & 12 \\ 6 & 8 & 5 \end{pmatrix} = \begin{pmatrix} 40 & 34 & 24 \\ 12 & 16 & 10 \end{pmatrix}.$$

Note that

$$2\begin{pmatrix} 20 & 17 & 12 \\ 6 & 8 & 5 \end{pmatrix} = \begin{pmatrix} 20 & 17 & 12 \\ 6 & 8 & 5 \end{pmatrix} + \begin{pmatrix} 20 & 17 & 12 \\ 6 & 8 & 5 \end{pmatrix}.$$

In general, the product of a real number (scalar) k and a matrix $((a_{ij}))$, denoted by $k((a_{ij}))$ or by $((a_{ij}))k$, is called the *scalar multiple* of the matrix $((a_{ij}))$ by k. The **scalar multiple** $k((a_{ij}))$ is defined as a matrix wherein the elements are products of k and the corresponding elements of $((a_{ij}))$. Since the multiplication of real numbers is commutative,

$$k((a_{ij})) = ((a_{ij}))k = ((ka_{ij})). \tag{4-6}$$

Notice that for any real number k and any real matrix $A = ((a_{ij}))$, the matrices A and kA are of the same order. In particular, if $k = -1$, then

$$A + (-1)A = ((a_{ij})) + ((-a_{ij})) = 0.$$

Thus, $(-1)A$ is the *additive inverse* of A. If B and A are any two matrices of the same order, the **difference** $B - A$ is defined by the relation

$$B - A = B + (-1)A. \tag{4-7}$$

In general, the scalars may be considered as scalar coefficients, and any algebraic sum of scalar multiples of matrices of the same order satisfies certain laws. For example, if k and l are scalars and A and B are matrices of the same order, then

$$kA + lA = (k + l)A; \tag{4-8}$$

$$klA = k(lA) = l(kA) = (kl)A; \tag{4-9}$$

and

$$k(A + B) = kA + kB. \tag{4-10}$$

Furthermore, if $kA = 0$, then either $k = 0$ or A is a null matrix.

EXAMPLE 4 / Find $3A - 2B$ where

$$A = \begin{pmatrix} 1 & 2 \\ 3 & 0 \end{pmatrix} \quad \text{and} \quad B = \begin{pmatrix} 1 & 3 \\ 0 & -4 \end{pmatrix}.$$

$$3A - 2B = 3\begin{pmatrix} 1 & 2 \\ 3 & 0 \end{pmatrix} - 2\begin{pmatrix} 1 & 3 \\ 0 & -4 \end{pmatrix} = 3\begin{pmatrix} 1 & 2 \\ 3 & 0 \end{pmatrix} + (-2)\begin{pmatrix} 1 & 3 \\ 0 & -4 \end{pmatrix}$$

$$= \begin{pmatrix} 3(1) & 3(2) \\ 3(3) & 3(0) \end{pmatrix} + \begin{pmatrix} -2(1) & -2(3) \\ -2(0) & -2(-4) \end{pmatrix}$$

$$= \begin{pmatrix} 3 & 6 \\ 9 & 0 \end{pmatrix} + \begin{pmatrix} -2 & -6 \\ 0 & 8 \end{pmatrix} = \begin{pmatrix} 1 & 0 \\ 9 & 8 \end{pmatrix}.$$

EXERCISES

1. Construct a square matrix $((a_{ij}))$ of order three where $a_{ij} = i^2 + 2j - 3$.
2. Construct a matrix $((a_{ij}))$ of order 3 by 2 where $a_{ij} = i^2 - ij$.
3. In the square matrix $((a_{ij}))$ of order two describe the position of the elements for which (a) $i = 2$; (b) $j = 1$; (c) $i = j$.
4. If
$$A = \begin{pmatrix} 2 & -1 & 5 \\ 3 & 2 & 1 \end{pmatrix} \text{ and } B = \begin{pmatrix} -2 & 0 & 3 \\ 1 & -4 & -1 \end{pmatrix},$$
then find (a) $A + B$; (b) $A - B$; (c) $A + 3B$.

5. Verify the associative law of addition of matrices (4-5) for
$$A = \begin{pmatrix} 3 & 1 & 1 \\ -2 & 5 & 0 \end{pmatrix}, \quad B = \begin{pmatrix} 1 & -3 & 0 \\ 0 & 1 & 4 \end{pmatrix}, \text{ and } C = \begin{pmatrix} 2 & 7 & -1 \\ -2 & 1 & 3 \end{pmatrix}.$$

6. Find the additive inverse of the matrix

   (a) $\begin{pmatrix} 2 & 5 & 3 \\ 0 & 2 & -1 \end{pmatrix}$; (b) $\begin{pmatrix} a & b \\ c & d \end{pmatrix}$.

7. Solve the matrix equation
$$\begin{pmatrix} a_{11} & a_{12} \\ a_{21} & a_{22} \end{pmatrix} + \begin{pmatrix} 3 & 1 \\ -4 & 0 \end{pmatrix} = \begin{pmatrix} -3 & 2 \\ 1 & 0 \end{pmatrix}.$$

8. Prove that the set of real matrices of order m by n is a real vector space under the operations of matrix addition and the multiplication of a matrix by a real number.

9. Prove that the set of square real matrices of order two is a real vector space of dimension 4 by determining 4 matrices that form a basis for the vector space.

4-2 / MATRIX MULTIPLICATION

Consider a system of linear equations such as
$$\begin{cases} 2x - y + 2z = 1 \\ x + 2y - 4z = 3 \\ 3x - y + z = 0. \end{cases} \quad (4\text{-}11)$$

This system may be represented by a single matrix equation:

$$\begin{pmatrix} 2x - y + 2z \\ x + 2y - 4z \\ 3x - y + z \end{pmatrix} = \begin{pmatrix} 1 \\ 3 \\ 0 \end{pmatrix}. \qquad (4\text{-}12)$$

The coefficients of x, y, and z may be obtained either from (4-11) or (4-12). In both cases the solution depends upon these coefficients. The **matrix of coefficients** is

$$\begin{pmatrix} 2 & -1 & 2 \\ 1 & 2 & -4 \\ 3 & -1 & 1 \end{pmatrix}.$$

The coefficients of each variable are positioned in a column, and coefficients of the variables of each equation are located in a row. It is customary and convenient to think of this matrix of coefficients as an operator that acts upon a column matrix of the variables:

$$\begin{pmatrix} 2 & -1 & 2 \\ 1 & 2 & -4 \\ 3 & -1 & 1 \end{pmatrix} \begin{pmatrix} x \\ y \\ z \end{pmatrix} = \begin{pmatrix} 2x - y + 2z \\ x + 2y - 4z \\ 3x - y + z \end{pmatrix}. \qquad (4\text{-}13)$$

Thus, the system of equations (4-11) may be represented by the single matrix equation

$$\begin{pmatrix} 2 & -1 & 2 \\ 1 & 2 & -4 \\ 3 & -1 & 1 \end{pmatrix} \begin{pmatrix} x \\ y \\ z \end{pmatrix} = \begin{pmatrix} 1 \\ 3 \\ 0 \end{pmatrix}. \qquad (4\text{-}14)$$

Use of the matrix of coefficients as an operator in (4-13) requires the introduction of *matrix multiplication*. Notice that the element $2x - y + 2z$ may be obtained from the matrices

$$\begin{pmatrix} 2 & -1 & 2 \\ 1 & 2 & -4 \\ 3 & -1 & 1 \end{pmatrix} \text{ and } \begin{pmatrix} x \\ y \\ z \end{pmatrix}$$

by summing the products of the elements of row one of the matrix of coefficients and the corresponding elements of the column matrix of variables, taken in order; that is,

$$(2)(x) + (-1)(y) + (2)(z) = 2x - y + 2z.$$

Similarly, the element $x + 2y - 4z$ may be obtained by summing the products of the elements of row two of the matrix of coefficients and the corresponding elements of the column matrix of variables, taken in order; that is,

$$(1)(x) + (2)(y) + (-4)(z) = x + 2y - 4z.$$

In a similar manner, using the elements of row three of the matrix of coefficients, we obtain

$$(3)(x) + (-1)(y) + (1)(z) = 3x - y + z.$$

In general, the **product** AB of two matrices A and B is defined to be a matrix C such that the element in the ith row and jth column of C is obtained by summing the products of the elements of the ith row of A and the corresponding elements of the jth column of B, taken in order. Notice that the number of columns of A must be the same as the number of rows of B. If $A = ((a_{ij}))$ is a matrix of order m by n and $B = ((b_{ij}))$ is a matrix of order n by r, then $C = ((c_{ij}))$ is a matrix of order m by r where

$$c_{ij} = a_{i1}b_{1j} + a_{i2}b_{2j} + \cdots + a_{in}b_{nj}. \tag{4-15}$$

We may use summation notation and write (4-15) as

$$c_{ij} = \sum_{k=1}^{n} a_{ik}b_{kj}. \tag{4-16}$$

When the number of columns of a matrix A is equal to the number of rows of a matrix B, the product AB exists, and the matrices A and B are said to be **conformable** for the product AB. Two matrices can be multiplied only when they are conformable. In the product AB, B is sometimes spoken of as being **premultiplied** by A, and A as being **postmultiplied** by B. Even if the product AB exists, the product BA may not exist since matrices A and B may not be conformable for this product. This illustrates an important property of matrix multiplication, namely that, in general, it is not commutative.

EXAMPLE 1 / Find the products AB and BA, if they exist, where

$$A = \begin{pmatrix} 2 & 3 \\ 1 & -4 \end{pmatrix} \quad \text{and} \quad B = \begin{pmatrix} 3 & -2 & 2 \\ 1 & 0 & -1 \end{pmatrix}.$$

Matrices A and B are conformable for the product AB since the number of columns of A is equal to the number of rows of B. Hence, AB exists. Furthermore, AB is of order 2 by 3 since A is of order 2 by 2 and B is of order 2 by 3. By definition, the general element of AB is given as $c_{ij} = a_{i1}b_{1j} + a_{i2}b_{2j}$. Then

$$c_{11} = (2)(3) + (3)(1), \quad c_{12} = (2)(-2) + (3)(0),$$
$$c_{13} = (2)(2) + (3)(-1),$$
$$c_{21} = (1)(3) + (-4)(1), \quad c_{22} = (1)(-2) + (-4)(0),$$
$$c_{23} = (1)(2) + (-4)(-1);$$

that is,

$$c_{11} = 9, \quad c_{12} = -4, \quad c_{13} = 1,$$
$$c_{21} = -1, \quad c_{22} = -2, \quad c_{23} = 6.$$

Therefore,

$$AB = \begin{pmatrix} 9 & -4 & 1 \\ -1 & -2 & 6 \end{pmatrix}.$$

The product BA does not exist since the matrices B and A are not conformable for this product.

If the elements of any row or column of a matrix are considered to represent the components of a vector, then for every pair of values (i, j) the element c_{ij} of (4-15) is the scalar product of the ith row vector of A and the jth column vector of B. A matrix consisting of a single row is sometimes called a **row matrix** or **row vector**; a matrix consisting of a single column is sometimes called a **column matrix** or **column vector**.

EXAMPLE 2 / Find the matrix products AB and BA of the row vector $A = (1 \quad 2 \quad 3)$ and the column vector

$$B = \begin{pmatrix} -2 \\ 4 \\ 1 \end{pmatrix}.$$

Since A is of order 1 by 3 and B is of order 3 by 1, the matrices are conformable regardless of the order in which they are considered. Hence, the products AB and BA both exist:

$$AB = ((1)(-2) + (2)(4) + (3)(1)) = (9);$$

$$BA = \begin{pmatrix} (-2)(1) & (-2)(2) & (-2)(3) \\ (4)(1) & (4)(2) & (4)(3) \\ (1)(1) & (1)(2) & (1)(3) \end{pmatrix} = \begin{pmatrix} -2 & -4 & -6 \\ 4 & 8 & 12 \\ 1 & 2 & 3 \end{pmatrix}.$$

Note that the product AB may be considerd a matrix whose only element represents the scalar product of two vectors whose components are the elements of A and B, respectively.

EXAMPLE 3 / Prove that $C(A + B) = CA + CB$ where

$$A = \begin{pmatrix} 1 & 2 \\ 3 & 0 \end{pmatrix}, \quad B = \begin{pmatrix} 2 & -1 \\ 3 & 4 \end{pmatrix}, \quad \text{and} \quad C = \begin{pmatrix} 2 & -2 \\ 1 & 3 \\ 4 & -1 \end{pmatrix}.$$

$$C(A + B) = \begin{pmatrix} 2 & -2 \\ 1 & 3 \\ 4 & -1 \end{pmatrix} \left[\begin{pmatrix} 1 & 2 \\ 3 & 0 \end{pmatrix} + \begin{pmatrix} 2 & -1 \\ 3 & 4 \end{pmatrix} \right]$$

$$= \begin{pmatrix} 2 & -2 \\ 1 & 3 \\ 4 & -1 \end{pmatrix} \begin{pmatrix} 3 & 1 \\ 6 & 4 \end{pmatrix} = \begin{pmatrix} -6 & -6 \\ 21 & 13 \\ 6 & 0 \end{pmatrix};$$

$$CA + CB = \begin{pmatrix} 2 & -2 \\ 1 & 3 \\ 4 & -1 \end{pmatrix} \begin{pmatrix} 1 & 2 \\ 3 & 0 \end{pmatrix} + \begin{pmatrix} 2 & -2 \\ 1 & 3 \\ 4 & -1 \end{pmatrix} \begin{pmatrix} 2 & -1 \\ 3 & 4 \end{pmatrix}$$

$$= \begin{pmatrix} -4 & 4 \\ 10 & 2 \\ 1 & 8 \end{pmatrix} + \begin{pmatrix} -2 & -10 \\ 11 & 11 \\ 5 & -8 \end{pmatrix} = \begin{pmatrix} -6 & -6 \\ 21 & 13 \\ 6 & 0 \end{pmatrix};$$

hence, $C(A + B) = CA + CB$.

Example 3 is an illustration of a general theorem of matrix algebra.

THEOREM 4.1 / *The multiplication of matrices is distributive with respect to addition.*

PROOF / Let $A = ((a_{ij}))$ and $B = ((b_{ij}))$ be matrices of order m by n, and let $C = ((c_{ij}))$ be a matrix of order k by m. Then $A + B$, CA, CB, and $C(A + B)$ exist. The elements of the ith row of C are

$$c_{i1}, c_{i2}, \ldots, c_{im},$$

and the elements of the jth column of $A + B$ are

$$a_{1j} + b_{1j}, a_{2j} + b_{2j}, \ldots, a_{mj} + b_{mj}.$$

Therefore, the ijth element of $C(A + B)$ is

$$c_{i1}(a_{1j} + b_{1j}) + c_{i2}(a_{2j} + b_{2j}) + \cdots + c_{im}(a_{mj} + b_{mj});$$

that is,

$$(c_{i1}a_{1j} + c_{i2}a_{2j} + \cdots + c_{im}a_{mj}) + (c_{i1}b_{1j} + c_{i2}b_{2j} + \cdots + c_{im}b_{mj}),$$

the sum of the ijth elements of CA and CB, respectively. Hence,

$$C(A + B) = CA + CB. \tag{4-17}$$

Equation (4-17) represents the *left-hand distributive property* of matrix multiplication. The *right-hand distributive property*

$$(A + B)C = AC + BC \tag{4-18}$$

also is valid, provided that $A + B$, AC, BC, and $(A + B)C$ exist. Note that $C(A + B)$ and $(A + B)C$ generally are not equal.

EXAMPLE 4 / Prove that $A(BC) = (AB)C$ where

$$A = \begin{pmatrix} 1 & 2 \\ -1 & 3 \end{pmatrix}, \quad B = \begin{pmatrix} 1 & 0 & -1 \\ 2 & 1 & 0 \end{pmatrix}, \quad \text{and} \quad C = \begin{pmatrix} 1 & -1 \\ 3 & 2 \\ 2 & 1 \end{pmatrix}.$$

$$A(BC) = \begin{pmatrix} 1 & 2 \\ -1 & 3 \end{pmatrix} \left[\begin{pmatrix} 1 & 0 & -1 \\ 2 & 1 & 0 \end{pmatrix} \begin{pmatrix} 1 & -1 \\ 3 & 2 \\ 2 & 1 \end{pmatrix} \right]$$

$$= \begin{pmatrix} 1 & 2 \\ -1 & 3 \end{pmatrix} \begin{pmatrix} -1 & -2 \\ 5 & 0 \end{pmatrix} = \begin{pmatrix} 9 & -2 \\ 16 & 2 \end{pmatrix};$$

$$(AB)C = \left[\begin{pmatrix} 1 & 2 \\ -1 & 3 \end{pmatrix} \begin{pmatrix} 1 & 0 & -1 \\ 2 & 1 & 0 \end{pmatrix} \right] \begin{pmatrix} 1 & -1 \\ 3 & 2 \\ 2 & 1 \end{pmatrix}$$

$$= \begin{pmatrix} 5 & 2 & -1 \\ 5 & 3 & 1 \end{pmatrix} \begin{pmatrix} 1 & -1 \\ 3 & 2 \\ 2 & 1 \end{pmatrix} = \begin{pmatrix} 9 & -2 \\ 16 & 2 \end{pmatrix};$$

hence, $A(BC) = (AB)C$.

Example 4 is an illustration of one of the most important theorems of matrix algebra.

THEOREM 4.2 / *The multiplication of matrices is associative.*

PROOF / Let $A = ((a_{ij}))$, $B = ((b_{ij}))$, and $C = ((c_{ij}))$ be matrices of order k by m, m by n, and n by p, respectively. Then the products AB, BC, $A(BC)$, and $(AB)C$ exist. The elements of the ith row of A are $a_{i1}, a_{i2}, \ldots, a_{im}$, and the elements of the jth column of BC are $b_{11}c_{1j} + b_{12}c_{2j} + \cdots + b_{1n}c_{nj}$, $b_{21}c_{1j} + b_{22}c_{2j} + \cdots + b_{2n}c_{nj}, \ldots,$ $b_{m1}c_{1j} + b_{m2}c_{2j} + \cdots + b_{mn}c_{nj}$. Therefore, the ijth element of $A(BC)$ is

$$a_{i1}(b_{11}c_{1j} + b_{12}c_{2j} + \cdots + b_{1n}c_{nj}) +$$
$$a_{i2}(b_{21}c_{1j} + b_{22}c_{2j} + \cdots + b_{2n}c_{nj}) +$$
$$\cdots + a_{im}(b_{m1}c_{1j} + b_{m2}c_{2j} + \cdots + b_{mn}c_{nj});$$

that is,

$$(a_{i1}b_{11} + a_{i2}b_{21} + \cdots + a_{im}b_{m1})c_{1j} +$$
$$(a_{i1}b_{12} + a_{i2}b_{22} + \cdots + a_{im}b_{m2})c_{2j} +$$
$$\cdots + (a_{i1}b_{1n} + a_{i2}b_{2n} + \cdots + a_{im}b_{mn})c_{nj},$$

which represents the ijth element of $(AB)C$. Hence,

$$A(BC) = (AB)C. \tag{4-19}$$

Consider the matrices

$$A = \begin{pmatrix} 1 & 0 \\ 1 & 0 \end{pmatrix} \quad \text{and} \quad B = \begin{pmatrix} 0 & 0 \\ 1 & 1 \end{pmatrix}.$$

Note that $AB = 0$, but neither A nor B is a null matrix. In matrix algebra it is possible to have **zero divisors**. These are nonzero elements whose product is the zero element. In the algebra of real numbers it can be shown that zero divisors do not exist; that is, if a and b are real numbers and $ab = 0$, then $a = 0$ or $b = 0$.

EXERCISES

1. Find AB and BA, if they exist, where

$$A = \begin{pmatrix} 3 & 4 & 0 \\ -1 & 0 & 2 \end{pmatrix} \quad \text{and} \quad B = \begin{pmatrix} 6 & -1 & 2 \\ 0 & 1 & 5 \\ -1 & 3 & 4 \end{pmatrix}.$$

2. Use the matrices given to verify that the multiplication of square matrices generally is not commutative:

$$A = \begin{pmatrix} -1 & 2 \\ 3 & 1 \end{pmatrix} \quad \text{and} \quad B = \begin{pmatrix} 0 & 2 \\ 1 & 4 \end{pmatrix}.$$

3. Verify the associative property of matrix multiplication (4-19) for

$$A = \begin{pmatrix} 2 & 1 \\ -1 & 0 \end{pmatrix}, \quad B = \begin{pmatrix} 3 & 1 \\ 2 & -2 \end{pmatrix}, \quad \text{and} \quad C = \begin{pmatrix} 0 & 1 \\ 2 & 1 \end{pmatrix}.$$

4. Verify the right-hand distributive property (4-18) for

$$A = \begin{pmatrix} 1 & 2 & 3 \\ -1 & 0 & 4 \end{pmatrix}, \quad B = \begin{pmatrix} 2 & 0 & 2 \\ 1 & 3 & -1 \end{pmatrix}, \quad \text{and} \quad C = \begin{pmatrix} 2 & 1 \\ -1 & 0 \\ 3 & 2 \end{pmatrix}.$$

5. Find

$$\begin{pmatrix} \cos\theta & \sin\theta \\ -\sin\theta & \cos\theta \end{pmatrix} \begin{pmatrix} \cos\theta & -\sin\theta \\ \sin\theta & \cos\theta \end{pmatrix}.$$

6. Determine conditions on m, n, and p such that the product AB of matrix A of order m by n and matrix B of order n by p is a square matrix.

7. Given

$$A = \begin{pmatrix} 1 & 3 \\ -2 & 0 \end{pmatrix} \quad \text{and} \quad B = \begin{pmatrix} 2 & 1 \\ -1 & 2 \end{pmatrix},$$

show that $(A + B)(A - B) \neq A^2 - B^2$.

8. Determine a necessary and sufficient condition for
(a) $(A + B)(A + B) = A^2 + B^2$; (b) $(A + B)(A - B) = A^2 - B^2$.

9. Find all matrices A such that
(a) $\begin{pmatrix} 3 & 0 \\ 5 & 0 \end{pmatrix} A = \begin{pmatrix} 2 & 0 & 0 \\ 1 & 0 & 0 \end{pmatrix}$; (b) $\begin{pmatrix} 2 & 1 \\ 3 & 2 \end{pmatrix} A = A \begin{pmatrix} 2 & 1 \\ 3 & 2 \end{pmatrix}$.

10. Prove that $AB = BA$ where

$$A = \begin{pmatrix} r & s \\ -s & r \end{pmatrix} \quad \text{and} \quad B = \begin{pmatrix} m & n \\ -n & m \end{pmatrix}.$$

11. Find A^3 if $A^3 = A(AA)$ and

$$A = \begin{pmatrix} 2 & 3 \\ 1 & 0 \end{pmatrix}.$$

12. If

$$A = \begin{pmatrix} 1 & 1 \\ 1 & 0 \end{pmatrix},$$

discuss the nature of A^n where n is a positive integer. (*Note:* The sequence 1, 1, 2, 3, 5, 8, 13, 21, 34, ..., where each term is the sum of the two preceding terms, is called a *Fibonacci sequence*.)

13. If $AC = CA$ and $BC = CB$, prove that $C(AB + BA) = (AB + BA)C$.

14. Construct the multiplication table for the set of *Pauli matrices*:

$$A = \begin{pmatrix} 1 & 0 \\ 0 & 1 \end{pmatrix}, \quad B = \begin{pmatrix} 0 & 1 \\ -1 & 0 \end{pmatrix}, \quad C = \begin{pmatrix} 0 & -1 \\ 1 & 0 \end{pmatrix}, \quad D = \begin{pmatrix} -1 & 0 \\ 0 & -1 \end{pmatrix},$$

$$E = \begin{pmatrix} i & 0 \\ 0 & -i \end{pmatrix}, \quad F = \begin{pmatrix} -i & 0 \\ 0 & i \end{pmatrix}, \quad G = \begin{pmatrix} 0 & -i \\ -i & 0 \end{pmatrix}, \quad H = \begin{pmatrix} 0 & i \\ i & 0 \end{pmatrix}.$$

The Pauli matrices are used in the study of atomic physics.

4-3 / DIAGONAL MATRICES

The elements a_{ij} where $i = j$ of a square matrix $((a_{ij}))$ are called the **diagonal elements** of $((a_{ij}))$ and are said to be on the **main diagonal** or **principal diagonal**. A square matrix of the form

$$A = \begin{pmatrix} a_{11} & 0 & \cdots & 0 \\ 0 & a_{22} & \cdots & 0 \\ \cdots & \cdots & \cdots & \cdots \\ 0 & 0 & \cdots & a_{nn} \end{pmatrix} \qquad (4\text{-}20)$$

is called a **diagonal matrix**; that is, a diagonal matrix is a square matrix $((a_{ij}))$ where $a_{ij} = 0$ if $i \neq j$ for all pairs (i, j). For example,

$$\begin{pmatrix} 3 & 0 \\ 0 & -1 \end{pmatrix}, \quad \begin{pmatrix} 2 & 0 & 0 \\ 0 & 5 & 0 \\ 0 & 0 & 0 \end{pmatrix}, \quad \text{and} \quad \begin{pmatrix} 3 & 0 & 0 \\ 0 & 3 & 0 \\ 0 & 0 & 3 \end{pmatrix} \qquad (4\text{-}21)$$

are diagonal matrices, as is any null matrix of order n. If all the a_{ii}'s of (4-20) are equal, then the diagonal matrix is called a **scalar matrix**. The third matrix of (4-21) is an example of a scalar matrix.

EXAMPLE 1 / Determine the effect of the premultiplication and the postmultiplication of any square matrix of order two by (a) a conformable diagonal matrix; (b) a conformable scalar matrix.

Let
$$A = \begin{pmatrix} a & b \\ c & d \end{pmatrix}$$
be any square matrix of order two,
$$D = \begin{pmatrix} k_1 & 0 \\ 0 & k_2 \end{pmatrix}$$
be a conformable diagonal matrix, and
$$S = \begin{pmatrix} k & 0 \\ 0 & k \end{pmatrix}$$
be a conformable scalar matrix.

(a) By the definition of matrix multiplication,
$$DA = \begin{pmatrix} k_1 & 0 \\ 0 & k_2 \end{pmatrix}\begin{pmatrix} a & b \\ c & d \end{pmatrix} = \begin{pmatrix} k_1 a & k_1 b \\ k_2 c & k_2 d \end{pmatrix};$$
that is, premultiplication of matrix A by D results in a matrix where each element of the ith row equals the product of the corresponding element of A and the diagonal element in the ith row of D. Similarly,
$$AD = \begin{pmatrix} a & b \\ c & d \end{pmatrix}\begin{pmatrix} k_1 & 0 \\ 0 & k_2 \end{pmatrix} = \begin{pmatrix} k_1 a & k_2 b \\ k_1 c & k_2 d \end{pmatrix};$$
that is, postmultiplication of matrix A by D results in a matrix where each element of the ith column equals the product of the corresponding element of A and the diagonal element in the ith column of D.

(b)
$$SA = \begin{pmatrix} k & 0 \\ 0 & k \end{pmatrix}\begin{pmatrix} a & b \\ c & d \end{pmatrix} = \begin{pmatrix} ka & kb \\ kc & kd \end{pmatrix};$$
$$AS = \begin{pmatrix} a & b \\ c & d \end{pmatrix}\begin{pmatrix} k & 0 \\ 0 & k \end{pmatrix} = \begin{pmatrix} ka & kb \\ kc & kd \end{pmatrix};$$

$SA = AS = kA$. In general, the product of any matrix A and a conformable scalar matrix S with diagonal elements k is equivalent to the scalar multiple kA.

A scalar matrix $((a_{ii}))$ where $a_{ii} = 1$ for all values of i is called an **identity matrix** or a **unit matrix**. An identity matrix of order n will be denoted by I and has the property that for every square matrix A of order n

$$AI = IA = A. \qquad (4\text{-}22)$$

Assume that I is the only matrix with this property (4-22). Note that if A is a matrix of order m by n, the premultiplicative identity matrix is of order m by m while the postmultiplicative identity matrix is of order n by n.

The identity matrix is often denoted by $((\delta_{ij}))$ where the Kronecker delta symbol δ_{ij} is defined by

$$\delta_{ij} \begin{cases} = 0, & \text{when } i \neq j \\ = 1, & \text{when } i = j. \end{cases} \qquad (4\text{-}23)$$

EXAMPLE 2 / Show that $AB = BA = I$ where

$$A = \begin{pmatrix} 1 & -1 & 1 \\ 0 & 1 & 0 \\ 2 & 0 & 3 \end{pmatrix} \quad \text{and} \quad B = \begin{pmatrix} 3 & 3 & -1 \\ 0 & 1 & 0 \\ -2 & -2 & 1 \end{pmatrix}.$$

$$AB = \begin{pmatrix} 3+0-2 & 3-1-2 & -1+0+1 \\ 0+0+0 & 0+1+0 & 0+0+0 \\ 6+0-6 & 6+0-6 & -2+0+3 \end{pmatrix} = \begin{pmatrix} 1 & 0 & 0 \\ 0 & 1 & 0 \\ 0 & 0 & 1 \end{pmatrix};$$

$$BA = \begin{pmatrix} 3+0-2 & -3+3+0 & 3+0-3 \\ 0+0+0 & 0+1+0 & 0+0+0 \\ -2+0+2 & 2-2+0 & -2+0+3 \end{pmatrix} = \begin{pmatrix} 1 & 0 & 0 \\ 0 & 1 & 0 \\ 0 & 0 & 1 \end{pmatrix};$$

hence, $AB = BA = I$.

EXERCISES

1. Construct a matrix $((a_{ij}))$ of order 3 by 4 where $a_{ij} = 3i + \delta_{ij}j^2$.

2. Show that $AB = BA = I$ where

$$A = \begin{pmatrix} 3 & -4 & 2 \\ -2 & 1 & 0 \\ -1 & -1 & 1 \end{pmatrix} \text{ and } B = \begin{pmatrix} 1 & 2 & -2 \\ 2 & 5 & -4 \\ 3 & 7 & -5 \end{pmatrix}.$$

3. Show that $A^2 = I$ where

$$A = \begin{pmatrix} -1 & -1 & -1 \\ 0 & 1 & 0 \\ 0 & 0 & 1 \end{pmatrix}.$$

4. Prove that the multiplication of any two diagonal matrices of the same order is commutative.

4-4 / SPECIAL REAL MATRICES

A special type of matrix, which will be used extensively in Chapter 7, is a symmetric matrix. Symmetric matrices play an important role in many branches of mathematics as well as in other sciences. A matrix $((a_{ij}))$ is called a **symmetric matrix** if, and only if, $a_{ij} = a_{ji}$ for all pairs (i, j). The matrix

$$\begin{pmatrix} 3 & 1 & 0 \\ 1 & 2 & -2 \\ 0 & -2 & 4 \end{pmatrix}$$

is an example of a symmetric matrix of order three.

A matrix $((a_{ij}))$ is called a **skew-symmetric matrix** or **anti-symmetric matrix** if, and only if, $a_{ij} = -a_{ji}$ for all pairs (i, j). The matrix

$$\begin{pmatrix} 0 & -2 \\ 2 & 0 \end{pmatrix}$$

is an example of a skew-symmetric matrix of order two.

A matrix must necessarily be a square matrix to be symmetric or skew-symmetric. Furthermore, the diagonal elements of a skew-symmetric matrix must be zero since $a_{ii} = -a_{ii}$ if, and only if, $a_{ii} = 0$.

EXAMPLE 1 / Determine which of the following matrices are symmetric matrices and which are skew-symmetric matrices:

$$A = \begin{pmatrix} 3 & 0 \\ 0 & 2 \end{pmatrix}, \qquad B = \begin{pmatrix} 3 & 4 \\ -4 & 1 \end{pmatrix}, \quad C = \begin{pmatrix} 2 & -1 \\ -1 & 1 \end{pmatrix},$$

$$D = \begin{pmatrix} 0 & 2 \\ -2 & 0 \end{pmatrix}, \qquad E = \begin{pmatrix} 0 & 0 \\ 1 & 0 \end{pmatrix}, \quad F = (3),$$

$$G = \begin{pmatrix} 0 & 1 & -2 \\ -1 & 0 & 3 \\ 2 & -3 & 0 \end{pmatrix}, \; H = \begin{pmatrix} 3 & 2 \\ 2 & 1 \\ 1 & 0 \end{pmatrix}, \quad J = \begin{pmatrix} 0 & 0 \\ 0 & 0 \end{pmatrix}.$$

Matrices A, C, F, and J are symmetric matrices, while D, G, and J are skew-symmetric matrices. Matrices B, E, and H are neither symmetric nor skew-symmetric. Note that the null matrix J of order two may be considered both symmetric and skew-symmetric. Null matrices of any order n are the only matrices that have this property.

Symmetric and skew-symmetric matrices may be discussed in terms of **transposition,** the operation of interchanging the rows and columns of a given matrix. The matrix A^T that is obtained by transposition from a matrix A is called the **transpose** of A. If $A = ((a_{ij}))$, then $A^T = ((a_{ji}))$. Note that the transpose of a column vector is a row vector. In general, the transpose of a matrix of order m by n is a matrix of order n by m. If $a_{ij} = a_{ji}$ for all pairs (i, j) for matrix A, then A is a symmetric matrix and $A^T = A$. If $a_{ij} = -a_{ji}$ for all pairs (i, j) for matrix A, then A is a skew-symmetric matrix and $A^T = -A$.

The following theorems are true for matrices in general.

THEOREM 4.3 / *The transposition operation is reflexive; that is, $(A^T)^T = A$.*

PROOF / Let $A = ((a_{ij}))$ and $A^T = ((b_{ij}))$. Then $b_{ij} = a_{ji}$ for all pairs (i, j). The transpose of A^T is $((b_{ji}))$ and thus $((a_{ij}))$. Hence, $(A^T)^T = A$.

THEOREM 4.4 / *The transpose of the sum (difference) of two matrices is equal to the sum (difference) of their transposes; that is, $(A + B)^T = A^T + B^T$ and $(A - B)^T = A^T - B^T$.*

PROOF / Let $A = ((a_{ij}))$ and $B = ((b_{ij}))$ be any two matrices of the same order. Then $A + B = ((c_{ij}))$ where $c_{ij} = a_{ij} + b_{ij}$ for all pairs (i, j);

$$A^T = ((a_{ji}));$$
$$B^T = ((b_{ji}));$$
$$(A + B)^T = ((c_{ji}))$$
$$= ((a_{ji} + b_{ji}))$$
$$= A^T + B^T.$$

In a similar manner, it can be shown that $(A - B)^T = A^T - B^T$.

THEOREM 4.5 / *The transpose of the product of two matrices is equal to the product of their transposes in reverse order; that is, $(AB)^T = B^T A^T$.*

PROOF / Let $A = ((a_{ij}))$ and $B = ((b_{ij}))$ be matrices of order k by m and m by n, respectively. Let $AB = ((c_{ij}))$. Then $c_{ij} = a_{i1}b_{1j} + a_{i2}b_{2j} + \cdots + a_{im}b_{mj}$ and is the element of the jth row and ith column of $((c_{ij}))^T$; that is, $(AB)^T$.

The elements $b_{1j}, b_{2j}, \ldots, b_{mj}$ of the jth column of B are the elements of the jth row of B^T. The elements $a_{i1}, a_{i2}, \ldots, a_{im}$ of the ith row of A are the elements of the ith column of A^T. The element of the jth row and ith column of $B^T A^T$ is $b_{1j}a_{i1} + b_{2j}a_{i2} + \cdots + b_{mj}a_{im}$; that is, $a_{i1}b_{1j} + a_{i2}b_{2j} + \cdots + a_{im}b_{mj}$. Hence, $(AB)^T = B^T A^T$.

THEOREM 4.6 / *The product of any matrix and its transpose is a symmetric matrix; that is, $(AA^T)^T = AA^T$.*

PROOF /

$$(AA^T)^T = (A^T)^T A^T \quad \text{by Theorem 4.5}$$
$$= AA^T \quad \text{by Theorem 4.3.}$$

Hence, AA^T is a symmetric matrix.

Note that if A is a matrix of order m by n, then AA^T is a symmetric matrix of order m; $A^T A$ is a symmetric matrix of order n.

The following theorems are true for square matrices.

THEOREM 4.7 / *The sum of any matrix and its transpose is a symmetric matrix; that is,* $(A + A^T)^T = A + A^T$.

PROOF /

$$(A + A^T)^T = A^T + (A^T)^T \quad \text{by Theorem 4.4}$$
$$= A^T + A \quad \text{by Theorem 4.3}$$
$$= A + A^T \quad \text{since the addition of matrices is commutative.}$$

Hence, $A + A^T$ is a symmetric matrix.

THEOREM 4.8 / *The difference of any matrix and its transpose is a skew-symmetric matrix; that is,* $(A - A^T)^T = -(A - A^T)$.

PROOF /

$$(A - A^T)^T = A^T - (A^T)^T \quad \text{by Theorem 4.4}$$
$$= A^T - A \quad \text{by Theorem 4.3}$$
$$= -(A - A^T) \quad \text{by the definitions and properties of scalar multiplication, matrix subtraction, and matrix addition.}$$

Hence, $A - A^T$ is a skew-symmetric matrix.

THEOREM 4.9 / *If A and B are symmetric matrices, then AB is a symmetric matrix if, and only if, $AB = BA$.*

PROOF / Let A and B be symmetric matrices of the same order such that AB is a symmetric matrix. Then

$$AB = (AB)^T \quad \text{since } AB \text{ is a symmetric matrix}$$
$$= B^T A^T \quad \text{by Theorem 4.5}$$
$$= BA \quad \text{since } A \text{ and } B \text{ are symmetric matrices.}$$

Hence, $AB = BA$.

If $AB = BA$, where A and B are symmetric matrices, then

$$(AB)^T = (BA)^T \quad \text{since } AB = BA$$
$$= A^T B^T \quad \text{by Theorem 4.5}$$
$$= AB \quad \text{since } A \text{ and } B \text{ are symmetric matrices.}$$

Hence, AB is a symmetric matrix.

EXAMPLE 2 / Verify Theorem 4.5 where

$$A = \begin{pmatrix} 2 & 3 \\ 0 & -1 \end{pmatrix} \quad \text{and} \quad B = \begin{pmatrix} 1 & 5 \\ 2 & 4 \end{pmatrix}.$$

$$AB = \begin{pmatrix} 2 & 3 \\ 0 & -1 \end{pmatrix}\begin{pmatrix} 1 & 5 \\ 2 & 4 \end{pmatrix} = \begin{pmatrix} 8 & 22 \\ -2 & -4 \end{pmatrix}; \quad (AB)^T = \begin{pmatrix} 8 & -2 \\ 22 & -4 \end{pmatrix};$$

$$B^T A^T = \begin{pmatrix} 1 & 2 \\ 5 & 4 \end{pmatrix}\begin{pmatrix} 2 & 0 \\ 3 & -1 \end{pmatrix} = \begin{pmatrix} 8 & -2 \\ 22 & -4 \end{pmatrix}; \quad (AB)^T = B^T A^T.$$

EXERCISES

1. Determine which of the following matrices are
(a) symmetric matrices;   (b) skew-symmetric matrices:

$$A = \begin{pmatrix} 2 & 3 \\ 4 & 5 \end{pmatrix}, \quad B = \begin{pmatrix} 3 & 1 \\ -1 & 3 \end{pmatrix}, \quad C = \begin{pmatrix} 2 & 2 \\ 2 & 2 \end{pmatrix},$$

$$D = \begin{pmatrix} 5 & 1 \\ 1 & 2 \end{pmatrix}, \quad E = \begin{pmatrix} 5 & 2 & 1 \\ 4 & 2 & 4 \\ 1 & 2 & 3 \end{pmatrix}, \quad F = \begin{pmatrix} 0 & -3 \\ 3 & 0 \end{pmatrix},$$

$$G = \begin{pmatrix} 1 & 1 & 1 \\ 2 & 2 & 2 \\ 3 & 3 & 3 \end{pmatrix}, \quad H = \begin{pmatrix} 0 \\ 0 \\ 0 \end{pmatrix}, \quad J = \begin{pmatrix} 0 & 0 & 0 \\ 0 & 0 & 0 \\ 0 & 0 & 0 \end{pmatrix}.$$

2. Determine the maximum number of distinct elements in any symmetric matrix of order n.

3. Determine the maximum number of distinct elements in any skew-symmetric matrix of order n.

4. Verify Theorem 4.5 for matrices A and B of Exercise 1.

5. Verify Theorem 4.6 for matrix E of Exercise 1.

6. Verify Theorem 4.7 for matrix A of Exercise 1.

7. Verify Theorem 4.8 for matrix E of Exercise 1.

8. Prove that every diagonal matrix is a symmetric matrix.

9. Prove that the square of a skew-symmetric matrix is a symmetric matrix.

10. Verify the results of Exercise 9 for the matrix

$$A = \begin{pmatrix} 0 & a & b \\ -a & 0 & -c \\ -b & c & 0 \end{pmatrix}.$$

11. Prove that if A and B are skew-symmetric matrices of the same order, then AB is a symmetric matrix if, and only if, $AB = BA$.

12. Prove that the set of symmetric matrices of order n is a subspace of the real vector space of square real matrices of order n.

4-5 / SPECIAL COMPLEX MATRICES

This book is concerned primarily with real matrices; however, some special complex matrices, which will be useful in Chapter 7 for proving a set of theorems about real symmetric matrices, are considered in this section. A **complex matrix** is a matrix whose elements are complex numbers. Since every real number is a complex number, every real matrix is a complex matrix, but not every complex matrix is a real matrix.

If A is a complex matrix, then $\bar{A}$ denotes the matrix obtained from A by replacing each element $z = a + bi$ with its conjugate $\bar{z} = a - bi$. The matrix $\bar{A}$ is called the **conjugate** of matrix A. For example, each of the matrices

$$\begin{pmatrix} 2+i & 3 \\ i & 5-2i \end{pmatrix} \text{ and } \begin{pmatrix} 2-i & 3 \\ -i & 5+2i \end{pmatrix}$$

is the conjugate of the other. Note that matrix A is a real matrix if, and only if, $A = \bar{A}$. The transpose of the conjugate of matrix A will be denoted by A^*; that is, $A^* = (\bar{A})^T$. If $A = ((a_{ij}))$, then $A^T = ((a_{ji}))$, $\bar{A} = ((\overline{a_{ij}}))$, and $(\bar{A})^T = ((\overline{a_{ji}})) = (\overline{A^T})$. Hence, the transpose of the conjugate of a matrix is equal to the conjugate of the transpose of the matrix.

A matrix A such that $A = A^*$ is called a **Hermitian matrix**; that is, a matrix $A = ((a_{ij}))$ is a Hermitian matrix if, and only if, $a_{ij} = \overline{a_{ji}}$ for all pairs (i, j). Since $a_{ii} = \overline{a_{ii}}$ only if a_{ii} is a real number, the diagonal elements of a Hermitian matrix are real numbers. If A is a real symmetric matrix, then $a_{ij} = a_{ji}$, and $a_{ij} = \overline{a_{ji}}$ for all pairs (i, j) since $a_{ji} = \overline{a_{ji}}$. Hence, every real symmetric matrix is a Hermitian matrix.

A matrix A such that $A = -A^*$ is called a **skew-Hermitian matrix**;

that is, a matrix $A = ((a_{ij}))$ is a skew-Hermitian matrix if, and only if, $a_{ij} = -\overline{a_{ji}}$ for all pairs (i, j). Every real skew-symmetric matrix is a skew-Hermitian matrix.

EXAMPLE 1 / Prove that A is a skew-Hermitian matrix where

$$A = \begin{pmatrix} 2i & 3 & i \\ -3 & 0 & -2-i \\ i & 2-i & i \end{pmatrix}.$$

$$\overline{A} = \begin{pmatrix} -2i & 3 & -i \\ -3 & 0 & -2+i \\ -i & 2+i & -i \end{pmatrix};$$

$$A^* = (\overline{A})^T = \begin{pmatrix} -2i & -3 & -i \\ 3 & 0 & 2+i \\ -i & -2+i & -i \end{pmatrix};$$

$$-A^* = \begin{pmatrix} 2i & 3 & i \\ -3 & 0 & -2-i \\ i & 2-i & i \end{pmatrix} = A.$$

Hence, A is a skew-Hermitian matrix.

EXAMPLE 2 / Prove that $\overline{AB} = \overline{A}\,\overline{B}$.

Let $A = ((a_{ij}))$ and $B = ((b_{ij}))$ be matrices of order m by n and n by p, respectively. Then the ijth element of AB is given by

$$\sum_{k=1}^{n} a_{ik}b_{kj},$$

and the ijth element of $\overline{AB}$ is given by

$$\sum_{k=1}^{n} \overline{a_{ik}b_{kj}}$$

since the conjugate of the sum of complex numbers is equal to the sum of their conjugates. Also, since the conjugate of the product of two complex numbers is equal to the product of their conjugates,

$$\sum_{k=1}^{n} \overline{a_{ik}b_{kj}} = \sum_{k=1}^{n} \overline{a_{ik}}\,\overline{b_{kj}}.$$

Hence, $\overline{AB} = \overline{A}\overline{B}$; that is, the conjugate of the product of two matrices is equal to the product of their conjugates.

EXAMPLE 3 / Prove that $(AB)^* = B^*A^*$.

$$\begin{aligned}(AB)^* &= (\overline{AB})^T && \text{by definition} \\ &= (\overline{A}\overline{B})^T && \text{by the results of Example 2} \\ &= (\overline{B})^T(\overline{A})^T && \text{by Theorem 4.5} \\ &= B^*A^* && \text{by definition.}\end{aligned}$$

Additional properties of Hermitian matrices and skew-Hermitian matrices are presented in the exercises.

EXERCISES

1. If

$$A = \begin{pmatrix} 1 & 3+2i \\ -i & 2-i \end{pmatrix},$$

find (a) $\overline{A}$; (b) A^*.

2. Determine which of the following matrices are
(a) Hermitian matrices;   (b) skew-Hermitian matrices:

$$A = \begin{pmatrix} 0 & -i \\ i & 0 \end{pmatrix}, \quad B = \begin{pmatrix} 2 & 3+i \\ -3-i & 5 \end{pmatrix}, \quad C = \begin{pmatrix} 0 & 0 \\ 0 & 0 \end{pmatrix},$$

$$D = \begin{pmatrix} 3 & 5+2i \\ 5-2i & 1 \end{pmatrix}, \quad E = \begin{pmatrix} 0 & 4+5i \\ -4+5i & 0 \end{pmatrix}, \quad F = \begin{pmatrix} 2 & 3 \\ -3 & 2 \end{pmatrix},$$

$$G = \begin{pmatrix} 0 & 2+i & i \\ 2+i & 0 & 1+3i \\ i & 1+3i & 0 \end{pmatrix}, \quad H = \begin{pmatrix} 0 & 3 & 2-i \\ -3 & 0 & -2i \\ -2-i & -2i & 0 \end{pmatrix}.$$

3. Prove that the diagonal elements of a skew-Hermitian matrix are either zeros or pure imaginary numbers.

Prove the following theorems for complex matrices.

4. $(A^*)^* = A$.

5. $(A + B)^* = A^* + B^*$.

6. If k is any complex number, then $(kA)^* = \bar{k}A^*$.

7. The product of any matrix and its transposed conjugate is a Hermitian matrix; that is, $AA^* = (AA^*)^*$.

8. The sum of any square matrix and its transposed conjugate is a Hermitian matrix; that is, $A + A^* = (A + A^*)^*$.

9. The difference of any square matrix and its transposed conjugate is a skew-Hermitian matrix; that is, $A - A^* = -(A - A^*)^*$.

10. Every square matrix can be expressed as the sum of a Hermitian matrix and a skew-Hermitian matrix. (*Hint:* See Exercises 8 and 9.)

11. Every Hermitian matrix can be expressed as $A + Bi$ where A is a real symmetric matrix and B is a real skew-symmetric matrix.

12. Every skew-Hermitian matrix can be expressed as $A + Bi$ where A is a real skew-symmetric matrix and B is a real symmetric matrix.

FIVE

INVERSES AND SYSTEMS OF MATRICES

5-1 / DETERMINANTS

Associated with each square matrix $A = ((a_{ij}))$ of order two is a function of the matrix called its *determinant* and denoted either by det A or by

$$\begin{vmatrix} a_{11} & a_{12} \\ a_{21} & a_{22} \end{vmatrix}.$$

The determinant function assigns to each square real matrix A of order two a unique real number $a_{11}a_{22} - a_{21}a_{12}$ called the *value of the determinant*; that is, det $A = a_{11}a_{22} - a_{21}a_{12}$. The value of the determinant of a matrix of order two may be remembered by the array

$$\begin{vmatrix} a_{11} & a_{12} \\ a_{21} & a_{22} \end{vmatrix} \begin{matrix} - \\ + \end{matrix} = a_{11}a_{22} - a_{21}a_{12}. \tag{5-1}$$

Note that the value of the determinant is the difference of the product of the diagonal elements and the product of the remaining two elements.

EXAMPLE 1 / Find the value of the determinant of matrix A where

$$A = \begin{pmatrix} 2 & -1 \\ 4 & 3 \end{pmatrix}.$$

$$\det A = \begin{vmatrix} 2 & -1 \\ 4 & 3 \end{vmatrix} = (2)(3) - (4)(-1) = 10.$$

A *determinant* of a matrix $A = ((a_{ij}))$ of order three is a function of the square matrix and is denoted by det A or by

$$\begin{vmatrix} a_{11} & a_{12} & a_{13} \\ a_{21} & a_{22} & a_{23} \\ a_{31} & a_{32} & a_{33} \end{vmatrix}.$$

The *value of the determinant* of A is defined as

$$a_{11}a_{22}a_{33} + a_{12}a_{23}a_{31} + a_{13}a_{21}a_{32}$$
$$- a_{31}a_{22}a_{13} - a_{32}a_{23}a_{11} - a_{33}a_{21}a_{12}.$$

The value of the determinant of a matrix of order three may be remembered by use of a particular scheme similar to that used for determinants of matrices of order two (5-1):

$$\begin{matrix} & -a_{31}a_{22}a_{13} & -a_{32}a_{23}a_{11} & -a_{33}a_{21}a_{12} \\ \begin{vmatrix} a_{11} & a_{12} & a_{13} \\ a_{21} & a_{22} & a_{23} \\ a_{31} & a_{32} & a_{33} \end{vmatrix} & \begin{matrix} a_{11} & a_{12} \\ a_{21} & a_{22} \\ a_{31} & a_{32} \end{matrix} & & \\ & +a_{11}a_{22}a_{33} & +a_{12}a_{23}a_{31} & +a_{13}a_{21}a_{32} \end{matrix} \quad (5\text{-}2)$$

In (5-2) the first two columns of the determinant are repeated at its right. The products of the three elements along the arrows running downward and to the right are noted, as well as the negative of the products of the three elements along the arrows running upward and to the right. The algebraic sum of these six products is the value of the determinant.

EXAMPLE 2 / Find the value of the determinant of matrix A where

$$A = \begin{pmatrix} 0 & 4 & 2 \\ 4 & -2 & -1 \\ 5 & 1 & 3 \end{pmatrix}.$$

Using the scheme of (5-2), we have

$$-(-20)-(0)-(48)$$

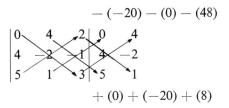

$$+(0)+(-20)+(8)$$

Hence, the value of the determinant of matrix A is $(0)+(-20)+(8)-(-20)-(0)-(48)$; that is, $\det A = -40$.

Note carefully that the method of diagonals employed for the evaluation of determinants of matrices of order three is not valid for determinants of matrices of higher order.

In general, a **determinant** of a matrix $A = ((a_{ij}))$ of order n is a function of the square matrix and is denoted by $\det A$ or by

$$\begin{vmatrix} a_{11} & a_{12} & \cdots & a_{1n} \\ a_{21} & a_{22} & \cdots & a_{2n} \\ \cdots & \cdots & \cdots & \cdots \\ a_{n1} & a_{n2} & \cdots & a_{nn} \end{vmatrix}.$$

The **value of the determinant** of a matrix of order n is defined as the sum of $n!$ terms of the form

$$(-1)^k a_{1i_1} a_{2i_2} \cdots a_{ni_n}.$$

Each term contains one and only one element from each row and one and only one element from each column; that is, the second subscripts $i_1, i_2, \ldots, i_n$ are equal to $1, 2, \ldots, n$, taken in some order. The exponent k represents the number of interchanges of two elements necessary for the second subscripts to be placed in the order $1, 2, \ldots, n$. For example, consider the term containing $a_{13}a_{21}a_{34}a_{42}$ in the evaluation of the determinant of a matrix of order four. The value of k is 3 since three interchanges of two elements are necessary for the second subscripts to be placed in the order 1, 2, 3, 4:

$$a_{13}a_{21}a_{34}a_{42} = a_{21}a_{13}a_{34}a_{42} = a_{21}a_{42}a_{34}a_{13} = a_{21}a_{42}a_{13}a_{34}.$$

Hence, the term containing the factor $a_{13}a_{21}a_{34}a_{42}$ has the additional factor $(-1)^3$; that is, -1.

Although the following theorems are valid for determinants of matrices of order n, their proofs will be considered only for determinants of matrices of order three. These theorems are extremely useful in the evaluation of determinants of matrices of order $n \geq 4$ since the values of such determinants are seldom computed from the definition. In addition, several of these theorems will be used to discuss certain concepts in matrix algebra.

THEOREM 5.1 / *The value of a determinant remains unchanged if corresponding rows and columns are interchanged; that is, $\det A = \det A^T$.*

PROOF / Consider the determinants

$$\det A = \begin{vmatrix} a_{11} & a_{12} & a_{13} \\ a_{21} & a_{22} & a_{23} \\ a_{31} & a_{32} & a_{33} \end{vmatrix} \quad \text{and} \quad \det A^T = \begin{vmatrix} a_{11} & a_{21} & a_{31} \\ a_{12} & a_{22} & a_{32} \\ a_{13} & a_{23} & a_{33} \end{vmatrix}.$$

Note that the elements of each row or column of $\det A$ correspond to the elements of the same numbered column or row, respectively, of $\det A^T$. By definition,

$$\det A = a_{11}a_{22}a_{33} + a_{12}a_{23}a_{31} + a_{13}a_{21}a_{32}$$
$$- a_{31}a_{22}a_{13} - a_{32}a_{23}a_{11} - a_{33}a_{21}a_{12}$$

and

$$\det A^T = a_{11}a_{22}a_{33} + a_{21}a_{32}a_{13} + a_{31}a_{12}a_{23}$$
$$- a_{13}a_{22}a_{31} - a_{23}a_{32}a_{11} - a_{33}a_{12}a_{21}.$$

By a simple rearrangement of factors and terms,

$$\det A = \det A^T.$$

The importance of Theorem 5.1 is that for every proved theorem concerning the rows of a determinant there exists a corresponding theorem concerning the columns. The corresponding theorem will not be stated in each instance, but the reader should interpret each theorem in terms of the corresponding theorem as well.

THEOREM 5.2 / *If any two rows of a determinant are interchanged, then the sign of the value of the determinant is changed.*

For example,

$$\begin{vmatrix} a_{11} & a_{12} & a_{13} \\ a_{31} & a_{32} & a_{33} \\ a_{21} & a_{22} & a_{23} \end{vmatrix} = - \begin{vmatrix} a_{11} & a_{12} & a_{13} \\ a_{21} & a_{22} & a_{23} \\ a_{31} & a_{32} & a_{33} \end{vmatrix}.$$

The proof of Theorem 5.2 is left to the reader as an exercise.

THEOREM 5.3 / *If every element of a row of a determinant is zero, then the value of the determinant is zero.*

PROOF / Consider the definition of the value of the determinant of a matrix of order three. Every term of this expression contains one and only one element from each row as a factor. Therefore, every term must contain a factor that is an element from the row of zeros. Hence, the value of the determinant is zero.

THEOREM 5.4 / *If every element of a row of a determinant is multiplied by the factor k, then the value of the determinant is multiplied by k.*

For example,

$$\begin{vmatrix} ka_{11} & ka_{12} & ka_{13} \\ a_{21} & a_{22} & a_{23} \\ a_{31} & a_{32} & a_{33} \end{vmatrix} = k \begin{vmatrix} a_{11} & a_{12} & a_{13} \\ a_{21} & a_{22} & a_{23} \\ a_{31} & a_{32} & a_{33} \end{vmatrix}.$$

PROOF / We have already mentioned that every term in the definition of the value of the determinant of a matrix of order three contains one and only one element from each row as a factor. Therefore, every term will contain the factor k once and only once; that is, the value of the determinant will be multiplied by k.

THEOREM 5.5 / *The value of a determinant remains unchanged if every element of a row is increased by a scalar multiple of the corresponding element of another row.*

For example,

$$\begin{vmatrix} a_{11} & a_{12} & a_{13} \\ a_{21} + ka_{31} & a_{22} + ka_{32} & a_{23} + ka_{33} \\ a_{31} & a_{32} & a_{33} \end{vmatrix} = \begin{vmatrix} a_{11} & a_{12} & a_{13} \\ a_{21} & a_{22} & a_{23} \\ a_{31} & a_{32} & a_{33} \end{vmatrix}.$$

The reader should be careful to note that the elements of the third row are left intact while a scalar multiple of these elements is added to each of the corresponding elements of the second row. The proof of Theorem 5.5 may be obtained by use of the definition of the value of the determinant of a matrix of order three and is left to the reader as an exercise.

The value of the determinant of a matrix of order three has been defined by the equation

$$\begin{vmatrix} a_{11} & a_{12} & a_{13} \\ a_{21} & a_{22} & a_{23} \\ a_{31} & a_{32} & a_{33} \end{vmatrix} = \begin{matrix} a_{11}a_{22}a_{33} + a_{12}a_{23}a_{31} + a_{13}a_{21}a_{32} \\ - a_{31}a_{22}a_{13} - a_{32}a_{23}a_{11} - a_{33}a_{21}a_{12}. \end{matrix} \qquad (5\text{-}3)$$

By a rearrangement of terms, the right-hand member of (5-3) may be written as

$$a_{11}(a_{22}a_{33} - a_{32}a_{23}) - a_{12}(a_{21}a_{33} - a_{31}a_{23}) + a_{13}(a_{21}a_{32} - a_{31}a_{22}),$$

and

$$a_{11}\begin{vmatrix} a_{22} & a_{23} \\ a_{32} & a_{33} \end{vmatrix} - a_{12}\begin{vmatrix} a_{21} & a_{23} \\ a_{31} & a_{33} \end{vmatrix} + a_{13}\begin{vmatrix} a_{21} & a_{22} \\ a_{31} & a_{32} \end{vmatrix}. \qquad (5\text{-}4)$$

The three determinants in (5-4) may be obtained from the original determinant by eliminating certain rows and columns. The pattern shown by (5-4) for expressing the value of the determinant of a matrix of order three may be generalized for determinants of matrices of higher order.

Consider the determinant of a square matrix $((a_{ij}))$ of order n. The determinant obtained from $\det((a_{ij}))$ by deleting the elements of the ith row and jth column is called the **minor** of the element a_{ij} and is denoted by M_{ij}. The minor, when prefixed by a positive or negative sign according to whether the sum of the position numbers of the row and column deleted from $\det((a_{ij}))$ is even or odd, respectively, is called the **cofactor** of the element a_{ij} and is denoted by the symbol A_{ij}. That is,

$$A_{ij} = (-1)^{i+j}M_{ij}. \qquad (5\text{-}5)$$

For example, the minor of a_{12} in

$$\det A = \begin{vmatrix} 2 & 1 & 3 \\ 4 & -1 & 2 \\ -2 & 0 & 1 \end{vmatrix}$$

is

$$M_{12} = \begin{vmatrix} 4 & 2 \\ -2 & 1 \end{vmatrix} = (4)(1) - (-2)(2) = 4 + 4 = 8,$$

and the cofactor is

$$A_{12} = (-1)^{1+2} M_{12} = (-1)(8) = -8.$$

THEOREM 5.6 / *The value of a determinant is equal to the sum of the products of the elements of any row and their cofactors.*

Theorem 5.6 is a key theorem for evaluating determinants of matrices of order n. The proof of Theorem 5.6 for determinants of matrices of order n may be found in most advanced linear algebra texts.

EXAMPLE 3 / Evaluate

$$\begin{vmatrix} 2 & -3 & 1 \\ 0 & 5 & 2 \\ -1 & -2 & 3 \end{vmatrix}$$

using the cofactors of the elements of the second column.

The cofactors of -3, 5, and -2, the elements of the second column, are denoted symbolically by A_{12}, A_{22}, and A_{32}, respectively. Now,

$$A_{12} = (-1)^{1+2} \begin{vmatrix} 0 & 2 \\ -1 & 3 \end{vmatrix} = -2; \quad A_{22} = (-1)^{2+2} \begin{vmatrix} 2 & 1 \\ -1 & 3 \end{vmatrix} = 7;$$

$$A_{32} = (-1)^{3+2} \begin{vmatrix} 2 & 1 \\ 0 & 2 \end{vmatrix} = -4.$$

Using Theorems 5.1 and 5.6, we obtain

$$\begin{vmatrix} 2 & -3 & 1 \\ 0 & 5 & 2 \\ -1 & -2 & 3 \end{vmatrix} = (-3)A_{12} + (5)A_{22} + (-2)A_{32}$$

$$= (-3)(-2) + (5)(7) + (-2)(-4) = 49.$$

The symbol $|A|$ is also used to denote the determinant of A. Be careful to note that $|A|$ does not mean the absolute value of A.

EXERCISES

In Exercises 1 through 6 find the value of the given determinant.

1. $\begin{vmatrix} 5 & 3 \\ 2 & 4 \end{vmatrix}$.

2. $\begin{vmatrix} 4 & -2 \\ 2 & 1 \end{vmatrix}$.

3. $\begin{vmatrix} 5 & 2 \\ 1 & 0 \end{vmatrix}$.

4. $\begin{vmatrix} 8 & 4 \\ 2 & 1 \end{vmatrix}$.

5. $\begin{vmatrix} 2 & 3 & 1 \\ 1 & 4 & -3 \\ -1 & 2 & 0 \end{vmatrix}$.

6. $\begin{vmatrix} -3 & 1 & 1 \\ 1 & -3 & 1 \\ 1 & 1 & -3 \end{vmatrix}$.

7. Prove that the value of the determinant of a matrix of order n is zero if two rows are identical.

8. Prove that for any determinant of a matrix of order three the sum of the products of the elements of any row and the cofactors of the corresponding elements of another row is zero. (*Note:* This result is true for the determinant of a matrix of any order n.)

9. Find $\det((ka_{ij}))$ if $((a_{ij}))$ is of order four and $\det((a_{ij})) = m$.

10. Prove that $\det AB = \det A \det B$ for any two square matrices A and B of order two. (*Note:* This result is true for square matrices A and B of any order n.)

11. Verify the results of Exercise 10 where

$$A = \begin{pmatrix} 3 & 2 \\ 5 & 1 \end{pmatrix} \text{ and } B = \begin{pmatrix} 1 & 6 \\ 2 & 9 \end{pmatrix}.$$

12. Matrix $((a_{ij}))$ of order n is called an **upper triangular matrix** if $a_{ij} = 0$ for every pair (i, j) such that $i > j$. Find $\det((a_{ij}))$ for such a matrix.

13. Show that the equation

$$\begin{vmatrix} 5 - \lambda & 1 \\ 2 & 3 - \lambda \end{vmatrix} = 0$$

is satisfied if λ is replaced by the matrix

$$A = \begin{pmatrix} 5 & 1 \\ 2 & 3 \end{pmatrix},$$

and each real number n is replaced by the scalar multiple nI where I is the identity matrix of order two.

14. Prove that

$$\begin{vmatrix} a & b & m & n \\ c & d & r & s \\ 0 & 0 & e & f \\ 0 & 0 & g & h \end{vmatrix} = \begin{vmatrix} a & b \\ c & d \end{vmatrix} \begin{vmatrix} e & f \\ g & h \end{vmatrix}.$$

15. Prove that

$$\begin{vmatrix} a+b & a & a \\ a & a+b & a \\ a & a & a+b \end{vmatrix} = b^2(3a+b).$$

16. Repeat Exercise 6 using the formula of Exercise 15.

17. Show that the vector product $\vec{a} \times \vec{b}$, where $\vec{a} = x_1\vec{i} + y_1\vec{j} + z_1\vec{k}$ and $\vec{b} = x_2\vec{i} + y_2\vec{j} + z_2\vec{k}$, can be expressed in determinant form as

$$\begin{vmatrix} \vec{i} & \vec{j} & \vec{k} \\ x_1 & y_1 & z_1 \\ x_2 & y_2 & z_2 \end{vmatrix}.$$

18. Use the results of Exercise 17 to find $\vec{a} \times \vec{b}$ where $\vec{a} = 3i - 2\vec{j} + \vec{k}$ and $\vec{b} = 2\vec{i} + 6\vec{j} + 5\vec{k}$.

19. Show that the scalar triple product $\vec{a} \cdot (\vec{b} \times \vec{c})$, where $\vec{a} = x_1\vec{i} + y_1\vec{j} + z_1\vec{k}$, $\vec{b} = x_2\vec{i} + y_2\vec{j} + z_2\vec{k}$, and $\vec{c} = x_3\vec{i} + y_3\vec{j} + z_3\vec{k}$, can be expressed in determinant form as

$$\begin{vmatrix} x_1 & y_1 & z_1 \\ x_2 & y_2 & z_2 \\ x_3 & y_3 & z_3 \end{vmatrix}.$$

20. Use the results of Exercise 19 to find $\vec{a} \cdot (\vec{b} \times \vec{c})$ where $\vec{a} = 3\vec{i} + 2\vec{k}$, $\vec{b} = \vec{i} + 2\vec{j} + \vec{k}$, and $\vec{c} = -\vec{j} + 4\vec{k}$.

21. Use the results of Exercise 19 to write the coordinate form of the equation of the plane through the points $A: (x_1, y_1, z_1)$, $B: (x_2, y_2, z_2)$, and $C: (x_3, y_3, z_3)$.

5-2 / INVERSE OF A MATRIX

This section will be concerned with the problem of finding a multiplicative inverse, if it exists, for any given square matrix. A **left multiplicative inverse** of a matrix A is a matrix B such that $BA = I$; a **right multiplicative inverse** of a matrix A is a matrix C such that $AC = I$. If a left and a right multiplicative inverse of a matrix A are equal, then the left (right) inverse is simply called a **multiplicative inverse** of A and is denoted by A^{-1}.

THEOREM 5.7 / *A left multiplicative inverse of a square matrix A is a multiplicative inverse of A.*

PROOF / Suppose $BA = I$; then

$A(BA) = AI$ by premultiplication by A,
$(AB)A = A$ since the multiplication of matrices is associative and $AI = A$,
$AB = I$ since I is the unique matrix such that $IA = A$.

Therefore, B is also a right multiplicative inverse of A. Hence, B is a multiplicative inverse of A; that is, $B = A^{-1}$.

Similarly, it can be shown that Theorem 5.8 is true. The proof is left to the reader as an exercise.

THEOREM 5.8 / *A right multiplicative inverse of a square matrix A is a multiplicative inverse of A.*

THEOREM 5.9 / *The multiplicative inverse, if it exists, of a square matrix A is unique.*

PROOF / Let A^{-1} and B be any two multiplicative inverses of the square matrix A. Since $A^{-1}A = I$ and $BA = I$, then

$$A^{-1}A = BA,$$
$$(A^{-1}A)A^{-1} = (BA)A^{-1},$$
$$A^{-1}(AA^{-1}) = B(AA^{-1}),$$
$$A^{-1}I = BI,$$
$$A^{-1} = B.$$

Hence, any two multiplicative inverses of matrix A are identically equal; that is, the multiplicative inverse, if it exists, of a square matrix is unique.

Not every square matrix has a multiplicative inverse. For example, the matrix

$$\begin{pmatrix} 1 & 2 \\ 0 & 0 \end{pmatrix}$$

does not have a multiplicative inverse. If

$$\begin{pmatrix} a & b \\ c & d \end{pmatrix}$$

were the multiplicative inverse of

$$\begin{pmatrix} 1 & 2 \\ 0 & 0 \end{pmatrix},$$

then

$$\begin{pmatrix} a & b \\ c & d \end{pmatrix} \begin{pmatrix} 1 & 2 \\ 0 & 0 \end{pmatrix} = \begin{pmatrix} 1 & 0 \\ 0 & 1 \end{pmatrix};$$

$$\begin{pmatrix} a & 2a \\ c & 2c \end{pmatrix} = \begin{pmatrix} 1 & 0 \\ 0 & 1 \end{pmatrix}. \tag{5-6}$$

This requires that four equations be satisfied: $a = 1$, $2a = 0$, $c = 0$, $2c = 1$. However, if $a = 1$, then $2a \neq 0$. Hence, values of a, b, c, and d that satisfy the matrix equation (5-6) do not exist, and the matrix

$$\begin{pmatrix} 1 & 2 \\ 0 & 0 \end{pmatrix}$$

does not have a multiplicative inverse.

EXAMPLE 1 / Find the multiplicative inverse, if it exists, of

$$\begin{pmatrix} 3 & 1 \\ 4 & 2 \end{pmatrix}.$$

If

$$\begin{pmatrix} 3 & 1 \\ 4 & 2 \end{pmatrix}$$

has a multiplicative inverse

$$\begin{pmatrix} a & b \\ c & d \end{pmatrix},$$

then

$$\begin{pmatrix} a & b \\ c & d \end{pmatrix} \begin{pmatrix} 3 & 1 \\ 4 & 2 \end{pmatrix} = \begin{pmatrix} 1 & 0 \\ 0 & 1 \end{pmatrix};$$

$$\begin{pmatrix} 3a + 4b & a + 2b \\ 3c + 4d & c + 2d \end{pmatrix} = \begin{pmatrix} 1 & 0 \\ 0 & 1 \end{pmatrix}.$$

Therefore,

$$\begin{cases} 3a + 4b = 1 \\ a + 2b = 0 \end{cases} \text{ and } \begin{cases} 3c + 4d = 0 \\ c + 2d = 1. \end{cases}$$

Solve these pairs of equations simultaneously and obtain $a = 1$, $b = -\frac{1}{2}$, $c = -2$, and $d = \frac{3}{2}$. Hence, the multiplicative inverse of

$$\begin{pmatrix} 3 & 1 \\ 4 & 2 \end{pmatrix} \text{ is } \begin{pmatrix} 1 & -\frac{1}{2} \\ -2 & \frac{3}{2} \end{pmatrix}.$$

EXAMPLE 2 / Find the form of the multiplicative inverse, if it exists, of the general square matrix of order two

$$\begin{pmatrix} a & b \\ c & d \end{pmatrix}.$$

If

$$\begin{pmatrix} a & b \\ c & d \end{pmatrix}$$

has a multiplicative inverse

$$\begin{pmatrix} w & x \\ y & z \end{pmatrix},$$

then
$$\begin{pmatrix} w & x \\ y & z \end{pmatrix} \begin{pmatrix} a & b \\ c & d \end{pmatrix} = \begin{pmatrix} 1 & 0 \\ 0 & 1 \end{pmatrix};$$
$$\begin{pmatrix} aw + cx & bw + dx \\ ay + cz & by + dz \end{pmatrix} = \begin{pmatrix} 1 & 0 \\ 0 & 1 \end{pmatrix}.$$

Therefore,
$$\begin{cases} aw + cx = 1 \\ bw + dx = 0 \end{cases} \text{ and } \begin{cases} ay + cz = 0 \\ by + dz = 1 \end{cases}.$$

If $ad - bc \neq 0$, the solutions of these pairs of equations are given as

$$w = \frac{d}{ad - bc}, \quad x = \frac{-b}{ad - bc},$$

$$y = \frac{-c}{ad - bc}, \quad z = \frac{a}{ad - bc}.$$

Hence, the multiplicative inverse of

$$\begin{pmatrix} a & b \\ c & d \end{pmatrix} \text{ is } \begin{pmatrix} \dfrac{d}{ad-bc} & \dfrac{-b}{ad-bc} \\ \dfrac{-c}{ad-bc} & \dfrac{a}{ad-bc} \end{pmatrix}$$

providing that $ad - bc \neq 0$. Notice that the multiplicative inverse of the square matrix of order two

$$\begin{pmatrix} a & b \\ c & d \end{pmatrix}$$

exists if

$$\begin{vmatrix} a & b \\ c & d \end{vmatrix} \neq 0.$$

Examples 1 and 2 illustrate the use of a method that involves considerable labor and is impractical for finding the multiplicative inverses of square matrices of orders greater than two. Now, a direct method for finding the multiplicative inverse, if it exists, of a square matrix of any order will be derived. If $A = ((a_{ij}))$, then by Theorem 5.6

$$\det A = \sum_{j=1}^{n} a_{ij} A_{ij}, \quad \text{for } i = 1, 2, \ldots, n. \tag{5-7}$$

By Exercise 8 of §5-1,

$$\sum_{j=1}^{n} a_{hj}A_{ij} = 0, \qquad \text{for } h, i = 1, 2, \ldots, n \text{ where } h \neq i. \tag{5-8}$$

Equations (5-7) and (5-8) may be expressed as the single equation

$$\sum_{j=1}^{n} a_{hj}A_{ij} = \delta_{hi} \det A, \qquad \text{for } h, i = 1, 2, \ldots, n. \tag{5-9}$$

The n^2 equations represented by (5-9) may be written in matrix form as

$$((a_{ij}))((A_{ij}))^T = ((\delta_{ij})) \det A. \tag{5-10}$$

For example, if $n = 3$ in equation (5-9), then equation (5-10) represents

$$\begin{pmatrix} a_{11} & a_{12} & a_{13} \\ a_{21} & a_{22} & a_{23} \\ a_{31} & a_{32} & a_{33} \end{pmatrix} \begin{pmatrix} A_{11} & A_{21} & A_{31} \\ A_{12} & A_{22} & A_{32} \\ A_{13} & A_{23} & A_{33} \end{pmatrix} = \begin{pmatrix} 1 & 0 & 0 \\ 0 & 1 & 0 \\ 0 & 0 & 1 \end{pmatrix} \det A.$$

Since $((a_{ij})) = A$ and $((\delta_{ij})) = I$, equation (5-10) may be written as

$$A((A_{ij}))^T = I \det A. \tag{5-11}$$

If $\det A \neq 0$, then

$$A \frac{((A_{ij}))^T}{\det A} = I. \tag{5-12}$$

Hence, the matrix $\dfrac{((A_{ij}))^T}{\det A}$ is the multiplicative inverse of A; that is,

$$A^{-1} = \frac{((A_{ij}))^T}{\det A} = \frac{1}{\det A}((A_{ij}))^T. \tag{5-13}$$

The multiplicative inverse, if it exists, of a square matrix is the product of the reciprocal of the determinant of that matrix and the transpose of the **matrix of cofactors**. That $\dfrac{1}{\det A}((A_{ij}))^T$ is not just the right inverse of A follows from Theorem 5.8.

Note that a necessary and sufficient condition for the multiplicative inverse of matrix A to exist is that $\det A \neq 0$. A square matrix A is said to be **nonsingular** if $\det A \neq 0$, and **singular** if $\det A = 0$.

It should be mentioned that if A is not a square matrix, then it is possible for A to have a left or a right multiplicative inverse, but not both. For example, consider

$$A = \begin{pmatrix} 1 & 0 & 0 \\ 0 & 1 & 0 \end{pmatrix}.$$

Then any matrix of the form

$$\begin{pmatrix} 1 & 0 \\ 0 & 1 \\ r & s \end{pmatrix},$$

where r and s are arbitrary scalars, is a right multiplicative inverse of A. A left multiplicative inverse

$$\begin{pmatrix} m & n \\ w & x \\ y & z \end{pmatrix}$$

does not exist since

$$\begin{pmatrix} m & n \\ w & x \\ y & z \end{pmatrix} \begin{pmatrix} 1 & 0 & 0 \\ 0 & 1 & 0 \end{pmatrix} = \begin{pmatrix} m & n & 0 \\ w & x & 0 \\ y & z & 0 \end{pmatrix} \neq \begin{pmatrix} 1 & 0 & 0 \\ 0 & 1 & 0 \\ 0 & 0 & 1 \end{pmatrix}$$

for any values of m, n, w, x, y, and z.

EXAMPLE 3 / Find the multiplicative inverse, if it exists, of A where

$$A = \begin{pmatrix} 1 & 2 & 3 \\ 1 & 3 & 5 \\ 1 & 5 & 12 \end{pmatrix}.$$

Using Theorem 5.6 and the cofactors of the elements of the first row, we have

$$\det A = \begin{vmatrix} 3 & 5 \\ 5 & 12 \end{vmatrix} - 2 \begin{vmatrix} 1 & 5 \\ 1 & 12 \end{vmatrix} + 3 \begin{vmatrix} 1 & 3 \\ 1 & 5 \end{vmatrix} = 11 - 2(7) + 3(2) = 3.$$

Since $\det A \neq 0$, then A^{-1} exists. The cofactors of the elements of A are

$$A_{11} = \begin{vmatrix} 3 & 5 \\ 5 & 12 \end{vmatrix} = 11, \quad A_{12} = -\begin{vmatrix} 1 & 5 \\ 1 & 12 \end{vmatrix} = -7, \quad A_{13} = \begin{vmatrix} 1 & 3 \\ 1 & 5 \end{vmatrix} = 2,$$

$$A_{21} = -\begin{vmatrix} 2 & 3 \\ 5 & 12 \end{vmatrix} = -9, \quad A_{22} = \begin{vmatrix} 1 & 3 \\ 1 & 12 \end{vmatrix} = 9, \quad A_{23} = -\begin{vmatrix} 1 & 2 \\ 1 & 5 \end{vmatrix} = -3,$$

$$A_{31} = \begin{vmatrix} 2 & 3 \\ 3 & 5 \end{vmatrix} = 1, \quad A_{32} = -\begin{vmatrix} 1 & 3 \\ 1 & 5 \end{vmatrix} = -2, \quad A_{33} = \begin{vmatrix} 1 & 2 \\ 1 & 3 \end{vmatrix} = 1.$$

Replacing each element of A by its cofactor, we obtain the matrix

$$((A_{ij})) = \begin{pmatrix} 11 & -7 & 2 \\ -9 & 9 & -3 \\ 1 & -2 & 1 \end{pmatrix}.$$

Hence,

$$A^{-1} = \frac{((A_{ij}))^T}{\det A} = \frac{1}{3}\begin{pmatrix} 11 & -9 & 1 \\ -7 & 9 & -2 \\ 2 & -3 & 1 \end{pmatrix} = \begin{pmatrix} \frac{11}{3} & -3 & \frac{1}{3} \\ -\frac{7}{3} & 3 & -\frac{2}{3} \\ \frac{2}{3} & -1 & \frac{1}{3} \end{pmatrix}.$$

EXAMPLE 4 / Solve the system of linear equations

$$\begin{cases} 5x - 2y = 12 \\ x + 2y = 0. \end{cases}$$

This system of linear equations may be expressed in matrix form as

$$\begin{pmatrix} 5 & -2 \\ 1 & 2 \end{pmatrix}\begin{pmatrix} x \\ y \end{pmatrix} = \begin{pmatrix} 12 \\ 0 \end{pmatrix}.$$

The multiplicative inverse of

$$\begin{pmatrix} 5 & -2 \\ 1 & 2 \end{pmatrix} \text{ is } \begin{pmatrix} \frac{1}{6} & \frac{1}{6} \\ -\frac{1}{12} & \frac{5}{12} \end{pmatrix}.$$

Premultiply both sides of the matrix equation representing the system of linear equations by this multiplicative inverse to obtain

$$\begin{pmatrix} \frac{1}{6} & \frac{1}{6} \\ -\frac{1}{12} & \frac{5}{12} \end{pmatrix}\begin{pmatrix} 5 & -2 \\ 1 & 2 \end{pmatrix}\begin{pmatrix} x \\ y \end{pmatrix} = \begin{pmatrix} \frac{1}{6} & \frac{1}{6} \\ -\frac{1}{12} & \frac{5}{12} \end{pmatrix}\begin{pmatrix} 12 \\ 0 \end{pmatrix}$$

$$\begin{pmatrix} 1 & 0 \\ 0 & 1 \end{pmatrix}\begin{pmatrix} x \\ y \end{pmatrix} = \begin{pmatrix} 2 \\ -1 \end{pmatrix}$$

$$\begin{pmatrix} x \\ y \end{pmatrix} = \begin{pmatrix} 2 \\ -1 \end{pmatrix}.$$

Hence, $x = 2$ and $y = -1$.

In subsequent sections other methods of determining the multiplicative inverse of a nonsingular matrix will be examined.

EXERCISES

In Exercises 1 through 6 determine the multiplicative inverse, if it exists, of the given matrix.

1. $\begin{pmatrix} 3 & 2 \\ 5 & 4 \end{pmatrix}.$

2. $\begin{pmatrix} 3 & 6 \\ 1 & 2 \end{pmatrix}.$

3. $\begin{pmatrix} \cos\theta & -\sin\theta \\ \sin\theta & \cos\theta \end{pmatrix}.$

4. $\begin{pmatrix} 2 & 3 & 4 \\ 0 & 5 & 6 \\ 0 & 0 & 1 \end{pmatrix}.$

5. $\begin{pmatrix} 1 & 2 & 3 \\ 2 & 4 & 5 \\ 3 & 5 & 6 \end{pmatrix}.$

6. $\begin{pmatrix} 0 & a & b \\ -a & 0 & c \\ -b & -c & 0 \end{pmatrix}.$

In Exercises 7 and 8 solve the system of simultaneous equations by matrix methods.

7. $\begin{cases} 2x + y = 4 \\ 3x + 4y = 1. \end{cases}$

8. $\begin{cases} x + 3y + 3z = 2 \\ x + 3y + 4z = 3 \\ x + 4y + 3z = 1. \end{cases}$

9. Find a left multiplicative inverse of

$$\begin{pmatrix} 1 & 1 \\ 3 & 4 \\ 0 & 0 \end{pmatrix}.$$

Show that a right multiplicative inverse does not exist.

10. Prove that $(A^{-1})^{-1} = A$.

11. Prove that if $AB = 0$ and $B \neq 0$, then A is a singular matrix.

12. Prove that $(AB)^{-1} = B^{-1}A^{-1}$.

13. Verify the results of Exercise 12 for

$$A = \begin{pmatrix} 6 & 2 \\ 5 & 2 \end{pmatrix} \quad \text{and} \quad B = \begin{pmatrix} 1 & 0 \\ -1 & 2 \end{pmatrix}.$$

14. Prove that if $AB = AC$ and A is a nonsingular matrix, then $B = C$.

15. Prove that the multiplicative inverse of a nonsingular symmetric matrix is a symmetric matrix.

16. Prove that if $AB = BA$, then $A^{-1}B^{-1} = B^{-1}A^{-1}$.

17. Prove that $(A^T)^{-1} = (A^{-1})^T$ for any nonsingular matrix A.

18. Verify the results of Exercise 17 for

$$A = \begin{pmatrix} 6 & 3 \\ 8 & 6 \end{pmatrix}.$$

5-3 / SYSTEMS OF MATRICES

The set of square matrices of a given order forms, under the operations of matrix addition and multiplication, a model of a particular abstract mathematical system called a ring. A **ring** is a mathematical system consisting of a set R of elements $a, b, c, \ldots$, an equivalence relation denoted by $=$, and two well-defined binary operations $+$ and $\times$ called "addition" and "multiplication," respectively, which satisfy the following properties:

(1) *Closure:* If $a, b \in R$, then $a + b \in R$ and $a \times b \in R$.

(2) *Commutative:* If $a, b \in R$, then $a + b = b + a$.

(3) *Associative:* If $a, b, c \in R$, then $a + (b + c) = (a + b) + c$ and $a \times (b \times c) = (a \times b) \times c$.

(4) *Additive identity:* There exists an element $z \in R$ such that $a + z = z + a = a$ for every $a \in R$.

(5) *Additive inverse:* For each $a \in R$ there exists an element $(-a) \in R$ such that $a + (-a) = (-a) + a = z$.

(6) *Distributive:* If $a, b, c \in R$, then $a \times (b + c) = (a \times b) + (a \times c)$ and $(b + c) \times a = (b \times a) + (c \times a)$.

EXAMPLE 1 / Show that the set S of square real matrices of order n forms a ring.

By the definitions of the sum and product of two matrices, the set S is closed under matrix addition and multiplication (Property 1). The addition of matrices of S is commutative (Property 2) and associative (part of Property 3) since the addition of real numbers is commutative and associative. Property 4 is satisfied because the null matrix of order n is the additive identity element for the set S. Since for each $((a_{ij})) \in S$ there exists a matrix $((-a_{ij}))$ such that $((a_{ij})) + ((-a_{ij})) = 0$, Property 5 is satisfied. Property 6 and the second part of Property 3 were proved in §4-2. Therefore, the set S of square real matrices of order n forms a ring.

A one-to-one correspondence between the elements of a ring R and the elements of a ring S is called an **isomorphism** if sums and products of corresponding elements correspond; that is, if any two elements $a, b \in R$ and any two elements $a', b' \in S$ are such that a and b may be made to correspond to a' and b', respectively, denoted by $a \longleftrightarrow a'$ and $b \longleftrightarrow b'$, the correspondence is an isomorphism if $a + b \longleftrightarrow a' + b'$ and $ab \longleftrightarrow a'b'$. Whenever an isomorphism between the elements of a ring R and those of a ring S exists, the rings are abstractly identical and are said to be **isomorphic**. Only the notations used to represent the elements and the operations of the two rings differ.

Two very important subsets of the set of square real matrices of order two exist. The first is the set of scalar matrices of order two

$$\begin{pmatrix} k & 0 \\ 0 & k \end{pmatrix}.$$

This set of matrices is isomorphic to the set of real numbers. Both the set of scalar matrices and the set of real numbers along with the operations of addition and multiplication defined on the sets are examples of rings.

Consider the one-to-one correspondence between the matrices

$$\begin{pmatrix} k & 0 \\ 0 & k \end{pmatrix}$$

and the scalars k; that is,

$$\begin{pmatrix} k & 0 \\ 0 & k \end{pmatrix} \longleftrightarrow k.$$

In order to establish that the two systems are isomorphic, it is necessary to show that the sums and products of corresponding elements in the two systems are in the same one-to-one correspondence. Let

$$\begin{pmatrix} a & 0 \\ 0 & a \end{pmatrix} \longleftrightarrow a \quad \text{and} \quad \begin{pmatrix} b & 0 \\ 0 & b \end{pmatrix} \longleftrightarrow b.$$

Then

$$\begin{pmatrix} a & 0 \\ 0 & a \end{pmatrix} + \begin{pmatrix} b & 0 \\ 0 & b \end{pmatrix} = \begin{pmatrix} a+b & 0 \\ 0 & a+b \end{pmatrix} \longleftrightarrow a + b$$

and

$$\begin{pmatrix} a & 0 \\ 0 & a \end{pmatrix} \begin{pmatrix} b & 0 \\ 0 & b \end{pmatrix} = \begin{pmatrix} ab & 0 \\ 0 & ab \end{pmatrix} \longleftrightarrow ab.$$

Hence, the ring of matrices of the form

$$\begin{pmatrix} k & 0 \\ 0 & k \end{pmatrix}$$

is isomorphic to the ring of real numbers k.

The second important subset of the set of square real matrices of order two is the set of matrices of the form

$$\begin{pmatrix} x & y \\ -y & x \end{pmatrix}.$$

This set of matrices is isomorphic to the set of complex numbers $x + yi$. Both the set of matrices of this form and the set of complex numbers along with the operations of addition and multiplication defined on the sets are examples of rings.

Consider the one-to-one correspondence between the matrices

$$\begin{pmatrix} x & y \\ -y & x \end{pmatrix}$$

and the complex numbers $x + yi$; that is,

$$\begin{pmatrix} x & y \\ -y & x \end{pmatrix} \longleftrightarrow x + yi.$$

Let

$$\begin{pmatrix} a & b \\ -b & a \end{pmatrix} \longleftrightarrow a + bi \quad \text{and} \quad \begin{pmatrix} c & d \\ -d & c \end{pmatrix} \longleftrightarrow c + di.$$

Then

$$\begin{pmatrix} a & b \\ -b & a \end{pmatrix} + \begin{pmatrix} c & d \\ -d & c \end{pmatrix} = \begin{pmatrix} a+c & b+d \\ -(b+d) & a+c \end{pmatrix}$$

and $(a + bi) + (c + di) = (a + c) + (b + d)i$. Since

$$\begin{pmatrix} a+c & b+d \\ -(b+d) & a+c \end{pmatrix} \longleftrightarrow (a+c) + (b+d)i,$$

then

$$\begin{pmatrix} a & b \\ -b & a \end{pmatrix} + \begin{pmatrix} c & d \\ -d & c \end{pmatrix} \longleftrightarrow (a+bi) + (c+di),$$

which demonstrates that the correspondence is preserved under addition. For multiplication,

$$\begin{pmatrix} a & b \\ -b & a \end{pmatrix}\begin{pmatrix} c & d \\ -d & c \end{pmatrix} = \begin{pmatrix} ac - bd & ad + bc \\ -(ad + bc) & ac - bd \end{pmatrix}$$

and $(a + bi) \times (c + di) = (ac - bd) + (ad + bc)i$. Since

$$\begin{pmatrix} ac - bd & ad + bc \\ -(ad + bc) & ac - bd \end{pmatrix} \longleftrightarrow (ac - bd) + (ad + bc)i,$$

then

$$\begin{pmatrix} a & b \\ -b & a \end{pmatrix}\begin{pmatrix} c & d \\ -d & c \end{pmatrix} \longleftrightarrow (a + bi) \times (c + di),$$

which demonstrates that the correspondence is preserved under multiplication.

EXAMPLE 2 / Verify that the one-to-one correspondence between the matrices

$$\begin{pmatrix} x & y \\ -y & x \end{pmatrix}$$

and the complex numbers $x + yi$ is preserved under multiplication for the complex numbers $5 + i$ and $2 - 4i$.

Let

$$5 + i \longleftrightarrow \begin{pmatrix} 5 & 1 \\ -1 & 5 \end{pmatrix} \quad \text{and} \quad 2 - 4i \longleftrightarrow \begin{pmatrix} 2 & -4 \\ 4 & 2 \end{pmatrix}.$$

Then

$$(5 + i) \times (2 - 4i) = [(5)(2) - (1)(-4)] + [(5)(-4) + (1)(2)]i$$
$$= 14 - 18i,$$

and

$$\begin{pmatrix} 5 & 1 \\ -1 & 5 \end{pmatrix}\begin{pmatrix} 2 & -4 \\ 4 & 2 \end{pmatrix} = \begin{pmatrix} 14 & -18 \\ 18 & 14 \end{pmatrix}.$$

Since

$$14 - 18i \longleftrightarrow \begin{pmatrix} 14 & -18 \\ 18 & 14 \end{pmatrix},$$

then
$$(5+i) \times (2-4i) \longleftrightarrow \begin{pmatrix} 5 & 1 \\ -1 & 5 \end{pmatrix} \begin{pmatrix} 2 & -4 \\ 4 & 2 \end{pmatrix}.$$

One reason why the algebra of matrices has become a valuable algebra for mathematicians to study is that many important mathematical systems are isomorphic to sets of matrices. Consider now one other such mathematical system which is of historical importance in the development of vector algebra.

Under the definitions of addition and multiplication of quaternions, it can be shown that the set of quaternions forms a ring. Furthermore, the ring of quaternions $a + bi + cj + dk$ can be shown to be isomorphic to two different sets of matrices: the set of square real matrices of order four of the form

$$\begin{pmatrix} a & b & c & d \\ -b & a & -d & c \\ -c & d & a & -b \\ -d & -c & b & a \end{pmatrix}$$

and the set of square complex matrices of order two of the form

$$\begin{pmatrix} a+bi & c+di \\ -c+di & a-bi \end{pmatrix}.$$

EXAMPLE 3 / Verify that the one-to-one correspondence between the matrices

$$\begin{pmatrix} a & b & c & d \\ -b & a & -d & c \\ -c & d & a & -b \\ -d & -c & b & a \end{pmatrix}$$

and the quaternions $a + bi + cj + dk$ is preserved under multiplication for the quaternions $2 + 3i + j + 4k$ and $1 - i + 3j - 2k$.

Let

$$2 + 3i + j + 4k \longleftrightarrow \begin{pmatrix} 2 & 3 & 1 & 4 \\ -3 & 2 & -4 & 1 \\ -1 & 4 & 2 & -3 \\ -4 & -1 & 3 & 2 \end{pmatrix}$$

and

$$1 - i + 3j - 2k \longleftrightarrow \begin{pmatrix} 1 & -1 & 3 & -2 \\ 1 & 1 & 2 & 3 \\ -3 & -2 & 1 & 1 \\ 2 & -3 & -1 & 1 \end{pmatrix}.$$

Then

$$\begin{aligned}
&(2 + 3i + j + 4k)(1 - i + 3j - 2k) \\
&= [(2)(1) - (3)(-1) - (1)(3) - (4)(-2)] \\
&\quad + [(2)(-1) + (3)(1) + (1)(-2) - (4)(3)]i \\
&\quad + [(2)(3) - (3)(-2) + (1)(1) + (4)(-1)]j \\
&\quad + [(2)(-2) + (3)(3) - (1)(-1) + (4)(1)]k \\
&= 10 - 13i + 9j + 10k,
\end{aligned}$$

and

$$\begin{pmatrix} 2 & 3 & 1 & 4 \\ -3 & 2 & -4 & 1 \\ -1 & 4 & 2 & -3 \\ -4 & -1 & 3 & 2 \end{pmatrix} \begin{pmatrix} 1 & -1 & 3 & -2 \\ 1 & 1 & 2 & 3 \\ -3 & -2 & 1 & 1 \\ 2 & -3 & -1 & 1 \end{pmatrix}$$

$$= \begin{pmatrix} 10 & -13 & 9 & 10 \\ 13 & 10 & -10 & 9 \\ -9 & 10 & 10 & 13 \\ -10 & -9 & -13 & 10 \end{pmatrix}.$$

Since

$$10 - 13i + 9j + 10k \longleftrightarrow \begin{pmatrix} 10 & -13 & 9 & 10 \\ 13 & 10 & -10 & 9 \\ -9 & 10 & 10 & 13 \\ -10 & -9 & -13 & 10 \end{pmatrix},$$

then the one-to-one correspondence is preserved under multiplication.

EXERCISES

Determine which of the following sets of elements form rings considering the operations as ordinary addition and multiplication unless otherwise stated.

1. The set of natural numbers.
2. The set of rational numbers.
3. The set of even integers.
4. The set of matrices

$$\begin{pmatrix} 1 & 0 \\ 0 & 1 \end{pmatrix}, \begin{pmatrix} -1 & 0 \\ 0 & -1 \end{pmatrix}, \text{ and } \begin{pmatrix} 0 & 0 \\ 0 & 0 \end{pmatrix}$$

under matrix addition and multiplication.

5. The set of square real matrices of order two of the form

$$\begin{pmatrix} a & b \\ 0 & 0 \end{pmatrix}$$

under matrix addition and multiplication.

In Exercises 6 through 9 verify that the one-to-one correspondence between the matrices

$$\begin{pmatrix} x & y \\ -y & x \end{pmatrix}$$

and the complex numbers $x + yi$ is preserved under the indicated operation for the given complex numbers.

6. $(5 + 2i) + (5 - 2i)$.
7. $(5 - 3i) + (2 - 4i)$.
8. $(3 + 4i) \times (2 - i)$.
9. $(i) \times (i)$.

10. Show that the reciprocal of the complex number $a + bi$ corresponds to the inverse of the matrix

$$\begin{pmatrix} a & b \\ -b & a \end{pmatrix}.$$

Consider an operation $\odot$ on the set S of nonsingular matrices of order two such that if $A, B \in S$, $A \odot B = AB - BA$. In Exercises 11 and 12 prove the indicated properties of $\odot$.

11. $A \odot B = -B \odot A$.
12. $A \odot (B \odot C) \neq (A \odot B) \odot C$.

In Exercises 13 and 14 verify that the one-to-one correspondence between the complex matrices

$$\begin{pmatrix} a + bi & c + di \\ -c + di & a - bi \end{pmatrix}$$

and the quaternions $a + bi + cj + dk$ is preserved under multiplication for the given quaternions.

13. $5i + j + k$ and $3 + i + 2j + 4k$.   **14.** $i + k$ and $3 + j$.

5-4 / LINEAR ALGEBRAS

In this section we define an abstract mathematical system that has the properties of both a ring and a real vector space.

A set V of elements (vectors) is called a **real linear algebra** provided that

(1) V is a real vector space under the operations of addition $+$ and multiplication by a real number;

(2) V is a ring under the operations of addition $+$ and multiplication $\times$;

(3) for each $a, b \in V$ and $m \in R$,

$$(ma) \times b = a \times (mb) = m(a \times b).$$

Note that if V is a real vector space, then the properties of a ring that relate to the addition of the elements (vectors) are satisfied. Thus, our definition of a real linear algebra is somewhat redundant. It is, however, stated in a most convenient form.

EXAMPLE / Prove that the set S of square real matrices of order n is a real linear algebra.

Under the operations of matrix addition and the multiplication of a matrix by a real number, the set S is a real vector space (Exercise 8 of §4-1). Hence, Property (1) of a linear algebra is satisfied. Under the operations of matrix addition and matrix multiplication, the set S is a ring (Example 1 of §5-3). Hence, Property (2) of a linear algebra is satisfied. Now, let $A, B \in V$ where $A = ((a_{ij}))$ and $B = ((b_{ij}))$; let $m \in R$. Since $mA = ((ma_{ij}))$ and $mB = ((mb_{ij}))$ by equation (4-6), it follows that the ijth element of $(mA)B$ is

$$(ma_{i1})b_{1j} + (ma_{i2})b_{2j} + \cdots + (ma_{in})b_{nj};$$

the ijth element of $A(mB)$ is

$$a_{i1}(mb_{1j}) + a_{i2}(mb_{2j}) + \cdots + a_{in}(mb_{nj}).$$

Since $AB = ((a_{i_1}b_{1j} + a_{i_2}b_{2j} + \cdots + a_{in}b_{nj}))$, the ijth element of $m(AB)$ is

$$m(a_{i_1}b_{1j}) + m(a_{i_2}b_{2j}) + \cdots + m(a_{in}b_{nj}).$$

Since the multiplication of real numbers is associative and commutative, the ijth elements of $(mA)B$, $A(mB)$, and $m(AB)$ are equal. Hence,

$$(mA)B = A(mB) = m(AB),$$

and Property (3) of a linear algebra is satisfied. Therefore, the set S of square real matrices of order n is a real linear algebra.

It can be shown that the set C of complex numbers and the set Q of quaternions are also real linear algebras (Exercises 2 and 3). The set R of real numbers is a trivial example of a real linear algebra (Exercise 1). The importance of the example of this section lies in the fact that *any real linear algebra V for which the vector space V is of finite dimension is isomorphic to some set of square real matrices*; that is, a one-to-one correspondence between the elements of the real linear algebra and the elements of some set of square real matrices exists such that the sums and products of corresponding elements correspond, and the real multiples of corresponding elements correspond.

EXERCISES

In Exercises 1 through 3 prove that each set is a real linear algebra.

1. The set R of real numbers.
2. The set C of complex numbers.
3. The set Q of quaternions.

In Exercises 4 through 7 determine whether or not each set is a real linear algebra.

4. Let S be the set of square matrices of order n whose elements are integers.
5. Let S be the set of ordered triples of real numbers. Consider the following definitions of addition, multiplication, and multiplication by a real number:
If $(a, b, c), (d, e, f) \in S$, then

$$(a, b, c) + (d, e, f) = (a + d, b + e, c + f).$$

If (a, b, c), $(d, e, f) \in S$, then

$$(a, b, c) \times (d, e, f) = (ad + cf, ae + bd + bf + ce, af + cd).$$

If $(a, b, c) \in S$ and $m \in R$, then

$$m(a, b, c) = (ma, mb, mc).$$

6. Let S be the set of three-dimensional vectors $a\vec{i} + b\vec{j} + c\vec{k}$, where a, b, and c are real numbers, under the operations of vector addition, "scalar product," and multiplication of a vector by a real number.

7. Let S be the set of three-dimensional vectors $a\vec{i} + b\vec{j} + c\vec{k}$, where a, b, and c are real numbers, under the operations of vector addition, "vector product," and multiplication of a vector by a real number.

5-5 / RANK OF A MATRIX

Before the discussion concerning the concept of the rank of a matrix, a brief review of the concept of linear dependence is in order.

Consider a set of functions, vectors, or elements $e_1, e_2, \ldots, e_n$. These functions, vectors, or elements $e_1, e_2, \ldots, e_n$ are said to be *linearly dependent* if there exists a set of scalars $k_1, k_2, \ldots, k_n$, not all zero, such that $k_1 e_1 + k_2 e_2 + \cdots + k_n e_n = 0$. If they are not linearly dependent, then they are said to be *linearly independent*.

EXAMPLE 1 / Show that the functions $f(x, y, z)$, $g(x, y, z)$, and $h(x, y, z)$ are linearly dependent where $f(x, y, z) = x - y + z$, $g(x, y, z) = x + y + 2z$, and $h(x, y, z) = 3x + y + 5z$.

If $f(x, y, z)$, $g(x, y, z)$, and $h(x, y, z)$ are linearly dependent functions, then there exists a set of scalars k_1, k_2, and k_3, not all zero, such that $k_1 f(x, y, z) + k_2 g(x, y, z) + k_3 h(x, y, z) = 0$.

Consider $k_1(x - y + z) + k_2(x + y + 2z) + k_3(3x + y + 5z) = 0$. Then $(k_1 + k_2 + 3k_3)x + (-k_1 + k_2 + k_3)y + (k_1 + 2k_2 + 5k_3)z = 0$, and

$$\begin{cases} k_1 + k_2 + 3k_3 = 0 \\ -k_1 + k_2 + k_3 = 0 \\ k_1 + 2k_2 + 5k_3 = 0. \end{cases}$$

Solve simultaneously by adding the second equation to the first and third equations and obtain

$$\begin{cases} 2k_2 + 4k_3 = 0 \\ 3k_2 + 6k_3 = 0; \end{cases}$$

$k_2 = -2k_3$ and $k_1 = -k_3$. Let $k_3 = -1$. Then $k_2 = 2, k_1 = 1$, and $1(x - y + z) + 2(x + y + 2z) - 1(3x + y + 5z) = 0$. Hence, $f(x, y, z), g(x, y, z)$, and $h(x, y, z)$ are linearly dependent functions.

Given any matrix A of order m by n, **square submatrices** of order r may be obtained by selecting the elements in any r rows and r columns of the matrix. The determinants of these square submatrices of order r are called **r-rowed minors** of the matrix A. The order of the largest square submatrix whose determinant has a nonzero value is called the **rank** of the matrix. The rank of any matrix other than a null matrix cannot be zero, that of a matrix of order m by n cannot exceed either m or n, and that of a null matrix is defined to be zero. In general, the rank of a matrix is equal to the largest number of linearly independent row vectors and column vectors since the value of the determinant of a square matrix is zero if the row vectors or column vectors are linearly dependent.

EXAMPLE 2 / Determine the rank of matrix A where

$$A = \begin{pmatrix} 4 & 2 & -1 & 3 \\ 0 & 5 & -1 & 2 \\ 12 & -4 & -1 & 5 \end{pmatrix}.$$

Since matrix A is of order 3 by 4, the rank of A cannot exceed three. Each of the 3-rowed minors of A, however, has value zero; that is,

$$\begin{vmatrix} 4 & 2 & -1 \\ 0 & 5 & -1 \\ 12 & -4 & -1 \end{vmatrix} = \begin{vmatrix} 4 & 2 & 3 \\ 0 & 5 & 2 \\ 12 & -4 & 5 \end{vmatrix}$$

$$= \begin{vmatrix} 4 & -1 & 3 \\ 0 & -1 & 2 \\ 12 & -1 & 5 \end{vmatrix} = \begin{vmatrix} 2 & -1 & 3 \\ 5 & -1 & 2 \\ -4 & -1 & 5 \end{vmatrix} = 0.$$

Therefore, the rank of A cannot be three. Since the 2-rowed minor

$$\begin{vmatrix} 4 & 2 \\ 0 & 5 \end{vmatrix}$$

does not have value zero, the rank of A is two.

The difficulty in determining the rank of a matrix of high order may be reduced by considering three **elementary row transformations** on the matrix:

(i) interchanging the corresponding elements of two rows;
(ii) multiplication of the elements of any row by a nonzero scalar;
(iii) addition of a nonzero scalar multiple of the elements of any row to the corresponding elements of another row.

From Theorems 5.2, 5.4, and 5.5, it is evident that the rank of a matrix is not changed by an application of any of the elementary row transformations since under (i) only the signs of the values of some r-rowed minors are changed; under (ii) the values of some r-rowed minors are multiplied by a nonzero scalar; and under (iii) the value of every r-rowed minor is unchanged. These elementary row transformations may be applied in any sequence until the first nonzero element that appears in each row is equal to 1 and is positioned at the right of the first nonzero element of the preceding row. The matrix is then said to be in **echelon form**, and its rank may be easily determined.

EXAMPLE 3 / Determine the rank of matrix A where

$$A = \begin{pmatrix} 0 & 2 & 4 & 6 \\ 3 & -1 & 4 & -2 \\ 6 & -1 & 10 & -1 \end{pmatrix}.$$

Interchange the corresponding elements of rows one and two:

$$\begin{pmatrix} 3 & -1 & 4 & -2 \\ 0 & 2 & 4 & 6 \\ 6 & -1 & 10 & -1 \end{pmatrix};$$

multiply the elements of row one by -2 and add the products to the corresponding elements of row three:

$$\begin{pmatrix} 3 & -1 & 4 & -2 \\ 0 & 2 & 4 & 6 \\ 0 & 1 & 2 & 3 \end{pmatrix};$$

multiply the elements of row two by $-\frac{1}{2}$ and add the products to the corresponding elements of row three:

$$\begin{pmatrix} 3 & -1 & 4 & -2 \\ 0 & 2 & 4 & 6 \\ 0 & 0 & 0 & 0 \end{pmatrix};$$

multiply the elements of rows one and two by $\frac{1}{3}$ and $\frac{1}{2}$, respectively:

$$\begin{pmatrix} 1 & -\frac{1}{3} & \frac{4}{3} & -\frac{2}{3} \\ 0 & 1 & 2 & 3 \\ 0 & 0 & 0 & 0 \end{pmatrix}.$$

The matrix A is now in echelon form, and it is immediately evident that every 3-rowed minor has value zero. The rank of A is two since

$$\begin{vmatrix} 1 & -\frac{1}{3} \\ 0 & 1 \end{vmatrix} \neq 0.$$

Each of the elementary row transformations on a matrix may be viewed as a premultiplication of the matrix by a conformable matrix obtained by performing that elementary row transformation on the identity matrix. For example, if the matrix whose rank is to be determined is of order 3 by n, then the first elementary row transformation may be accomplished by premultiplying the given matrix by one of the matrices

$$\begin{pmatrix} 0 & 0 & 1 \\ 0 & 1 & 0 \\ 1 & 0 & 0 \end{pmatrix}, \quad \begin{pmatrix} 0 & 1 & 0 \\ 1 & 0 & 0 \\ 0 & 0 & 1 \end{pmatrix}, \quad \text{or} \quad \begin{pmatrix} 1 & 0 & 0 \\ 0 & 0 & 1 \\ 0 & 1 & 0 \end{pmatrix}. \quad (5\text{-}14)$$

The second elementary row transformation also may be accomplished by premultiplying the given matrix by one of the matrices

$$\begin{pmatrix} k & 0 & 0 \\ 0 & 1 & 0 \\ 0 & 0 & 1 \end{pmatrix}, \quad \begin{pmatrix} 1 & 0 & 0 \\ 0 & k & 0 \\ 0 & 0 & 1 \end{pmatrix}, \quad \text{or} \quad \begin{pmatrix} 1 & 0 & 0 \\ 0 & 1 & 0 \\ 0 & 0 & k \end{pmatrix}; \quad (5\text{-}15)$$

furthermore, the third elementary row transformation may be accomplished by premultiplying the given matrix by one of the matrices

$$\begin{pmatrix} 1 & k & 0 \\ 0 & 1 & 0 \\ 0 & 0 & 1 \end{pmatrix}, \quad \begin{pmatrix} 1 & 0 & k \\ 0 & 1 & 0 \\ 0 & 0 & 1 \end{pmatrix}, \quad \begin{pmatrix} 1 & 0 & 0 \\ k & 1 & 0 \\ 0 & 0 & 1 \end{pmatrix},$$
$$\begin{pmatrix} 1 & 0 & 0 \\ 0 & 1 & k \\ 0 & 0 & 1 \end{pmatrix}, \quad \begin{pmatrix} 1 & 0 & 0 \\ 0 & 1 & 0 \\ k & 0 & 1 \end{pmatrix}, \quad \text{or} \quad \begin{pmatrix} 1 & 0 & 0 \\ 0 & 1 & 0 \\ 0 & k & 1 \end{pmatrix}.$$
(5-16)

The matrices that permit the performance of elementary row transformations are called **elementary row transformation matrices.** For example, the matrices of (5-14), (5-15), and (5-16) are elementary row transformation matrices of order three. Note that every elementary row transformation matrix is nonsingular.

It is possible to show that every nonsingular matrix is a product of elementary row transformation matrices. Then every nonsingular matrix A has a nonsingular inverse A^{-1} which is equal to a product of elementary row transformation matrices. The product of the elementary row transformation matrices transforming A into I is A^{-1}. Note carefully that the elementary row transformation matrices used must be multiplied in the reverse order of their application.

EXAMPLE 4 / Determine the inverse of A where

$$A = \begin{pmatrix} 3 & 1 \\ 4 & 2 \end{pmatrix}.$$

Since A is nonsingular, perform the elementary row transformations on A until the identity matrix I is obtained. The product of the matrices representing the transformations will be equal to A^{-1}. For example, multiply the elements of row one by $-\frac{4}{3}$ and add the products to the corresponding elements of row two:

$$\begin{pmatrix} 1 & 0 \\ -\frac{4}{3} & 1 \end{pmatrix} \begin{pmatrix} 3 & 1 \\ 4 & 2 \end{pmatrix} = \begin{pmatrix} 3 & 1 \\ 0 & \frac{2}{3} \end{pmatrix};$$

multiply the elements of row one by $\frac{1}{3}$, then multiply the elements of row two by $\frac{3}{2}$:

$$\begin{pmatrix} 1 & 0 \\ 0 & \frac{3}{2} \end{pmatrix} \begin{pmatrix} \frac{1}{3} & 0 \\ 0 & 1 \end{pmatrix} \begin{pmatrix} 3 & 1 \\ 0 & \frac{2}{3} \end{pmatrix} = \begin{pmatrix} 1 & \frac{1}{3} \\ 0 & 1 \end{pmatrix};$$

5-5 / RANK OF A MATRIX

multiply the elements of row two by $-\frac{1}{3}$ and add the products to the corresponding elements of row one:

$$\begin{pmatrix} 1 & -\frac{1}{3} \\ 0 & 1 \end{pmatrix} \begin{pmatrix} 1 & \frac{1}{3} \\ 0 & 1 \end{pmatrix} = \begin{pmatrix} 1 & 0 \\ 0 & 1 \end{pmatrix}.$$

Hence,

$$A^{-1} = \begin{pmatrix} 1 & -\frac{1}{3} \\ 0 & 1 \end{pmatrix} \begin{pmatrix} 1 & 0 \\ 0 & \frac{3}{2} \end{pmatrix} \begin{pmatrix} \frac{1}{3} & 0 \\ 0 & 1 \end{pmatrix} \begin{pmatrix} 1 & 0 \\ -\frac{4}{3} & 1 \end{pmatrix} = \begin{pmatrix} 1 & -\frac{1}{2} \\ -2 & \frac{3}{2} \end{pmatrix}.$$

EXERCISES

1. Show that the functions $2x + z$, $x + y$, and $2y - z$ are linearly dependent.

2. Show that the functions $x + y - z$, $2x + y + 3z$, and $x + 2y + 4z$ are linearly independent.

In Exercises 3 through 5 determine the rank of the given matrix.

3. $\begin{pmatrix} 4 & 2 & 3 \\ 8 & 5 & 2 \\ 12 & -4 & 5 \end{pmatrix}.$
4. $\begin{pmatrix} 2 & 3 & 3 \\ 3 & 6 & 12 \\ 2 & 4 & 8 \end{pmatrix}.$
5. $\begin{pmatrix} 3 & 4 & 0 & 2 \\ 6 & 8 & 0 & 4 \\ 1 & 0 & 1 & 1 \end{pmatrix}.$

6. Prove that the rank of any skew-symmetric matrix cannot be one.

7. Determine the maximum number of elementary row transformations necessary to transform a square matrix of order n to echelon form.

8. State the general forms of the elementary row transformation matrices of order two.

In Exercises 9 through 11 express the inverse of the given matrix as a product of elementary row transformation matrices.

9. $\begin{pmatrix} 6 & 2 \\ 2 & 1 \end{pmatrix}.$
10. $\begin{pmatrix} 1 & 2 \\ -2 & 3 \end{pmatrix}.$
11. $\begin{pmatrix} 1 & -2 & 0 \\ -2 & 3 & 1 \\ -5 & 9 & 3 \end{pmatrix}.$

12. Express the matrix given in Exercise 9 as a product of elementary row transformation matrices.

5-6 / SYSTEMS OF LINEAR EQUATIONS

Consider a system of m linear equations in n variables $x_1, x_2, \ldots, x_n$:

$$\begin{cases} a_{11}x_1 + a_{12}x_2 + \cdots + a_{1n}x_n = c_1 \\ a_{21}x_1 + a_{22}x_2 + \cdots + a_{2n}x_n = c_2 \\ \ldots \\ a_{m1}x_1 + a_{m2}x_2 + \cdots + a_{mn}x_n = c_m. \end{cases} \qquad (5\text{-}17)$$

The system of (5-17) may be written in matrix form as

$$\begin{pmatrix} a_{11} & a_{12} & \cdots & a_{1n} \\ a_{21} & a_{22} & \cdots & a_{2n} \\ \cdots & \cdots & \cdots & \cdots \\ a_{m1} & a_{m2} & \cdots & a_{mn} \end{pmatrix} \begin{pmatrix} x_1 \\ x_2 \\ \cdots \\ x_n \end{pmatrix} = \begin{pmatrix} c_1 \\ c_2 \\ \cdots \\ c_m \end{pmatrix}. \qquad (5\text{-}18)$$

The matrix $((a_{ij}))$ of order m by n is called the **matrix of coefficients**. The matrix composed of the mn elements a_{ij} plus an additional column whose elements are the constants c_i is called the **augmented matrix** of the system; that is,

$$\begin{pmatrix} a_{11} & a_{12} & \cdots & a_{1n} & c_1 \\ a_{21} & a_{22} & \cdots & a_{2n} & c_2 \\ \cdots & \cdots & \cdots & \cdots & \cdots \\ a_{m1} & a_{m2} & \cdots & a_{mn} & c_m \end{pmatrix} \qquad (5\text{-}19)$$

is the augmented matrix of (5-17). Note that the rank of the augmented matrix is either equal to or one greater than the rank of the matrix of coefficients.

A solution, if it exists, to a system of linear equations (5-17) may be determined by using the elementary row transformations to transform the augmented matrix into echelon form. The system of linear equations represented by the transformed augmented matrix will be an equivalent system to (5-17); "equivalent" is used in the sense that the solution set is not changed.

If the system of linear equations (5-17) is such that the equations are all satisfied simultaneously by at least one set of values of the variables, then it is said to be **consistent**. The system is said to be **inconsistent** if the equations are not satisfied simultaneously by any set of values of the variables.

EXAMPLE 1 / Solve the system of linear equations by performing the elementary row transformations on the augmented matrix:

$$\begin{cases} 2x - y + z = 1 \\ x + y - z = 2 \\ 3x - y + z = 0. \end{cases}$$

The augmented matrix is

$$\begin{pmatrix} 2 & -1 & 1 & 1 \\ 1 & 1 & -1 & 2 \\ 3 & -1 & 1 & 0 \end{pmatrix}.$$

Interchange the corresponding elements of rows one and two:

$$\begin{pmatrix} 1 & 1 & -1 & 2 \\ 2 & -1 & 1 & 1 \\ 3 & -1 & 1 & 0 \end{pmatrix};$$

multiply the elements of row one by -2 and add the products to the corresponding elements of row two:

$$\begin{pmatrix} 1 & 1 & -1 & 2 \\ 0 & -3 & 3 & -3 \\ 3 & -1 & 1 & 0 \end{pmatrix};$$

multiply the elements of row one by -3 and add the products to the corresponding elements of row three:

$$\begin{pmatrix} 1 & 1 & -1 & 2 \\ 0 & -3 & 3 & -3 \\ 0 & -4 & 4 & -6 \end{pmatrix};$$

multiply the elements of row two by $-\frac{1}{3}$:

$$\begin{pmatrix} 1 & 1 & -1 & 2 \\ 0 & 1 & -1 & 1 \\ 0 & -4 & 4 & -6 \end{pmatrix};$$

multiply the elements of row two by 4 and add the products to the corresponding elements of row three:

$$\begin{pmatrix} 1 & 1 & -1 & 2 \\ 0 & 1 & -1 & 1 \\ 0 & 0 & 0 & -2 \end{pmatrix};$$

multiply the elements of row three by $-\frac{1}{2}$:

$$\begin{pmatrix} 1 & 1 & -1 & 2 \\ 0 & 1 & -1 & 1 \\ 0 & 0 & 0 & 1 \end{pmatrix}.$$

The transformed augmented matrix is now in echelon form and represents the system of linear equations

$$\begin{cases} x + y - z = 2 \\ y - z = 1 \\ 0 = 1. \end{cases}$$

Since no values of x, y, and z exist such that 0 equals 1, the system is inconsistent.

Note that in Example 1 the rank of the augmented matrix is three while the rank of the matrix of coefficients is two.

EXAMPLE 2 / Solve the system of linear equations by performing the elementary row transformations on the augmented matrix:

$$\begin{cases} x + 3y + z = 6 \\ 3x - 2y - 8z = 7 \\ 4x + 5y - 3z = 17. \end{cases}$$

The augmented matrix is

$$\begin{pmatrix} 1 & 3 & 1 & 6 \\ 3 & -2 & -8 & 7 \\ 4 & 5 & -3 & 17 \end{pmatrix}.$$

Multiply the elements of row one by -3 and add the products to the corresponding elements of row two:

$$\begin{pmatrix} 1 & 3 & 1 & 6 \\ 0 & -11 & -11 & -11 \\ 4 & 5 & -3 & 17 \end{pmatrix};$$

multiply the elements of row one by -4 and add the products to the corresponding elements of row three:

$$\begin{pmatrix} 1 & 3 & 1 & 6 \\ 0 & -11 & -11 & -11 \\ 0 & -7 & -7 & -7 \end{pmatrix};$$

multiply the elements of row two by $-\frac{1}{11}$:

$$\begin{pmatrix} 1 & 3 & 1 & 6 \\ 0 & 1 & 1 & 1 \\ 0 & -7 & -7 & -7 \end{pmatrix};$$

multiply the elements of row two by 7 and add the products to the corresponding elements of row three:

$$\begin{pmatrix} 1 & 3 & 1 & 6 \\ 0 & 1 & 1 & 1 \\ 0 & 0 & 0 & 0 \end{pmatrix}.$$

The transformed augmented matrix is now in echelon form and represents the system of linear equations

$$\begin{cases} x + 3y + z = 6 \\ y + z = 1 \\ 0 = 0. \end{cases}$$

Hence, there exist infinitely many solutions of the form $x = 3 + 2z$ and $y = 1 - z$. The value of the variable z may be assigned arbitrarily, thus fixing the values of x and y.

Note that in Example 2 the rank of both the augmented matrix and the matrix of coefficients is two. In general, a necessary and sufficient condition for a system of linear equations to be consistent is that the augmented matrix and the matrix of coefficients be of the same rank.

For the reader's information, some general statements about systems of

m linear equations in n variables follow. Although the statements may be proved, they will be assumed throughout this book. If the system (5-17) is consistent in the case of $m > n$ (more equations than variables), the equations are necessarily linearly dependent. Furthermore, if the rank of the augmented matrix is n, a unique solution exists; however, if the rank r of the augmented matrix is less than n, a solution may be obtained by choosing arbitrary values for $n - r$ of the variables and proceeding to determine values of the remaining variables; that is, infinitely many solutions exist. Note that when $m < n$, values for at least $n - m$ variables must be chosen arbitrarily if the system is consistent.

A system of *linear homogeneous equations* is of the form

$$\begin{cases} a_{11}x_1 + a_{12}x_2 + \cdots + a_{1n}x_n = 0 \\ a_{21}x_1 + a_{22}x_2 + \cdots + a_{2n}x_n = 0 \\ \cdots \\ a_{m1}x_1 + a_{m2}x_2 + \cdots + a_{mn}x_n = 0. \end{cases} \quad (5\text{-}20)$$

The system (5-20) may be expressed in matrix form as $AX = 0$, where A is the matrix of coefficients, X is the column matrix of variables $x_1, x_2, \ldots, x_n$, and 0 is the null column matrix of order m by 1. Such a system necessarily is consistent since it has the *trivial* solution $x_1 = x_2 = \cdots = x_n = 0$. If a *nontrivial* solution exists, infinitely many solutions exist; that is, if $t_1, t_2, \ldots, t_n$ represents a nontrivial solution, then $kt_1, kt_2, \ldots, kt_n$ for any real number k is also a solution. This property of a system of linear homogeneous equations may be proved by direct substitution in (5-20).

EXERCISES

In Exercises 1 and 2 solve the system of linear equations by performing the elementary row transformations on the augmented matrix.

1. $\begin{cases} 2x - 4y + 5z = 10 \\ 2x - 11y + 10z = 36 \\ 4x - y + 5z = -6. \end{cases}$

2. $\begin{cases} x + 2y - z = 2 \\ x - 2y - z = 5 \\ 2x + 4y - 2z = 7. \end{cases}$

3. Determine k such that the system of linear equations is consistent:

$$\begin{cases} 2x + y - z = 12 \\ x - y - 2z = -3 \\ 3y + 3z = k. \end{cases}$$

4. Prove that a system of n linear homogeneous equations in n variables has a nontrivial solution if, and only if, the rank of the matrix of coefficients is less than n.

In Exercises 5 and 6 determine whether or not the system of linear homogeneous equations has nontrivial solutions. (See Exercise 4.)

5. $\begin{cases} 3x + 8y + 2z = 0 \\ 2x + y + 3z = 0 \\ -5x - y + z = 0. \end{cases}$

6. $\begin{cases} x + 3y - 5z = 0 \\ 3x - y + 5z = 0 \\ 3x + 2y - z = 0. \end{cases}$

SIX

TRANSFORMATIONS OF THE PLANE

6-1 / MAPPINGS

In elementary mathematics a function may be considered as a rule that associates with each element of some set A an element of some set B. Except in advanced mathematics, the sets A and B are usually subsets of the set of real numbers. For example, if A consists of the set of real numbers x and B consists of the set of nonnegative real numbers y, then $y = x^2$ represents a function; that is, to each $x \in A$ the rule $y = x^2$ associates an element $y \in B$ where y is the square of x. In this section some additional terminology for the concept of a function is introduced. This terminology is based on geometric concepts and lends itself to the description of the problems to be discussed in this chapter.

A **single-valued mapping** T of a set A *into* a set B is a rule that associates with each element $a \in A$ a unique element $b \in B$. The element b is called the **image** of a under the mapping T and is denoted by $b = T(a)$. For example, consider two sets, $A = \{m, n\}$ and $B = \{x, y, z\}$. Let T be a rule that associates x and z with m and n, respectively; then T is a single-valued mapping of A into B such that $x = T(m)$ and $z = T(n)$. This mapping is illustrated geometrically by Figure 6-1.

Figure 6-2 illustrates a situation which cannot be described as a single-valued mapping of A into B since no element of set A may have more than one image.

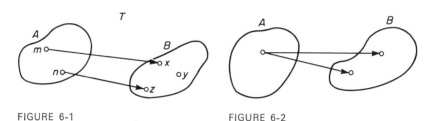

FIGURE 6-1   FIGURE 6-2

If T is a single-valued mapping of A into B such that each element of B is the image of some element of A, then T is a single-valued mapping of A **onto** B. In Figure 6-1, T is a mapping of A into B, but is not a mapping of A onto B. Consider two sets, $A = \{1, 2, 3\}$ and $B = \{a, b, c\}$. Let T be a rule that associates a, b, and c with 2, 1, and 3, respectively (Figure 6-3); then T is a single-valued mapping of A onto B. Every mapping of A onto B is necessarily a mapping of A into B; however, the converse is not true.

As another example of a single-valued mapping of A onto B, let $A = \{1, 2, 3\}$, $B = \{m, n\}$, and T be a rule such that $T(1) = m$, $T(2) = m$, and $T(3) = n$ (Figure 6-4). Notice that while each element of B is the image of some element of A, m is the image of both 1 and 2. The mapping T is nevertheless a single-valued mapping of A onto B. A distinction exists, however, between the two single-valued mappings illustrated in Figures 6-3 and 6-4. The mapping in Figure 6-3 is called a **one-to-one mapping** of A onto B and is a special type of onto mapping. Every one-to-one mapping of A onto B is a single-valued mapping of A onto B, but not every single-valued mapping of A onto B is a one-to-one mapping of A onto B. The mapping in Figure 6-4 is not a one-to-one mapping of A onto B.

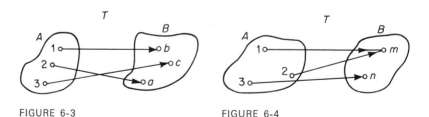

FIGURE 6-3   FIGURE 6-4

If T is a one-to-one mapping of A onto B, then each element of B is the image of exactly one element of A. Hence, it is possible to define an inverse mapping of B onto A, denoted by T^{-1}. The **inverse mapping** T^{-1} is such that it associates with each element $b \in B$ the element $a \in A$ that has b as its image under T; that is, $T^{-1}(T(a)) = a$ for all $a \in A$.

EXAMPLE 1 / Let N be the set of integers and $n \in N$. Let T be a single-valued mapping of N into N. Determine if T is a mapping of N onto N where (a) $T(n) = n + 1$; (b) $T(n) = 2n + 1$.

(a) T is a mapping of N onto N since each integer is the unique image of another integer one unit less. Furthermore, T is a one-to-one mapping of N onto N since each integer is the image of one and only one other integer.

(b) T is not a mapping of N onto N since no even integer is the image of another integer under the mapping; that is, $2n + 1$ cannot be even for any integer n.

EXAMPLE 2 / Let T be a single-valued mapping of A into B where $A = \{a, b\}$ and $B = \{x, y\}$. Define the possible mappings.

There are four possible mappings T of A into B:
(i) $T(a) = x$ and $T(b) = x$;
(ii) $T(a) = x$ and $T(b) = y$;
(iii) $T(a) = y$ and $T(b) = x$;
(iv) $T(a) = y$ and $T(b) = y$.

Note that the mappings T described in (ii) and (iii) are one-to-one mappings of A onto B.

Some important mappings of the set of points on a plane into itself are illustrated in the remaining sections of this chapter.

EXERCISES

1. Let R be the set of real numbers and $x \in R$. Let T be a single-valued mapping of R into R. Determine if T is a mapping of R onto R where

(a) $T(x) = x + 3$;   (b) $T(x) = 2$;
(c) $T(x) = x^3$;   (d) $T(x) = x^3 - x$.

2. Let A and B be sets with the same finite number of elements. Prove that T is a one-to-one mapping of A onto B if T is a single-valued mapping of A onto B.

3. Determine the possible number of mappings T in Exercise 2 if A and B are sets with n elements.

4. Determine the possible number of single-valued mappings T of a set A into itself if A is a set with n elements.

5. Let $A = \{1, 2, 3\}$ and $B = \{a, b\}$. Let T be a single-valued mapping of A into B such that $T(1) = b$, $T(2) = a$, and $T(3) = b$. Determine the single-valued mapping T^{-1}, if it exists.

6-2 / ROTATIONS

It is sometimes necessary or desirable to consider a mapping of the set of points on a plane into itself in order to simplify a problem under examination. Such a mapping is called a **point transformation** or a **transformation of the plane**; that is, a transformation of the plane is a rule by which each point P on the plane is transformed or *mapped* onto a point P' on the plane. A transformation of the plane may be considered as a movement of the points on the plane to the position of their respective image points. Each transformation of the plane may be described algebraically by means of a set of equations relating the rectangular Cartesian coordinates of each point on the plane to the coordinates of its image point. There exist various types of transformations of the plane such as rotations, reflections, dilations, magnifications, shears, projections, and inversions, among others. In this section and in subsequent sections some of the most important transformations are considered and represented in matrix form.

Consider a **rotation of the plane** about the origin through an angle θ, retaining the same rectangular Cartesian coordinate reference system. Let $P: (x, y)$ be any point on the plane. Under a rotation of the plane about the origin, each point P is mapped onto a point $P': (x', y')$ as shown in Figure 6-5. Note that $|\overrightarrow{OP}| = |\overrightarrow{OP'}|$. Then

$$\overrightarrow{OP} = x\vec{i} + y\vec{j}$$
$$= |\overrightarrow{OP}| \cos \phi \vec{i} + |\overrightarrow{OP}| \sin \phi \vec{j}$$
$$= |\overrightarrow{OP'}| \cos \phi \vec{i} + |\overrightarrow{OP'}| \sin \phi \vec{j},$$

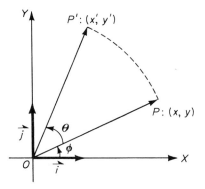

FIGURE 6-5

and
$$\overrightarrow{OP'} = x'\vec{i} + y'\vec{j}$$
$$= |\overrightarrow{OP'}|\cos(\theta + \phi)\vec{i} + |\overrightarrow{OP'}|\sin(\theta + \phi)\vec{j}$$
$$= |\overrightarrow{OP'}|(\cos\theta\cos\phi - \sin\theta\sin\phi)\vec{i}$$
$$+ |\overrightarrow{OP'}|(\sin\theta\cos\phi + \cos\theta\sin\phi)\vec{j}$$
$$= (x\cos\theta - y\sin\theta)\vec{i} + (x\sin\theta + y\cos\theta)\vec{j}.$$

Hence,
$$\begin{cases} x' = x\cos\theta - y\sin\theta \\ y' = x\sin\theta + y\cos\theta. \end{cases} \quad (6\text{-}1)$$

The relationship between the coordinates of each point P and its image point P' may be expressed in matrix form as

$$\begin{pmatrix} x' \\ y' \end{pmatrix} = \begin{pmatrix} \cos\theta & -\sin\theta \\ \sin\theta & \cos\theta \end{pmatrix} \begin{pmatrix} x \\ y \end{pmatrix}. \quad (6\text{-}2)$$

The matrix

$$\begin{pmatrix} \cos\theta & -\sin\theta \\ \sin\theta & \cos\theta \end{pmatrix} \quad (6\text{-}3)$$

of the rotation transformation defined by (6-2) is called a **rotation matrix.**

The rotation matrix (6-3) may be considered as an *operator* that maps each point (x, y) on the plane onto its image point

$$(x\cos\theta - y\sin\theta, \ x\sin\theta + y\cos\theta)$$

when the plane is rotated about the origin through an angle θ. When considered as an operator, a matrix of the form (6-3) may be used to represent, describe, or characterize a rotation of the plane about the origin through an angle θ. Note that the determinant of every rotation matrix is equal to 1. Furthermore, either the row or column vectors of a rotation matrix may be considered to be a pair of orthogonal unit vectors.

EXAMPLE 1 / Determine the coordinates of the image point of $P: (5, \sqrt{3})$ under a rotation of the plane about the origin through an angle of 30° (Figure 6-6).

Since the angle of rotation θ is 30°, $\cos \theta = \sqrt{3}/2$ and $\sin \theta = \frac{1}{2}$. Therefore, by (6-3), the rotation matrix for the transformation of the plane is

$$\begin{pmatrix} \frac{\sqrt{3}}{2} & -\frac{1}{2} \\ \frac{1}{2} & \frac{\sqrt{3}}{2} \end{pmatrix}.$$

Hence, by (6-2), the image point (x', y') of $P: (5, \sqrt{3})$ is given as

$$\begin{pmatrix} x' \\ y' \end{pmatrix} = \begin{pmatrix} \frac{\sqrt{3}}{2} & -\frac{1}{2} \\ \frac{1}{2} & \frac{\sqrt{3}}{2} \end{pmatrix} \begin{pmatrix} 5 \\ \sqrt{3} \end{pmatrix} = \begin{pmatrix} 2\sqrt{3} \\ 4 \end{pmatrix};$$

that is, the coordinates of the image point of $P: (5, \sqrt{3})$ under a rotation of the plane about the origin through an angle of 30° are $(2\sqrt{3}, 4)$.

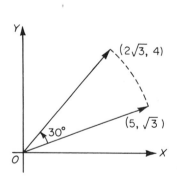

FIGURE 6-6

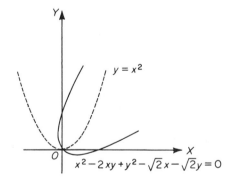

FIGURE 6-7

EXAMPLE 2 / Determine the equation satisfied by the set of image points of the locus of $x^2 - 2xy + y^2 - \sqrt{2}x - \sqrt{2}y = 0$ under a rotation of the plane about the origin through an angle of 45° (Figure 6-7).

Since the angle of rotation θ is 45°, $\cos\theta = \sqrt{2}/2$ and $\sin\theta = \sqrt{2}/2$. The image point (x', y') of each point (x, y) is given by the matrix equation (6-2) as

$$\begin{pmatrix} x' \\ y' \end{pmatrix} = \begin{pmatrix} \frac{\sqrt{2}}{2} & -\frac{\sqrt{2}}{2} \\ \frac{\sqrt{2}}{2} & \frac{\sqrt{2}}{2} \end{pmatrix} \begin{pmatrix} x \\ y \end{pmatrix}.$$

Then

$$\begin{pmatrix} x \\ y \end{pmatrix} = \begin{pmatrix} \frac{\sqrt{2}}{2} & \frac{\sqrt{2}}{2} \\ -\frac{\sqrt{2}}{2} & \frac{\sqrt{2}}{2} \end{pmatrix} \begin{pmatrix} x' \\ y' \end{pmatrix}.$$

Hence, replacing x by $\sqrt{2}x'/2 + \sqrt{2}y'/2$ and y by $-\sqrt{2}x'/2 + \sqrt{2}y'/2$ in $x^2 - 2xy + y^2 - \sqrt{2}x - \sqrt{2}y = 0$, we find that the equation satisfied by the set of image points becomes

$$(\tfrac{\sqrt{2}}{2}x' + \tfrac{\sqrt{2}}{2}y')^2 - 2(\tfrac{\sqrt{2}}{2}x' + \tfrac{\sqrt{2}}{2}y')(-\tfrac{\sqrt{2}}{2}x' + \tfrac{\sqrt{2}}{2}y')$$
$$+ (-\tfrac{\sqrt{2}}{2}x' + \tfrac{\sqrt{2}}{2}y')^2 - \sqrt{2}(\tfrac{\sqrt{2}}{2}x' + \tfrac{\sqrt{2}}{2}y')$$
$$- \sqrt{2}(-\tfrac{\sqrt{2}}{2}x' + \tfrac{\sqrt{2}}{2}y') = 0.$$

Simplifying, we obtain $y' = (x')^2$; that is, $y = x^2$ since (x', y') are the coordinates of each image point with reference to the xy-axes.

A point is called a **fixed point** under a transformation of the plane if it is mapped onto itself. Note that when $\theta \neq 360°k$ for any integer k, the origin is the only fixed point under a rotation of the plane about the origin through an angle θ. When $\theta = 360°k$ for some integer k, every point on the plane is a fixed point, and the rotation matrix (6-3) is equal to the identity matrix of order two. A rotation of the plane about the origin represented by an identity matrix is called the **identity transformation**.

Occasionally it is necessary to apply one transformation of the plane after another. If T_1 and T_2 are matrices representing transformations, then the **product transformation** T_1 followed by T_2 is denoted by T_2T_1 and defined such that

$$T_2T_1\begin{pmatrix} x \\ y \end{pmatrix} = T_2\left[T_1\begin{pmatrix} x \\ y \end{pmatrix}\right]. \tag{6-4}$$

A rotation of the plane about the origin through an angle θ is a one-to-one mapping of the set of points on the plane onto itself. Hence, an inverse transformation exists for each rotation. Consider a rotation of the plane about the origin through an angle $(-\theta)$. By (6-3), the rotation matrix of this transformation is

$$\begin{pmatrix} \cos(-\theta) & -\sin(-\theta) \\ \sin(-\theta) & \cos(-\theta) \end{pmatrix};$$

that is,

$$\begin{pmatrix} \cos\theta & \sin\theta \\ -\sin\theta & \cos\theta \end{pmatrix} \tag{6-5}$$

since $\cos(-\theta) = \cos\theta$ and $\sin(-\theta) = -\sin\theta$. Then, since

$$\begin{pmatrix} \cos\theta & \sin\theta \\ -\sin\theta & \cos\theta \end{pmatrix} \begin{pmatrix} \cos\theta & -\sin\theta \\ \sin\theta & \cos\theta \end{pmatrix} = \begin{pmatrix} 1 & 0 \\ 0 & 1 \end{pmatrix},$$

a rotation of the plane about the origin through an angle $(-\theta)$ is the inverse transformation of a rotation of the plane about the origin through an angle θ.

EXERCISES

In Exercises 1 through 6 determine the image point of the point P under a rotation of the plane about the origin through the angle θ.

1. $P: (2, -3);\ \theta = 90°$.
2. $P: (-5, -2);\ \theta = 180°$.
3. $P: (\sqrt{3}, 1);\ \theta = 30°$.
4. $P: (\sqrt{3}, 1);\ \theta = -60°$.
5. $P: (1, 2);\ \theta = 45°$.
6. $P: (2, \sqrt{3});\ \theta = 30°$.

7. Determine the equation satisfied by the set of image points of the locus of $x^2 - xy + y^2 = 16$ under a rotation of the plane about the origin through an angle of 45°.

8. Determine the equation satisfied by the set of image points of the locus of $16x^2 + 24xy + 9y^2 - 60x + 80y = 0$ under a rotation of the plane about the origin through an acute angle $\theta = \arctan \frac{4}{3}$.

9. Determine the equation satisfied by the set of image points of the locus of $x^2 + y^2 = r^2$ under a rotation of the plane about the origin through an angle θ.

10. Determine whether or not the matrix

$$\begin{pmatrix} -\frac{\sqrt{3}}{2} & \frac{1}{2} \\ \frac{1}{2} & \frac{\sqrt{3}}{2} \end{pmatrix}$$

may represent a rotation of the plane about the origin.

11. Prove that the multiplication of rotation matrices of the form (6-3) is (a) closed; (b) commutative.

12. Determine a rotation matrix that maps the points $(3, 4)$ and $(1, -2)$ onto the points $(-4, 3)$ and $(2, 1)$, respectively.

13. A geometric property that is unchanged under a given transformation is called an **invariant** under that transformation. Prove that the distance between two points on a plane is invariant under a rotation of the plane about the origin.

6-3 / REFLECTIONS, DILATIONS, AND MAGNIFICATIONS

In this section some additional types of transformations of the plane are presented.

Consider the transformations of the plane represented by the matrices

$$\begin{pmatrix} -1 & 0 \\ 0 & 1 \end{pmatrix} \text{ and } \begin{pmatrix} 1 & 0 \\ 0 & -1 \end{pmatrix}. \qquad (6\text{-}6)$$

These matrices map each point (x, y) on the plane onto the points $(-x, y)$ and $(x, -y)$, respectively, and represent one-to-one mappings of a plane onto itself that are called **reflections of the plane**. Each point on the plane is mapped onto its "mirror image" with respect to one of the coordinate axes by these **reflection matrices** (6-6). Note that under the reflection described by

$$\begin{pmatrix} -1 & 0 \\ 0 & 1 \end{pmatrix}$$

the points on the y-axis are fixed points; under the reflection described by

$$\begin{pmatrix} 1 & 0 \\ 0 & -1 \end{pmatrix}$$

the points on the x-axis are fixed points.

Another pair of matrices that represent reflections of the plane is given by

$$\begin{pmatrix} 0 & 1 \\ 1 & 0 \end{pmatrix} \text{ and } \begin{pmatrix} 0 & -1 \\ -1 & 0 \end{pmatrix}. \tag{6-7}$$

The matrix

$$\begin{pmatrix} 0 & 1 \\ 1 & 0 \end{pmatrix}$$

maps each point (x, y) onto (y, x); that is, each point on the plane is mapped onto its mirror image with respect to the line $y = x$. The matrix

$$\begin{pmatrix} 0 & -1 \\ -1 & 0 \end{pmatrix}$$

maps each point (x, y) onto $(-y, -x)$; that is, each point on the plane is mapped onto its mirror image with respect to the line $y = -x$.

It is interesting and important to note that the product of any two of the reflection matrices of (6-6) and (6-7) is a rotation matrix. In general, the product of any two reflections of the plane with respect to intersecting lines passing through the origin is a rotation of the plane about the origin. Also note that the determinant of each of the reflection matrices in (6-6) and (6-7) is equal to -1.

The reflection matrices of (6-7) may be considered as products of the reflection matrices of (6-6) and rotation matrices of the form (6-3). For example,

$$\begin{pmatrix} 0 & 1 \\ 1 & 0 \end{pmatrix} = \begin{pmatrix} \frac{\sqrt{2}}{2} & -\frac{\sqrt{2}}{2} \\ \frac{\sqrt{2}}{2} & \frac{\sqrt{2}}{2} \end{pmatrix} \begin{pmatrix} 1 & 0 \\ 0 & -1 \end{pmatrix} \begin{pmatrix} \frac{\sqrt{2}}{2} & \frac{\sqrt{2}}{2} \\ -\frac{\sqrt{2}}{2} & \frac{\sqrt{2}}{2} \end{pmatrix}; \tag{6-8}$$

$$\begin{pmatrix} 0 & -1 \\ -1 & 0 \end{pmatrix} = \begin{pmatrix} \frac{\sqrt{2}}{2} & \frac{\sqrt{2}}{2} \\ -\frac{\sqrt{2}}{2} & \frac{\sqrt{2}}{2} \end{pmatrix} \begin{pmatrix} 1 & 0 \\ 0 & -1 \end{pmatrix} \begin{pmatrix} \frac{\sqrt{2}}{2} & -\frac{\sqrt{2}}{2} \\ \frac{\sqrt{2}}{2} & \frac{\sqrt{2}}{2} \end{pmatrix}. \tag{6-9}$$

Reflection matrices other than those of (6-6) and (6-7) exist. These reflection matrices, which map each point on the plane onto its mirror image with respect to a line l, can be shown to be equal to

$$T^{-1}RT, \tag{6-10}$$

where T is a transformation matrix that maps the line l onto the x-axis and R is the matrix that represents a reflection of the plane with respect to the x-axis (Figure 6-8).

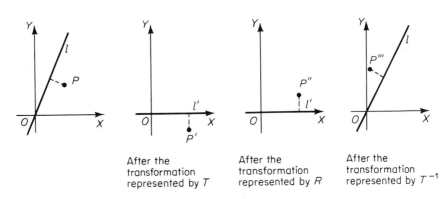

FIGURE 6-8

In (6-8) and (6-9), transformation (6-10) has been applied in expressing the matrices that represent reflections of the plane with regard to the lines $y = x$ and $y = -x$, respectively. In (6-8), T is a matrix representing a rotation of the plane about the origin through an angle $\theta = -45°$, and in (6-9), T is a matrix representing a rotation of the plane about the origin through an angle $\theta = 45°$.

If T is a transformation matrix that maps the line l onto the y-axis, then in (6-10) R is the matrix that represents a reflection of the plane with respect to the y-axis. Another application of (6-10) is considered in the next example.

EXAMPLE 1 / Determine the reflection matrix F that maps each point (x, y) on the plane onto its mirror image with respect to the line $y = \sqrt{3}\, x$. Determine the image of the point $P: (\sqrt{3}, 1)$ under the reflection of the plane represented by F (Figure 6-9).

A rotation of the plane about the origin through an angle $\theta = 30°$

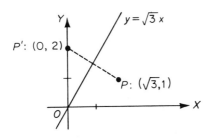

FIGURE 6-9

6-3 / REFLECTIONS, DILATIONS, AND MAGNIFICATIONS

maps the line $y = \sqrt{3}\,x$ onto the y-axis. Such a rotation of the plane may be represented by the matrix

$$T = \begin{pmatrix} \frac{\sqrt{3}}{2} & -\frac{1}{2} \\ \frac{1}{2} & \frac{\sqrt{3}}{2} \end{pmatrix}, \quad \text{whose inverse} \quad T^{-1} = \begin{pmatrix} \frac{\sqrt{3}}{2} & \frac{1}{2} \\ -\frac{1}{2} & \frac{\sqrt{3}}{2} \end{pmatrix}.$$

The matrix R representing a reflection of the plane with respect to the y-axis is

$$\begin{pmatrix} -1 & 0 \\ 0 & 1 \end{pmatrix}.$$

Hence, by (6-10), the reflection matrix F that maps each point (x, y) on the plane onto its mirror image with respect to the line $y = \sqrt{3}\,x$ is given as

$$F = T^{-1}RT = \begin{pmatrix} \frac{\sqrt{3}}{2} & \frac{1}{2} \\ -\frac{1}{2} & \frac{\sqrt{3}}{2} \end{pmatrix} \begin{pmatrix} -1 & 0 \\ 0 & 1 \end{pmatrix} \begin{pmatrix} \frac{\sqrt{3}}{2} & -\frac{1}{2} \\ \frac{1}{2} & \frac{\sqrt{3}}{2} \end{pmatrix};$$

that is,

$$F = \begin{pmatrix} -\frac{1}{2} & \frac{\sqrt{3}}{2} \\ \frac{\sqrt{3}}{2} & \frac{1}{2} \end{pmatrix}.$$

The image of the point $P: (\sqrt{3}, 1)$ under the reflection of the plane represented by F is $P': (0, 2)$ since

$$\begin{pmatrix} -\frac{1}{2} & \frac{\sqrt{3}}{2} \\ \frac{\sqrt{3}}{2} & \frac{1}{2} \end{pmatrix} \begin{pmatrix} \sqrt{3} \\ 1 \end{pmatrix} = \begin{pmatrix} 0 \\ 2 \end{pmatrix}.$$

Another important class of transformations of the plane may be represented by the scalar matrices of order two

$$\begin{pmatrix} k & 0 \\ 0 & k \end{pmatrix},$$

where $k > 0$. The effect of such a transformation matrix is to map each point P on the plane onto a point P' on the plane such that $\overrightarrow{OP'} = k\overrightarrow{OP}$. If $k > 1$, the matrix represents a "uniform stretching" of the plane; if $0 < k < 1$, the matrix represents a "uniform compression" of the plane. A transformation of the plane represented by a scalar matrix of order two having diagonal elements greater than zero is called a **dilation of the plane**.

Transformations represented by diagonal matrices of order two

$$\begin{pmatrix} d_1 & 0 \\ 0 & d_2 \end{pmatrix},$$

where $d_1 > 0$ and $d_2 > 0$, are related closely to the dilations of the plane. Many plane geometric figures are distorted under such transformations. A transformation of the plane represented by a diagonal matrix of order two with diagonal elements greater than zero is called a **magnification of the plane**. The set of dilations of the plane is a subset of the set of magnifications of the plane. Note that a magnification of the plane is a one-to-one mapping of the set of points on the plane onto itself.

EXAMPLE 2 / Determine the effect of the dilation of the plane represented by

$$\begin{pmatrix} 4 & 0 \\ 0 & 4 \end{pmatrix}$$

upon the line $y = 3x + 2$.

Under the dilation of the plane, each point (x, y) on the plane is mapped onto (x', y') where

$$\begin{pmatrix} x' \\ y' \end{pmatrix} = \begin{pmatrix} 4 & 0 \\ 0 & 4 \end{pmatrix} \begin{pmatrix} x \\ y \end{pmatrix}.$$

Hence,

$$\begin{pmatrix} x \\ y \end{pmatrix} = \begin{pmatrix} \frac{1}{4} & 0 \\ 0 & \frac{1}{4} \end{pmatrix} \begin{pmatrix} x' \\ y' \end{pmatrix};$$

that is, $x = \frac{1}{4}x'$ and $y = \frac{1}{4}y'$. Therefore, under the given dilation of the plane, the line $y = 3x + 2$ is mapped onto the line $y' = 3x' + 8$; that is, $y = 3x + 8$. Note that the image of the given line is a line with the same slope.

EXAMPLE 3 / Determine the effect of the magnification of the plane represented by

$$\begin{pmatrix} 1 & 0 \\ 0 & 3 \end{pmatrix}$$

upon the ellipse $\frac{x^2}{9} + \frac{y^2}{1} = 1$ (Figure 6-10).

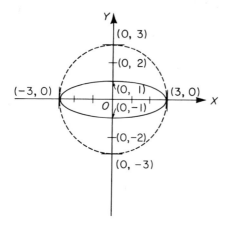

FIGURE 6-10

Under the magnification of the plane, each point (x, y) on the plane is mapped onto (x', y') where

$$\begin{pmatrix} x' \\ y' \end{pmatrix} = \begin{pmatrix} 1 & 0 \\ 0 & 3 \end{pmatrix} \begin{pmatrix} x \\ y \end{pmatrix}.$$

Hence,

$$\begin{pmatrix} x \\ y \end{pmatrix} = \begin{pmatrix} 1 & 0 \\ 0 & \frac{1}{3} \end{pmatrix} \begin{pmatrix} x' \\ y' \end{pmatrix};$$

that is, $x = x'$ and $y = \frac{1}{3}y'$. Therefore, under the magnification of the plane represented by

$$\begin{pmatrix} 1 & 0 \\ 0 & 3 \end{pmatrix},$$

the ellipse $\frac{x^2}{9} + \frac{y^2}{1} = 1$ is mapped onto the circle $\frac{(x')^2}{9} + \frac{(y')^2}{9} = 1$; that is, $x^2 + y^2 = 9$.

EXERCISES

In Exercises 1 through 3 determine the effect upon the coordinate plane of the transformation of the plane represented by the given matrix.

1. $\begin{pmatrix} 2 & 0 \\ 0 & 2 \end{pmatrix}.$
2. $\begin{pmatrix} -1 & 0 \\ 0 & -1 \end{pmatrix}.$
3. $\begin{pmatrix} 0 & 2 \\ 2 & 0 \end{pmatrix}.$

4. Prove that the products of the reflection matrices (6-6) represent rotations of the plane. Describe these rotations.

5. Prove that the distance between two points on a plane is invariant under a reflection of the plane with respect to the (a) x-axis; (b) y-axis.

6. Determine the reflection matrix F that maps each point (x, y) on the plane onto its mirror image with respect to the line $3x - 4y = 0$.

7. Prove that the multiplication of any dilation matrix and any rotation matrix of the form (6-3) is commutative.

8. Determine a matrix that represents the product of a reflection of the plane with respect to the line $y = x$ followed by a reflection of the plane with respect to the line $y = \sqrt{3}\, x$. Describe the resulting transformation matrix.

9. Find the dilation matrix that maps the points on a unit circle onto the points on a circle with radius (a) 4; (b) $\frac{1}{2}$; (c) r.

10. Determine the effect of the dilation of the plane represented by

$$\begin{pmatrix} 3 & 0 \\ 0 & 3 \end{pmatrix}$$

upon the triangle with vertices at $(0, 0)$, $(1, 2)$, and $(3, 1)$.

11. Determine the effect of the magnification of the plane represented by

$$\begin{pmatrix} \frac{1}{5} & 0 \\ 0 & \frac{1}{2} \end{pmatrix}$$

upon the line $2x + 5y = 10$.

12. Determine the effect of the magnification of the plane represented by

$$\begin{pmatrix} 2 & 0 \\ 0 & 1 \end{pmatrix}$$

upon the unit square with vertices at $(0, 0)$, $(1, 0)$, $(1, 1)$, and $(0, 1)$.

6-4 / OTHER TRANSFORMATIONS

The transformations of the plane discussed so far have been examples of one-to-one mappings of the plane onto itself. The following examples present for consideration some other interesting transformations of the plane, two of which are not one-to-one mappings.

EXAMPLE 1 / The transformation of the plane represented by a matrix of the form

$$\begin{pmatrix} 1 & k \\ 0 & 1 \end{pmatrix}$$

is called a **shear parallel to the x-axis**. Determine the effect of the shear parallel to the x-axis represented by

$$\begin{pmatrix} 1 & 3 \\ 0 & 1 \end{pmatrix}$$

upon a rectangle with vertices at (0, 0), (2, 0), (2, 1), and (0, 1) (Figure 6-11).

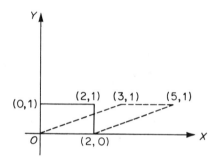

FIGURE 6-11

Under the shear parallel to the x-axis represented by

$$\begin{pmatrix} 1 & 3 \\ 0 & 1 \end{pmatrix},$$

each point $(x, 0)$ is a fixed point; that is, each point on the x-axis is mapped onto itself. Hence, the line segment with end points (0, 0) and (2, 0) is mapped onto itself. Each point $(x, 1)$ is mapped onto a point with coordinates $(x + 3, 1)$; that is, each point on the line $y = 1$ is mapped onto another point of the line and positioned three units to the right. Hence, the line segment with end points (0, 1) and (2, 1) is mapped onto the line segment with end points (3, 1) and (5, 1). Each point $(0, y)$ is mapped onto a point with coordinates $(3y, y)$; that is, each point on the y-axis is mapped onto a point on the line $3y = x$. As a result, the line segment with end points $(0, 0)$ and $(0, 1)$ is mapped onto the line segment with end points $(0, 0)$ and $(3, 1)$. Each point $(2, y)$ is mapped onto a point with coordinates $(2 + 3y, y)$; that is, each point on the line $x = 2$ is mapped onto a point on the line $2 + 3y = x$. The result is that the line segment with end points $(2, 0)$ and $(2, 1)$ is mapped onto the line segment with end points $(2, 0)$ and $(5, 1)$. Hence, the rectangle is mapped onto a parallelogram with vertices at $(0, 0)$, $(2, 0)$, $(5, 1)$, and $(3, 1)$.

EXAMPLE 2 / Determine the effect of the transformation of the plane represented by

$$\begin{pmatrix} 1 & 0 \\ 0 & 0 \end{pmatrix}.$$

Under the transformation of the plane, each point (x, y) on the plane is mapped onto a point with coordinates $(x, 0)$. The matrix represents a vertical projection of the points on the plane onto the x-axis. The transformation of the plane represented by

$$\begin{pmatrix} 1 & 0 \\ 0 & 0 \end{pmatrix}$$

is an example of a mapping of the set of points on the plane into, but not onto, itself.

EXAMPLE 3 / Determine the effect of the transformation of the plane represented by

$$\begin{pmatrix} 1 & 0 \\ 1 & 0 \end{pmatrix}.$$

Under the transformation of the plane, each point (x, y) on the plane is mapped onto a point with coordinates (x, x). The matrix represents a projection of the plane onto the line $y = x$. The transformation of the plane represented by

$$\begin{pmatrix} 1 & 0 \\ 1 & 0 \end{pmatrix}$$

is an example of a mapping of the set of points on the plane into, but not onto, itself.

The transformations of the plane represented by the matrices of Examples 2 and 3 are called **projections of the plane**. Other projections of the plane exist.

EXERCISES

In Exercises 1 through 3 determine the effect upon the coordinate plane of the transformation of the plane represented by the given matrix.

1. $\begin{pmatrix} 0 & 0 \\ 0 & 1 \end{pmatrix}$.
2. $\begin{pmatrix} 1 & 1 \\ 0 & 0 \end{pmatrix}$.
3. $\begin{pmatrix} 1 & -2 \\ 0 & 1 \end{pmatrix}$.

4. Determine the effect of the shear parallel to the *x*-axis represented by

$$\begin{pmatrix} 1 & 2 \\ 0 & 1 \end{pmatrix}$$

upon a rectangle with vertices at $(-1, -1)$, $(3, -1)$, $(3, 2)$, and $(-1, 2)$.

5. The transformation of the plane represented by a matrix of the form

$$\begin{pmatrix} 1 & 0 \\ k & 1 \end{pmatrix}$$

is called a **shear parallel to the *y*-axis**. (a) Determine the effect of the shear parallel to the *y*-axis represented by

$$\begin{pmatrix} 1 & 0 \\ 2 & 1 \end{pmatrix}$$

upon the unit circle $x^2 + y^2 = 1$. (b) Determine the effect of the shear parallel to the *y*-axis represented by

$$\begin{pmatrix} 1 & 0 \\ k & 1 \end{pmatrix}$$

upon a rectangle with vertices at $(0, 0)$, $(2, 0)$, $(2, 1)$, and $(0, 1)$.

6. Find a matrix that represents the product of a shear parallel to the

x-axis followed by a dilation of the plane that maps the line $y = 3x - 2$ onto the line $x = 6$.

7. Determine a transformation matrix that maps the points (1, 0) and (0, 1) onto (2, 1) and (3, 2), respectively.

8. Prove that the multiplication of a matrix representing a shear parallel to the x-axis and a rotation matrix of the form (6-3) generally is not commutative.

6-5 / LINEAR HOMOGENEOUS TRANSFORMATIONS

The transformations of the plane that have been discussed so far are examples of a general class of transformations called **linear homogeneous transformations of the plane**. These transformations always may be expressed in matrix form as

$$\begin{pmatrix} x' \\ y' \end{pmatrix} = \begin{pmatrix} a & b \\ c & d \end{pmatrix} \begin{pmatrix} x \\ y \end{pmatrix}; \qquad (6\text{-}11)$$

that is, under a linear homogeneous transformation of the plane, each point (x, y) is mapped onto its image point (x', y') where

$$\begin{cases} x' = ax + by \\ y' = cx + dy. \end{cases} \qquad (6\text{-}12)$$

If the matrix of the transformation

$$\begin{pmatrix} a & b \\ c & d \end{pmatrix}$$

is nonsingular, the transformation defined by (6-11) is called a **nonsingular linear homogeneous transformation of the plane**. For example, rotations of the plane about the origin, reflections of the plane with respect to a line through the origin, dilations of the plane, magnifications of the plane, and shears parallel to a coordinate axis are nonsingular linear homogeneous transformations of the plane. The projections of the plane considered in §6-4 are linear homogeneous transformations of the plane that are not nonsingular.

THEOREM 6.1 / *Every nonsingular linear homogeneous transformation of the plane is a one-to-one mapping of the set of points on the plane onto itself.*

PROOF / Let T be a nonsingular matrix of order two. Consider

$$\begin{pmatrix} x' \\ y' \end{pmatrix} = T \begin{pmatrix} x \\ y \end{pmatrix}.$$

Then, since T^{-1} exists,

$$T^{-1} \begin{pmatrix} x' \\ y' \end{pmatrix} = T^{-1} T \begin{pmatrix} x \\ y \end{pmatrix} = \begin{pmatrix} x \\ y \end{pmatrix};$$

that is, given any point $P': (x', y')$, it is possible to find the point $P: (x, y)$ such that P' is the image of P under the transformation represented by T. Hence, each point on the plane is the image of some point on the plane under the transformation represented by T, and T represents a mapping of the set of points on the plane onto itself.

Let $A: (x_1, y_1)$ and $B: (x_2, y_2)$ be any two points on the plane that have the same image point under the transformation represented by T; that is, let

$$\begin{pmatrix} x' \\ y' \end{pmatrix} = T \begin{pmatrix} x_1 \\ y_1 \end{pmatrix} \quad \text{and} \quad \begin{pmatrix} x' \\ y' \end{pmatrix} = T \begin{pmatrix} x_2 \\ y_2 \end{pmatrix}.$$

Then

$$T \begin{pmatrix} x_1 \\ y_1 \end{pmatrix} = T \begin{pmatrix} x_2 \\ y_2 \end{pmatrix}$$

$$T^{-1} T \begin{pmatrix} x_1 \\ y_1 \end{pmatrix} = T^{-1} T \begin{pmatrix} x_2 \\ y_2 \end{pmatrix}$$

$$\begin{pmatrix} x_1 \\ y_1 \end{pmatrix} = \begin{pmatrix} x_2 \\ y_2 \end{pmatrix}.$$

Since A and B are identical points if they have the same image point, T represents a one-to-one mapping of the set of points on the plane onto itself.

EXERCISES

In Exercises 1 through 4 prove that under a nonsingular linear homogeneous transformation of the plane:

1. The image of the origin is the origin.

2. The image of a line is a line.

3. The images of parallel lines are parallel lines.

4. The image of a square is a parallelogram.

5. Determine the effect upon the coordinate plane of the linear homogeneous transformation represented by

$$\begin{pmatrix} 6 & 3 \\ 2 & 1 \end{pmatrix}.$$

6. The set of points on the plane whose image points under a transformation of the plane are the origin is called the **null space** of the transformation. Find the null space of the transformation of the plane represented by

$$\begin{pmatrix} 4 & 2 \\ 2 & 1 \end{pmatrix}.$$

6-6 / ORTHOGONAL MATRICES

A square real matrix A for which $AA^T = I$ is called an **orthogonal matrix**. Then A^T is the right inverse of an orthogonal matrix A. Since A is a square matrix its left and right inverses are equal. Hence, $A^T A = I$. Furthermore, $A^T = A^{-1}$; that is, a matrix is orthogonal if, and only if, its transpose and inverse are equal.

Let a_{ij} be the general element of an orthogonal matrix A of order n. Since

$$\sum_{k=1}^{n} a_{ik} a_{jk} = \delta_{ij},$$

the row vectors of A are mutually orthogonal unit vectors; that is, since the sum of the products of the elements of the ith row of A and the corresponding elements of the jth column of A^T taken in order is 0 if $i \neq j$ and 1 if $i = j$, and since the jth column of A^T is the jth row of A, any two rows of A represent orthogonal vectors and any row vector is a unit vector. Similarly, it follows that the column vectors of an orthogonal matrix A are mutually orthogonal unit vectors since $A^T A = I$.

If A is an orthogonal matrix, $(A^{-1})(A^{-1})^T = A^T(A^T)^T = A^T A = I$; that is, the inverse (transpose) of an orthogonal matrix is an orthogonal matrix.

If two matrices A and B are orthogonal matrices, their product is an orthogonal matrix since $(AB)^T = B^T A^T = B^{-1} A^{-1} = (AB)^{-1}$.

If A is an orthogonal matrix, $\det AA^T = \det A \det A^T = \det A \det A = 1$, and $\det A = \pm 1$; that is, the determinant of an orthogonal matrix equals 1 or -1. If $\det A = 1$, the orthogonal matrix A is called a **proper orthogonal matrix**; if $\det A = -1$, the orthogonal matrix A is called an **improper orthogonal matrix**.

The transformation matrices that represent rotations of the plane about the origin are examples of proper orthogonal matrices. In each case, the row (column) vectors are mutually orthogonal unit vectors and the value of the determinant of the matrix is 1. The transformation matrices that represent reflections of the plane with respect to a line through the origin are examples of improper orthogonal matrices.

EXAMPLE 1 / Verify that the matrix A is a proper orthogonal matrix where

$$A = \begin{pmatrix} \frac{12}{13} & \frac{5}{13} \\ -\frac{5}{13} & \frac{12}{13} \end{pmatrix}.$$

$$AA^T = \begin{pmatrix} \frac{12}{13} & \frac{5}{13} \\ -\frac{5}{13} & \frac{12}{13} \end{pmatrix} \begin{pmatrix} \frac{12}{13} & -\frac{5}{13} \\ \frac{5}{13} & \frac{12}{13} \end{pmatrix} = \begin{pmatrix} 1 & 0 \\ 0 & 1 \end{pmatrix}.$$

Since $AA^T = I$, A is an orthogonal matrix. Furthermore, $\det A = 1$. Hence, A is a proper orthogonal matrix.

Since the matrices that represent rotations of the plane about the origin and reflections of the plane with respect to a line through the origin are orthogonal matrices, it is possible to present a concise proof which shows that the scalar product of two plane vectors is a scalar invariant under these transformations. That is, if $\vec{a}' = a_1' \vec{i} + a_2' \vec{j}$ and $\vec{b}' = b_1' \vec{i} + b_2' \vec{j}$ are the image vectors of $\vec{a} = a_1 \vec{i} + a_2 \vec{j}$ and $\vec{b} = b_1 \vec{i} + b_2 \vec{j}$, respectively, under a rotation of the plane about the origin or under a reflection of the plane with respect to a line through the origin, then $a_1 b_1 + a_2 b_2 = a_1' b_1' + a_2' b_2'$. Let R be either a rotation matrix or a reflection matrix. Then

$$(a_1' \quad a_2')^T = R(a_1 \quad a_2)^T$$

and

$$(a_1 \quad a_2)R^T = (a_1' \quad a_2').$$

Similarly,
$$R(b_1 \ b_2)^T = (b'_1 \ b'_2)^T.$$

Then, multiplying equals by equals, we obtain
$$(a_1 \ a_2)R^T R(b_1 \ b_2)^T = (a'_1 \ a'_2)(b'_1 \ b'_2)^T.$$

Since R is an orthogonal matrix, $R^T R = I$. Hence,
$$(a_1 \ a_2)(b_1 \ b_2)^T = (a'_1 \ a'_2)(b'_1 \ b'_2)^T$$
and
$$a_1 b_1 + a_2 b_2 = a'_1 b'_1 + a'_2 b'_2.$$

Since magnitudes of line segments, distances, and the measures of angles may be expressed in terms of the scalar product of two vectors, it immediately follows that these properties are invariants under rotations of the plane about the origin and under reflections of the plane with respect to a line through the origin.

EXAMPLE 2 / Verify that the scalar product of the vectors $\vec{a} = 2\vec{i} + \vec{j}$ and $\vec{b} = 3\vec{i} - \vec{j}$ is a scalar invariant under a rotation of the plane about the origin represented by

$$\begin{pmatrix} \frac{3}{5} & \frac{4}{5} \\ -\frac{4}{5} & \frac{3}{5} \end{pmatrix}.$$

Under the given rotation of the plane, $\vec{a}' = a'_1 \vec{i} + a'_2 \vec{j}$ where

$$\begin{pmatrix} a'_1 \\ a'_2 \end{pmatrix} = \begin{pmatrix} \frac{3}{5} & \frac{4}{5} \\ -\frac{4}{5} & \frac{3}{5} \end{pmatrix} \begin{pmatrix} 2 \\ 1 \end{pmatrix} = \begin{pmatrix} 2 \\ -1 \end{pmatrix}.$$

Similarly, $\vec{b}' = b'_1 \vec{i} + b'_2 \vec{j}$ where

$$\begin{pmatrix} b'_1 \\ b'_2 \end{pmatrix} = \begin{pmatrix} \frac{3}{5} & \frac{4}{5} \\ -\frac{4}{5} & \frac{3}{5} \end{pmatrix} \begin{pmatrix} 3 \\ -1 \end{pmatrix} = \begin{pmatrix} 1 \\ -3 \end{pmatrix}.$$

Then $\vec{a} \cdot \vec{b} = (2)(3) + (1)(-1) = 5$, and $\vec{a}' \cdot \vec{b}' = (2)(1) + (-1)(-3) = 5$. Hence, the scalar product $\vec{a} \cdot \vec{b}$ is a scalar invariant under the rotation of the plane about the origin represented by

$$\begin{pmatrix} \frac{3}{5} & \frac{4}{5} \\ -\frac{4}{5} & \frac{3}{5} \end{pmatrix}.$$

EXERCISES

In Exercises 1 through 4 determine whether the matrix is a proper orthogonal matrix or an improper orthogonal matrix.

1. $\begin{pmatrix} \frac{\sqrt{2}}{2} & \frac{\sqrt{2}}{2} \\ -\frac{\sqrt{2}}{2} & \frac{\sqrt{2}}{2} \end{pmatrix}.$

2. $\begin{pmatrix} \frac{\sqrt{3}}{2} & \frac{1}{2} \\ \frac{1}{2} & -\frac{\sqrt{3}}{2} \end{pmatrix}.$

3. $\begin{pmatrix} 2 & 0 \\ 1 & \frac{1}{2} \end{pmatrix}.$

4. $\begin{pmatrix} 3 & 5 \\ 1 & 2 \end{pmatrix}.$

5. Verify that the inverse of an orthogonal matrix A is an orthogonal matrix where

$$A = \begin{pmatrix} \frac{2}{3} & -\frac{2}{3} & \frac{1}{3} \\ \frac{1}{3} & \frac{2}{3} & \frac{2}{3} \\ \frac{2}{3} & \frac{1}{3} & -\frac{2}{3} \end{pmatrix}.$$

6. Verify that the product of two orthogonal matrices A and B is an orthogonal matrix where

$$A = \begin{pmatrix} -\frac{\sqrt{2}}{2} & -\frac{\sqrt{2}}{2} \\ \frac{\sqrt{2}}{2} & -\frac{\sqrt{2}}{2} \end{pmatrix} \quad \text{and} \quad B = \begin{pmatrix} \frac{3}{5} & \frac{4}{5} \\ \frac{4}{5} & -\frac{3}{5} \end{pmatrix}.$$

7. Verify that if A is an improper orthogonal matrix, then $\det(A + I) = 0$ where

$$A = \begin{pmatrix} \frac{12}{13} & \frac{5}{13} \\ \frac{5}{13} & -\frac{12}{13} \end{pmatrix}.$$

8. Prove that the product of two improper orthogonal matrices of the same order is a proper orthogonal matrix.

9. Determine all the orthogonal matrices of order two whose elements are 0's and 1's.

10. Prove that if $AB = BA$ and C is an orthogonal matrix, then the multiplication of $C^T AC$ and $C^T BC$ is commutative.

6-7 / TRANSLATIONS

The set of linear homogeneous transformations of the plane discussed in §6-5 is a subset of the set of **general linear transformations of the plane**.

Every general linear transformation of the plane may be described by the equations

$$\begin{cases} x' = ax + by + e \\ y' = cx + dy + f, \end{cases} \quad (6\text{-}13)$$

where (x', y') is the image of the point (x, y). It is not possible to express a general linear transformation of the plane described by (6-13) in matrix form where the matrix of the transformation is a square matrix of order two. To express (6-13) in matrix form it will be necessary to consider the use of homogeneous rectangular Cartesian coordinates.

On a coordinate plane, the **homogeneous coordinates** of a point with **nonhomogeneous coordinates** (x, y) are any three ordered scalars (x_1, x_2, x_3), where $x_3 \neq 0$, and for which $x = x_1/x_3$ and $y = x_2/x_3$. For example, the point on the plane with nonhomogeneous coordinates $(1, -3)$ may be represented by an infinite set of homogeneous coordinates of the form $(k, -3k, k)$ where k is any nonzero real number. One set of homogeneous coordinates for the point (x, y) is always of the form $(x, y, 1)$; all other sets of homogeneous coordinates are of the form (kx, ky, k) where $k \neq 0$. Generally, the context of a discussion will indicate whether homogeneous coordinates or nonhomogeneous coordinates are being considered.

Now, a general linear transformation of the plane described by (6-13) may be expressed in matrix form as

$$\begin{pmatrix} x' \\ y' \\ 1 \end{pmatrix} = \begin{pmatrix} a & b & e \\ c & d & f \\ 0 & 0 & 1 \end{pmatrix} \begin{pmatrix} x \\ y \\ 1 \end{pmatrix}, \quad (6\text{-}14)$$

where $(x', y', 1)$ are the homogeneous coordinates of the image of the point with homogeneous coordinates $(x, y, 1)$.

Note that every linear homogeneous transformation of the plane may be expressed in the matrix form

$$\begin{pmatrix} x' \\ y' \\ 1 \end{pmatrix} = \begin{pmatrix} a & b & 0 \\ c & d & 0 \\ 0 & 0 & 1 \end{pmatrix} \begin{pmatrix} x \\ y \\ 1 \end{pmatrix}. \quad (6\text{-}15)$$

For example, a rotation of the plane about the origin through an angle θ may be expressed as

$$\begin{pmatrix} x' \\ y' \\ 1 \end{pmatrix} = \begin{pmatrix} \cos\theta & -\sin\theta & 0 \\ \sin\theta & \cos\theta & 0 \\ 0 & 0 & 1 \end{pmatrix} \begin{pmatrix} x \\ y \\ 1 \end{pmatrix}.$$

6-7 / TRANSLATIONS

The matrix equation (6-15) represents the two equations of (6-12), which define a linear homogeneous transformation of the plane, and a trivial third equation, $1 = 1$, which does not place any conditions on the relationship between the coordinates of a point and those of its image point.

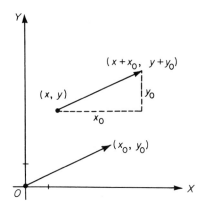

FIGURE 6-12

Consider a transformation of the plane whereby each point on the plane is mapped onto a point a fixed directed distance x_0 away parallel to the x-axis and another fixed directed distance y_0 away parallel to the y-axis (Figure 6-12). Such a transformation of the plane is called a **translation of the plane** and is defined by the equations

$$\begin{cases} x' = x + x_0 \\ y' = y + y_0, \end{cases} \quad (6\text{-}16)$$

where (x', y') is the image of the point (x, y). A translation of the plane may be expressed in matrix form as

$$\begin{pmatrix} x' \\ y' \\ 1 \end{pmatrix} = \begin{pmatrix} 1 & 0 & x_0 \\ 0 & 1 & y_0 \\ 0 & 0 & 1 \end{pmatrix} \begin{pmatrix} x \\ y \\ 1 \end{pmatrix}, \quad (6\text{-}17)$$

where $(x', y', 1)$ are the homogeneous coordinates of the image of the point with homogeneous coordinates $(x, y, 1)$. The matrix

$$\begin{pmatrix} 1 & 0 & x_0 \\ 0 & 1 & y_0 \\ 0 & 0 & 1 \end{pmatrix} \quad (6\text{-}18)$$

of the translation transformation defined by (6-17) is called a **translation matrix**. Every matrix representing a translation of the plane is of the form (6-18). Note that the value of the determinant of every translation matrix is equal to 1.

A translation of the plane is a one-to-one mapping of the set of points on the plane onto itself. Furthermore, there are no fixed points under the translation transformation (6-17) when x_0 and y_0 are not both zero; that is, each point is mapped onto another point. When $x_0 = y_0 = 0$, every point on the plane is a fixed point.

EXAMPLE 1 / Determine the coordinates of the image point of $P: (3, 2)$ under a translation of the plane that maps the origin onto $(4, -1)$.

The given translation of the plane may be described by the matrix equation

$$\begin{pmatrix} x' \\ y' \\ 1 \end{pmatrix} = \begin{pmatrix} 1 & 0 & 4 \\ 0 & 1 & -1 \\ 0 & 0 & 1 \end{pmatrix} \begin{pmatrix} x \\ y \\ 1 \end{pmatrix}$$

since

$$\begin{pmatrix} 4 \\ -1 \\ 1 \end{pmatrix} = \begin{pmatrix} 1 & 0 & 4 \\ 0 & 1 & -1 \\ 0 & 0 & 1 \end{pmatrix} \begin{pmatrix} 0 \\ 0 \\ 1 \end{pmatrix}.$$

Hence, the image of $P: (3, 2)$ is the point $P': (x', y')$ where

$$\begin{pmatrix} x' \\ y' \\ 1 \end{pmatrix} = \begin{pmatrix} 1 & 0 & 4 \\ 0 & 1 & -1 \\ 0 & 0 & 1 \end{pmatrix} \begin{pmatrix} 3 \\ 2 \\ 1 \end{pmatrix} = \begin{pmatrix} 7 \\ 1 \\ 1 \end{pmatrix};$$

that is, $P': (7, 1)$ is the image point of $P: (3, 2)$ under the translation of the plane that maps the origin onto $(4, -1)$.

EXAMPLE 2 / Determine the equation satisfied by the set of image points of the locus of $x^2 + y^2 + 4x + 6y + 9 = 0$ under a translation of the plane represented by

$$\begin{pmatrix} 1 & 0 & 2 \\ 0 & 1 & 3 \\ 0 & 0 & 1 \end{pmatrix}.$$

Under the given translation of the plane, each point (x, y) is mapped onto the point $(x + 2, y + 3)$. Hence, if x is replaced by $x - 2$ and y by $y - 3$ in $x^2 + y^2 + 4x + 6y + 9 = 0$, the equation satisfied by the set of image points becomes

$$(x - 2)^2 + (y - 3)^2 + 4(x - 2) + 6(y - 3) + 9 = 0,$$
$$x^2 - 4x + 4 + y^2 - 6y + 9 + 4x - 8 + 6y - 18 + 9 = 0,$$
$$x^2 + y^2 = 4;$$

that is, a circle with center at $(-2, -3)$ and a radius of 2 units is mapped onto a circle with center at the origin and a radius of 2 units (Figure 6-13).

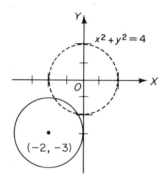

FIGURE 6-13

EXAMPLE 3 / Show that the image of a point $P: (x, y)$ under a rotation of the plane about the origin followed by a translation of the plane generally is not the image of the point under the same transformations considered in reverse order.

The matrices that represent a rotation of the plane about the origin and a translation of the plane are of the forms

$$\begin{pmatrix} \cos\theta & -\sin\theta & 0 \\ \sin\theta & \cos\theta & 0 \\ 0 & 0 & 1 \end{pmatrix} \quad \text{and} \quad \begin{pmatrix} 1 & 0 & x_0 \\ 0 & 1 & y_0 \\ 0 & 0 & 1 \end{pmatrix},$$

respectively, if they are considered to be operating on P with coordinates expressed in homogeneous form. Then the coordinates of the image of P under a rotation followed by a translation are expressed by

$$\begin{pmatrix} x' \\ y' \\ 1 \end{pmatrix} = \begin{pmatrix} 1 & 0 & x_0 \\ 0 & 1 & y_0 \\ 0 & 0 & 1 \end{pmatrix} \begin{pmatrix} \cos\theta & -\sin\theta & 0 \\ \sin\theta & \cos\theta & 0 \\ 0 & 0 & 1 \end{pmatrix} \begin{pmatrix} x \\ y \\ 1 \end{pmatrix}$$
$$= \begin{pmatrix} x\cos\theta - y\sin\theta + x_0 \\ x\sin\theta + y\cos\theta + y_0 \\ 1 \end{pmatrix};$$

that is, (x, y) is mapped onto

$$(x\cos\theta - y\sin\theta + x_0,\ x\sin\theta + y\cos\theta + y_0).$$

The coordinates of the image of P under the same transformations considered in reverse order are expressed by

$$\begin{pmatrix} x' \\ y' \\ 1 \end{pmatrix} = \begin{pmatrix} \cos\theta & -\sin\theta & 0 \\ \sin\theta & \cos\theta & 0 \\ 0 & 0 & 1 \end{pmatrix} \begin{pmatrix} 1 & 0 & x_0 \\ 0 & 1 & y_0 \\ 0 & 0 & 1 \end{pmatrix} \begin{pmatrix} x \\ y \\ 1 \end{pmatrix}$$
$$= \begin{pmatrix} x\cos\theta - y\sin\theta + x_0\cos\theta - y_0\sin\theta \\ x\sin\theta + y\cos\theta + x_0\sin\theta + y_0\cos\theta \\ 1 \end{pmatrix};$$

that is, (x, y) is mapped onto

$$(x\cos\theta - y\sin\theta + x_0\cos\theta - y_0\sin\theta,$$
$$x\sin\theta + y\cos\theta + x_0\sin\theta + y_0\cos\theta).$$

In general, the two image points

$$(x\cos\theta - y\sin\theta + x_0,\ x\sin\theta + y\cos\theta + y_0)$$

and

$$(x\cos\theta - y\sin\theta + x_0\cos\theta - y_0\sin\theta,$$
$$x\sin\theta + y\cos\theta + x_0\sin\theta + y_0\cos\theta)$$

will not be the same.

EXAMPLE 4 / Use the results of Example 3 to determine the image of the point $P: (-5, 2)$ under a rotation of the plane about the origin through an angle of 90° followed by a translation of the plane that maps the origin onto (2, 3).

Under the transformations taken in the given order, the image of each point (x, y) is

$$(x \cos \theta - y \sin \theta + x_0, \; x \sin \theta + y \cos \theta + y_0).$$

Since $\theta = 90°$, $x_0 = 2$, $y_0 = 3$, $x = -5$, and $y = 2$, the image of $P: (-5, 2)$ is $(-5 \cdot 0 - 2 \cdot 1 + 2, \; -5 \cdot 1 + 2 \cdot 0 + 3)$; that is, $(0, -2)$.

EXERCISES

1. Find the form of the homogeneous coordinates of the points having nonhomogeneous coordinates:
(a) $(1, -2)$;   (b) $(3, 0)$;
(c) $(0, 0)$;   (d) $(3, 4)$.

2. Find the nonhomogeneous coordinates of the points with homogeneous coordinates:
(a) $(1, 3, 1)$;   (b) $(-2, 0, -1)$;
(c) $(0, 4, 2)$;   (d) $(3, -6, 3)$.

3. Represent the following translations of the plane using matrices:
(a) $x' = x - 2, \; y' = y + 4$;   (b) $x' = x, \; y' = y + 1$.

4. Determine the inverse of the translation matrix (6-18).

In Exercises 5 through 7 determine the image point of the point P under a translation of the plane that maps the origin onto $(5, 2)$.

5. $(3, -4)$.   **6.** $(-5, -2)$.   **7.** $(5, 2)$.

8. Determine the equation satisfied by the set of image points of the locus of $xy + 4x - 3y - 13 = 0$ under a translation of the plane that maps the origin onto $(-3, 4)$.

9. Determine the equation satisfied by the set of image points of the locus of $3x^2 + 2y^2 - 6x + 8y - 7 = 0$ under a translation of the plane that maps the origin onto $(-1, 2)$.

10. Prove that the multiplication of translation matrices of the form (6-18) is (a) closed; (b) commutative.

11. Prove that the distance between two points on a plane is invariant under a translation of the plane.

12. Find the image of the point $P: (3 + \sqrt{2}, 4 - \sqrt{2})$ under a translation of the plane that maps the origin onto $(-3, -4)$ followed by a rotation of the plane about the new origin through an angle of $45°$.

13. Find the image of the point $P: \left(\dfrac{3 - 5\sqrt{3}}{2}, \dfrac{5 + \sqrt{3}}{2}\right)$ under a rotation of the plane about the origin through an angle of $30°$ followed by a translation of the plane that maps the origin onto $(2, -1)$.

6-8 / RIGID MOTION TRANSFORMATIONS

In §6-2 and §6-3 rotations of the plane about the origin and reflections of the plane with respect to a line through the origin are considered. By means of the translation transformation and a generalization of the concept expressed by (6-10), it is now possible to consider rotations of the plane about any point on the plane and reflections of the plane with respect to any line on the plane. The following three examples illustrate the procedure for determining the matrices that represent such transformations.

EXAMPLE 1 / Determine the matrix $\mathscr{R}$ that represents a rotation of the plane about the point $(1, 1, 1)$ through an angle of $45°$. Determine the image of the origin under the rotation of the plane (Figure 6-14).

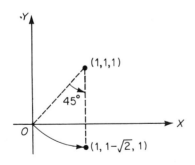

FIGURE 6-14

The translation of the plane that maps the point $(1, 1, 1)$ onto the origin is represented by the matrix

$$T = \begin{pmatrix} 1 & 0 & -1 \\ 0 & 1 & -1 \\ 0 & 0 & 1 \end{pmatrix}, \text{ whose inverse } T^{-1} = \begin{pmatrix} 1 & 0 & 1 \\ 0 & 1 & 1 \\ 0 & 0 & 1 \end{pmatrix}.$$

The matrix

$$R = \begin{pmatrix} \frac{\sqrt{2}}{2} & -\frac{\sqrt{2}}{2} & 0 \\ \frac{\sqrt{2}}{2} & \frac{\sqrt{2}}{2} & 0 \\ 0 & 0 & 1 \end{pmatrix}$$

represents a rotation of the plane about the origin through an angle $\theta = 45°$. Hence, by a generalization of the concept expressed by (6-10), the matrix $\mathcal{R}$ that represents a rotation of the plane about the point (1, 1, 1) through an angle of 45° is given as

$$\mathcal{R} = \begin{pmatrix} 1 & 0 & 1 \\ 0 & 1 & 1 \\ 0 & 0 & 1 \end{pmatrix} \begin{pmatrix} \frac{\sqrt{2}}{2} & -\frac{\sqrt{2}}{2} & 0 \\ \frac{\sqrt{2}}{2} & \frac{\sqrt{2}}{2} & 0 \\ 0 & 0 & 1 \end{pmatrix} \begin{pmatrix} 1 & 0 & -1 \\ 0 & 1 & -1 \\ 0 & 0 & 1 \end{pmatrix};$$

that is,

$$\mathcal{R} = \begin{pmatrix} \frac{\sqrt{2}}{2} & -\frac{\sqrt{2}}{2} & 1 \\ \frac{\sqrt{2}}{2} & \frac{\sqrt{2}}{2} & 1 - \sqrt{2} \\ 0 & 0 & 1 \end{pmatrix}.$$

The image of the origin under the rotation of the plane represented by $\mathcal{R}$ is $(1, 1 - \sqrt{2}, 1)$ since

$$\begin{pmatrix} \frac{\sqrt{2}}{2} & -\frac{\sqrt{2}}{2} & 1 \\ \frac{\sqrt{2}}{2} & \frac{\sqrt{2}}{2} & 1 - \sqrt{2} \\ 0 & 0 & 1 \end{pmatrix} \begin{pmatrix} 0 \\ 0 \\ 1 \end{pmatrix} = \begin{pmatrix} 1 \\ 1 - \sqrt{2} \\ 1 \end{pmatrix}.$$

EXAMPLE 2 / Determine the reflection matrix F that maps each point $(x, y, 1)$ on the plane onto its mirror image with respect to the line $y = 2$.

The translation of the plane that maps the line $y = 2$ onto the x-axis is represented by the matrix

$$T = \begin{pmatrix} 1 & 0 & 0 \\ 0 & 1 & -2 \\ 0 & 0 & 1 \end{pmatrix}, \text{ whose inverse } T^{-1} = \begin{pmatrix} 1 & 0 & 0 \\ 0 & 1 & 2 \\ 0 & 0 & 1 \end{pmatrix}.$$

The matrix

$$R = \begin{pmatrix} 1 & 0 & 0 \\ 0 & -1 & 0 \\ 0 & 0 & 1 \end{pmatrix}$$

represents a reflection of the plane with respect to the x-axis. Hence, by (6-10), the reflection matrix F that maps each point $(x, y, 1)$ on the plane onto its mirror image with respect to the line $y = 2$ is given as

$$F = T^{-1}RT = \begin{pmatrix} 1 & 0 & 0 \\ 0 & 1 & 2 \\ 0 & 0 & 1 \end{pmatrix} \begin{pmatrix} 1 & 0 & 0 \\ 0 & -1 & 0 \\ 0 & 0 & 1 \end{pmatrix} \begin{pmatrix} 1 & 0 & 0 \\ 0 & 1 & -2 \\ 0 & 0 & 1 \end{pmatrix};$$

that is,

$$F = \begin{pmatrix} 1 & 0 & 0 \\ 0 & -1 & 4 \\ 0 & 0 & 1 \end{pmatrix}.$$

EXAMPLE 3 / Determine the reflection matrix F that maps each point $(x, y, 1)$ on the plane onto its mirror image with respect to the line $x - y + 2 = 0$.

A translation of the plane represented by the matrix

$$\begin{pmatrix} 1 & 0 & 2 \\ 0 & 1 & 0 \\ 0 & 0 & 1 \end{pmatrix}$$

maps the line $x - y + 2 = 0$ onto the line $x - y = 0$; a rotation of the plane about the origin through an angle of $-45°$, represented by the matrix

$$\begin{pmatrix} \frac{\sqrt{2}}{2} & \frac{\sqrt{2}}{2} & 0 \\ -\frac{\sqrt{2}}{2} & \frac{\sqrt{2}}{2} & 0 \\ 0 & 0 & 1 \end{pmatrix},$$

maps the line $x - y = 0$ onto the x-axis. Hence, a product transformation that maps the line $x - y + 2 = 0$ onto the x-axis is represented by the matrix

$$T = \begin{pmatrix} \frac{\sqrt{2}}{2} & \frac{\sqrt{2}}{2} & 0 \\ -\frac{\sqrt{2}}{2} & \frac{\sqrt{2}}{2} & 0 \\ 0 & 0 & 1 \end{pmatrix} \begin{pmatrix} 1 & 0 & 2 \\ 0 & 1 & 0 \\ 0 & 0 & 1 \end{pmatrix} = \begin{pmatrix} \frac{\sqrt{2}}{2} & \frac{\sqrt{2}}{2} & \sqrt{2} \\ -\frac{\sqrt{2}}{2} & \frac{\sqrt{2}}{2} & -\sqrt{2} \\ 0 & 0 & 1 \end{pmatrix},$$

whose inverse

$$T^{-1} = \begin{pmatrix} 1 & 0 & -2 \\ 0 & 1 & 0 \\ 0 & 0 & 1 \end{pmatrix} \begin{pmatrix} \frac{\sqrt{2}}{2} & -\frac{\sqrt{2}}{2} & 0 \\ \frac{\sqrt{2}}{2} & \frac{\sqrt{2}}{2} & 0 \\ 0 & 0 & 1 \end{pmatrix} = \begin{pmatrix} \frac{\sqrt{2}}{2} & -\frac{\sqrt{2}}{2} & -2 \\ \frac{\sqrt{2}}{2} & \frac{\sqrt{2}}{2} & 0 \\ 0 & 0 & 1 \end{pmatrix}.$$

The matrix

$$R = \begin{pmatrix} 1 & 0 & 0 \\ 0 & -1 & 0 \\ 0 & 0 & 1 \end{pmatrix}$$

represents a reflection of the plane with respect to the x-axis. Hence, by (6-10), the reflection matrix F that maps each point $(x, y, 1)$ on the plane onto its mirror image with respect to the line $x - y + 2 = 0$ is given as

$$F = T^{-1}RT = \begin{pmatrix} \frac{\sqrt{2}}{2} & -\frac{\sqrt{2}}{2} & -2 \\ \frac{\sqrt{2}}{2} & \frac{\sqrt{2}}{2} & 0 \\ 0 & 0 & 1 \end{pmatrix} \begin{pmatrix} 1 & 0 & 0 \\ 0 & -1 & 0 \\ 0 & 0 & 1 \end{pmatrix} \begin{pmatrix} \frac{\sqrt{2}}{2} & \frac{\sqrt{2}}{2} & \sqrt{2} \\ -\frac{\sqrt{2}}{2} & \frac{\sqrt{2}}{2} & -\sqrt{2} \\ 0 & 0 & 1 \end{pmatrix};$$

that is,

$$F = \begin{pmatrix} 0 & 1 & -2 \\ 1 & 0 & 2 \\ 0 & 0 & 1 \end{pmatrix}.$$

Note that under the transformation of the plane represented by F every point on the line $x - y + 2 = 0$ is a fixed point; that is,

$$\begin{pmatrix} 0 & 1 & -2 \\ 1 & 0 & 2 \\ 0 & 0 & 1 \end{pmatrix} \begin{pmatrix} x \\ x+2 \\ 1 \end{pmatrix} = \begin{pmatrix} x \\ x+2 \\ 1 \end{pmatrix}.$$

The set of rotations of the plane about a point, reflections of the plane with respect to a line, and translations of the plane is called the set of **rigid motion transformations** because the distance between any two points on the plane is a scalar invariant under these transformations. (Euclidean geometry is sometimes characterized as a study of the properties of geometric figures which remain invariant under the rigid motion transformations.) The set of rigid motion transformations is a subset of the set of general linear transformations of the plane. As indicated in Figure 6-15, the set R of rotations of the plane about the origin and reflections of the plane with respect to a line through the origin is the intersection of the set M of rigid motion transformations and the set H of linear homogeneous transformations

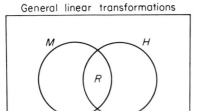

FIGURE 6-15

The set of ordered products of the rigid motion transformations of the plane may be represented by a matrix of the form

$$\begin{pmatrix} a_{11} & a_{12} & a_{13} \\ a_{21} & a_{22} & a_{23} \\ 0 & 0 & 1 \end{pmatrix}, \quad \text{where} \quad \begin{vmatrix} a_{11} & a_{12} \\ a_{21} & a_{22} \end{vmatrix} = \pm 1. \tag{6-19}$$

Furthermore,

$$\sum_{i=1}^{2} a_{ij}^2 = 1, \quad \text{for } j = 1, 2; \tag{6-20}$$

and

$$\sum_{i=1}^{2} a_{i1} a_{i2} = 0. \tag{6-21}$$

Two theorems which state that a general linear transformation of the plane is a rigid motion transformation if, and only if, the matrix representing the transformation satisfies conditions (6-20) and (6-21) will be proved now.

THEOREM 6.2 / *Let*

$$T = \begin{pmatrix} a_{11} & a_{12} & a_{13} \\ a_{21} & a_{22} & a_{23} \\ 0 & 0 & 1 \end{pmatrix}$$

represent a general linear transformation of the plane under which distance is a scalar invariant. Then

$$a_{11}^2 + a_{21}^2 = 1,$$
$$a_{12}^2 + a_{22}^2 = 1,$$

and

$$a_{11}a_{12} + a_{21}a_{22} = 0.$$

PROOF / Consider the points $O: (0, 0, 1)$ and $P: (1, 0, 1)$. Under the transformation represented by T, the image points of O and P are $O': (a_{13}, a_{23}, 1)$ and $P': (a_{11} + a_{13}, a_{21} + a_{23}, 1)$, respectively. Since $|\overrightarrow{O'P'}| = |\overrightarrow{OP}|$ under the transformation represented by T,

$$\sqrt{[(a_{11} + a_{13}) - a_{13}]^2 + [(a_{21} + a_{23}) - a_{23}]^2}$$
$$= \sqrt{(1 - 0)^2 + (0 - 0)^2},$$
$$\sqrt{a_{11}^2 + a_{21}^2} = \sqrt{1^2},$$
$$a_{11}^2 + a_{21}^2 = 1.$$

In a similar manner, choosing $O: (0, 0, 1)$ and $P: (0, 1, 1)$, we can show that

$$a_{12}^2 + a_{22}^2 = 1.$$

Now, consider the points $O: (0, 0, 1)$ and $P: (1, 1, 1)$. Under the transformation represented by T, the image points of O and P are $O': (a_{13}, a_{23}, 1)$ and $P': (a_{11} + a_{12} + a_{13}, a_{21} + a_{22} + a_{23}, 1)$, respectively. Again, since $|\overrightarrow{O'P'}| = |\overrightarrow{OP}|$ under the transformation represented by T,

$$\sqrt{(a_{11} + a_{12})^2 + (a_{21} + a_{22})^2} = \sqrt{1^2 + 1^2},$$
$$(a_{11} + a_{12})^2 + (a_{21} + a_{22})^2 = 2,$$
$$(a_{11}^2 + a_{21}^2) + (a_{12}^2 + a_{22}^2) + 2(a_{11}a_{12} + a_{21}a_{22}) = 2,$$
$$1 + 1 + 2(a_{11}a_{12} + a_{21}a_{22}) = 2,$$
$$a_{11}a_{12} + a_{21}a_{22} = 0.$$

THEOREM 6.3 / *Let*

$$T = \begin{pmatrix} a_{11} & a_{12} & a_{13} \\ a_{21} & a_{22} & a_{23} \\ 0 & 0 & 1 \end{pmatrix}$$

represent a general linear transformation of the plane such that $a_{11}^2 + a_{21}^2 = 1$, $a_{12}^2 + a_{22}^2 = 1$, and $a_{11}a_{12} + a_{21}a_{22} = 0$. Then distance is a scalar invariant under the transformation represented by T.

PROOF / Let $P_1: (x_1, y_1, 1)$ and $P_2: (x_2, y_2, 1)$ be any two points on the plane. Under the transformation represented by T, the image points of P_1 and P_2 are $P_1': (a_{11}x_1 + a_{12}y_1 + a_{13}, a_{21}x_1 + a_{22}y_1 + a_{23}, 1)$ and $P_2': (a_{11}x_2 + a_{12}y_2 + a_{13}, a_{21}x_2 + a_{22}y_2 + a_{23}, 1)$, respectively. Now,

$|\overrightarrow{P_1'P_2'}|$
$= \sqrt{[a_{11}(x_2 - x_1) + a_{12}(y_2 - y_1)]^2 + [a_{21}(x_2 - x_1) + a_{22}(y_2 - y_1)]^2}$
$= [(a_{11}^2 + a_{21}^2)(x_2 - x_1)^2 + 2(a_{11}a_{12} + a_{21}a_{22})(x_2 - x_1)(y_2 - y_1)$
$\quad + (a_{12}^2 + a_{22}^2)(y_2 - y_1)^2]^{1/2}$
$= \sqrt{(x_2 - x_1)^2 + (y_2 - y_1)^2}$
$= |\overrightarrow{P_1P_2}|;$

that is, the distance between the image points of P_1 and P_2 is equal to the distance between P_1 and P_2. Hence, distance is a scalar invariant under the transformation represented by T.

EXERCISES

1. Determine the matrix $\mathscr{R}$ that represents a rotation of the plane about the point $(1, \sqrt{3}, 1)$ through an angle of 60°. Determine the image of the origin under the rotation of the plane.

2. Determine the matrix $\mathscr{R}$ that represents a rotation of the plane about the point $(1, 0, 1)$ through an angle of 90°. Determine the image of the origin under the rotation of the plane.

3. Verify that the matrix of Exercise 1 is of the form (6-19) and satisfies the conditions (6-20) and (6-21).

4. Verify that the matrix of Exercise 2 is of the form (6-19) and satisfies the conditions (6-20) and (6-21).

5. Determine the reflection matrix F that maps each point $(x, y, 1)$ on the plane onto its mirror image with respect to the line (a) $x = 3$; (b) $x - \sqrt{3}y + 1 = 0$.

6. Verify that the product of a matrix representing a rotation of the plane

about the origin and a matrix representing a translation of the plane, taken in either order, is of the form (6-19) and satisfies the conditions (6-20) and (6-21).

7. Verify that the product of a matrix representing a rotation of the plane about the origin and a matrix representing a reflection of the plane about a coordinate axis, taken in either order, is of the form (6-19) and satisfies the conditions (6-20) and (6-21).

8. Prove that every matrix representing a general linear transformation of the plane under which distance is a scalar invariant can be expressed as the product of rotation, translation, and reflection matrices.

SEVEN

EIGENVALUES AND EIGENVECTORS

7-1 / CHARACTERISTIC FUNCTIONS

Associated with each square matrix $A = ((a_{ij}))$ of order n is a function

$$f(\lambda) = |A - \lambda I| = \begin{vmatrix} a_{11} - \lambda & a_{12} & \cdots & a_{1n} \\ a_{21} & a_{22} - \lambda & \cdots & a_{2n} \\ \cdots & \cdots & \cdots & \cdots \\ a_{n1} & a_{n2} & \cdots & a_{nn} - \lambda \end{vmatrix} \tag{7-1}$$

called the **characteristic function** of A. The equation

$$f(\lambda) = |A - \lambda I| = 0 \tag{7-2}$$

can be expressed in the polynomial form

$$c_0 \lambda^n + c_1 \lambda^{n-1} + \cdots + c_{n-1} \lambda + c_n = 0 \tag{7-3}$$

and is called the **characteristic equation** of matrix A.

EXAMPLE 1 / Find the characteristic equation of matrix A where

$$A = \begin{pmatrix} 1 & 2 & 0 \\ 2 & 2 & 2 \\ 0 & 2 & 3 \end{pmatrix}.$$

The characteristic equation of A is

$$\begin{vmatrix} 1-\lambda & 2 & 0 \\ 2 & 2-\lambda & 2 \\ 0 & 2 & 3-\lambda \end{vmatrix} = 0;$$

that is,

$$(1-\lambda)\begin{vmatrix} 2-\lambda & 2 \\ 2 & 3-\lambda \end{vmatrix} - 2\begin{vmatrix} 2 & 2 \\ 0 & 3-\lambda \end{vmatrix} = 0,$$

$$(1-\lambda)(\lambda^2 - 5\lambda + 2) - 2(6 - 2\lambda) = 0,$$

$$\lambda^3 - 6\lambda^2 + 3\lambda + 10 = 0.$$

In some instances the task of expressing the characteristic equation of a matrix in polynomial form may be simplified considerably by introducing the concept of the trace of a matrix. The sum of the diagonal elements of a matrix A is called the **trace** of A and is denoted by $tr(A)$. For example, the trace of matrix A in Example 1 is $1 + 2 + 3$; that is, 6. Let $t_1 = tr(A)$, $t_2 = tr(A^2), \ldots, t_n = tr(A^n)$. It can be shown that the coefficients of the characteristic equation are given by the equations:

$$c_0 = 1,$$
$$c_1 = -t_1,$$
$$c_2 = -\tfrac{1}{2}(c_1 t_1 + t_2), \qquad (7\text{-}4)$$
$$c_3 = -\tfrac{1}{3}(c_2 t_1 + c_1 t_2 + t_3),$$
$$\cdots$$
$$c_n = -\frac{1}{n}(c_{n-1} t_1 + c_{n-2} t_2 + \cdots + c_1 t_{n-1} + t_n).$$

Equations (7-4) make it possible to calculate the coefficients of the characteristic equation of a matrix A by summing the diagonal elements of the matrices of the form A^n. This numerical process is easily programmed on a large-scale digital computer, or for small values of n may be computed manually without difficulty.

EXAMPLE 2 / Find the characteristic equation of matrix A in Example 1 by using the equations of (7-4).

The characteristic equation of A is of the form $c_0\lambda^3 + c_1\lambda^2 + c_2\lambda + c_3 = 0$. Now,
$$t_1 = tr(A) = 1 + 2 + 3 = 6;$$
since
$$A^2 = \begin{pmatrix} 5 & 6 & 4 \\ 6 & 12 & 10 \\ 4 & 10 & 13 \end{pmatrix} \quad \text{and} \quad A^3 = \begin{pmatrix} 17 & 30 & 24 \\ 30 & 56 & 54 \\ 24 & 54 & 59 \end{pmatrix},$$
$$t_2 = tr(A^2) = 5 + 12 + 13 = 30;$$
$$t_3 = tr(A^3) = 17 + 56 + 59 = 132.$$

Then, using the equations of (7-4), we obtain
$$c_0 = 1,$$
$$c_1 = -6,$$
$$c_2 = -\tfrac{1}{2}[(-6)(6) + 30] = 3,$$
$$c_3 = -\tfrac{1}{3}[(3)(6) + (-6)(30) + 132] = 10.$$

Hence, $\lambda^3 - 6\lambda^2 + 3\lambda + 10 = 0$ is the characteristic equation of A.

The n roots $\lambda_1, \lambda_2, \ldots, \lambda_n$ of the characteristic equation (7-3) of a matrix A are called the **eigenvalues** of A.

EXAMPLE 3 / Determine the eigenvalues of the matrix A where
$$A = \begin{pmatrix} 3 & 1 \\ 2 & 2 \end{pmatrix}.$$

The characteristic equation of A is
$$\begin{vmatrix} 3 - \lambda & 1 \\ 2 & 2 - \lambda \end{vmatrix} = 0.$$

In expanded form this becomes $\lambda^2 - 5\lambda + 4 = 0$. The eigenvalues $\lambda_1 = 1$ and $\lambda_2 = 4$ are the roots of the characteristic equation.

EXAMPLE 4 / Prove that the trace of a matrix A of order n is equal to the sum of the n eigenvalues of A; that is,

$$tr(A) = \sum_{i=1}^{n} \lambda_i.$$

Let $A = ((a_{ij}))$. By definition,

$$tr(A) = \sum_{i=1}^{n} a_{ii}.$$

Consider the characteristic function of A expressed in factored form:

$$|A - \lambda I| = (-1)^n(\lambda - \lambda_1)(\lambda - \lambda_2) \cdots (\lambda - \lambda_n).$$

The coefficient of λ^{n-1} in $|A - \lambda I|$ is

$$(-1)^{n-1} \sum_{i=1}^{n} a_{ii};$$

the coefficient of λ^{n-1} in the factored form of the characteristic function is

$$(-1)^{n+1} \sum_{i=1}^{n} \lambda_i.$$

Therefore,

$$(-1)^{n-1} \sum_{i=1}^{n} a_{ii} = (-1)^{n+1} \sum_{i=1}^{n} \lambda_i$$

$$\sum_{i=1}^{n} a_{ii} = \sum_{i=1}^{n} \lambda_i;$$

that is,

$$tr(A) = \sum_{i=1}^{n} \lambda_i.$$

Many applications of matrix algebra in mathematics, physics, and engineering involve the concept of a set of nonzero vectors being mapped onto the null vector by means of the matrix $A - \lambda_i I$, where λ_i is an eigenvalue of matrix A. Any nonzero column vector, denoted by X_i, such that

$$(A - \lambda_i I)X_i = 0 \tag{7-5}$$

is called an **eigenvector** of matrix A. It is guaranteed that at least one eigenvector exists for each λ_i since equation (7-5) represents a system of n linear homogeneous equations which has a nontrivial solution $X_i \neq 0$ if, and only if, $|A - \lambda_i I| = 0$; that is, if, and only if, λ_i is an eigenvalue of A. Furthermore, note that any nonzero scalar multiple of an eigenvector associated with an eigenvalue is also an eigenvector associated with that eigenvalue.

EXAMPLE 5 / Determine a set of eigenvectors of the matrix A in Example 3.

Associated with $\lambda_1 = 1$ are the eigenvectors $(x_1 \quad x_2)^T$ for which

$$(A - I)(x_1 \quad x_2)^T = 0;$$

that is,

$$\begin{pmatrix} 2 & 1 \\ 2 & 1 \end{pmatrix} \begin{pmatrix} x_1 \\ x_2 \end{pmatrix} = \begin{pmatrix} 0 \\ 0 \end{pmatrix}.$$

It follows that $-2x_1 = x_2$. If x_1 is chosen as some convenient arbitrary scalar, say 1, x_2 becomes -2. Hence, $(1 \quad -2)^T$ is an eigenvector associated with the eigenvalue 1.

Similarly, associated with $\lambda_2 = 4$ are the eigenvectors $(x_1 \quad x_2)^T$ for which

$$(A - 4I)(x_1 \quad x_2)^T = 0;$$

that is,

$$\begin{pmatrix} -1 & 1 \\ 2 & -2 \end{pmatrix} \begin{pmatrix} x_1 \\ x_2 \end{pmatrix} = \begin{pmatrix} 0 \\ 0 \end{pmatrix}.$$

Hence, $x_1 = x_2$ and $(1 \quad 1)^T$ is an eigenvector associated with the eigenvalue 4. Therefore, one set of eigenvectors of the matrix A is $\{(1 \quad -2)^T, (1 \quad 1)^T\}$.

It should be noted that $(k \quad -2k)^T$ and $(k \quad k)^T$, where k is any nonzero scalar, represent the general forms of the eigenvectors of A.

The eigenvalues of a matrix are also called the **proper values**, the **latent values**, and the **characteristic values** of the matrix. The eigenvectors of a matrix are also called the **proper vectors**, the **latent vectors**, and the **characteristic vectors** of the matrix.

EXERCISES

In Exercises 1 through 6 determine the characteristic equation, the eigenvalues, and a set of eigenvectors of the given matrix.

1. $\begin{pmatrix} 5 & 3 \\ 2 & 4 \end{pmatrix}$. 2. $\begin{pmatrix} 1 & 2 \\ 4 & 3 \end{pmatrix}$. 3. $\begin{pmatrix} 2 & 0 \\ 0 & 0 \end{pmatrix}$.

4. $\begin{pmatrix} 2 & 0 & 0 \\ 0 & 1 & 0 \\ 0 & 0 & 3 \end{pmatrix}$. 5. $\begin{pmatrix} 2 & -2 & 3 \\ 1 & 1 & 1 \\ 1 & 3 & -1 \end{pmatrix}$. 6. $\begin{pmatrix} 3 & 0 & 2 \\ 0 & 1 & 2 \\ 2 & 2 & 2 \end{pmatrix}$.

7. Use the equations of (7-4) to find the characteristic equation of the matrix in (a) Exercise 1; (b) Exercise 5.

8. Verify the results of Example 4 for the matrix in (a) Exercise 1; (b) Exercise 6.

9. Prove that if an eigenvalue of A is zero, then $\det A = 0$.

10. Determine the necessary and sufficient conditions for a symmetric matrix of order two to have distinct eigenvalues.

11. If λ_1, λ_2, and λ_3 are the eigenvalues of a matrix A of order three, find the eigenvalues of (a) kA; (b) $A - kI$.

12. Prove that the eigenvalues of A and A^T are identical.

13. Prove that the eigenvalues of a diagonal matrix are equal to the diagonal elements.

7-2 / A GEOMETRIC INTERPRETATION OF EIGENVECTORS

Consider a magnification of the plane represented by the matrix A where

$$A = \begin{pmatrix} 3 & 0 \\ 0 & 2 \end{pmatrix}.$$

The eigenvalues of A are $\lambda_1 = 3$ and $\lambda_2 = 2$. Every eigenvector associated with λ_1 is of the form $(k \;\; 0)^T$, where k is any nonzero scalar, since

$$\begin{pmatrix} 3-3 & 0 \\ 0 & 2-3 \end{pmatrix} \begin{pmatrix} k \\ 0 \end{pmatrix} = \begin{pmatrix} 0 \\ 0 \end{pmatrix}.$$

Furthermore, the set of vectors of the form $(k\ \ 0)^T$ is such that

$$A(k\ \ 0)^T = \lambda_1(k\ \ 0)^T;$$

that is,

$$\begin{pmatrix} 3 & 0 \\ 0 & 2 \end{pmatrix}\begin{pmatrix} k \\ 0 \end{pmatrix} = 3\begin{pmatrix} k \\ 0 \end{pmatrix}.$$

Hence, the set of eigenvectors associated with $\lambda_1 = 3$ is mapped onto itself under the transformation represented by A, and the image of each eigenvector is a fixed scalar multiple of that eigenvector. The fixed scalar multiple is equal to the eigenvalue with which the set of eigenvectors is associated.

Similarly, every eigenvector associated with λ_2 is of the form $(0\ \ k)^T$, where k is any nonzero scalar. The set of vectors of the form $(0\ \ k)^T$ is such that

$$A(0\ \ k)^T = \lambda_2(0\ \ k)^T;$$

that is,

$$\begin{pmatrix} 3 & 0 \\ 0 & 2 \end{pmatrix}\begin{pmatrix} 0 \\ k \end{pmatrix} = 2\begin{pmatrix} 0 \\ k \end{pmatrix}.$$

Hence, the set of eigenvectors associated with $\lambda_2 = 2$ is mapped onto itself under the transformation represented by A, and the image of each eigenvector is a fixed scalar multiple of that eigenvector. The fixed scalar multiple is λ_2; that is, 2.

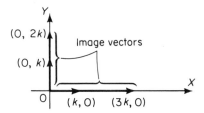

FIGURE 7-1

Note that the sets of vectors of the forms $(k\ \ 0)^T$ and $(0\ \ k)^T$ lie along the x-axis and y-axis, respectively (Figure 7-1). Under the magnification of the plane represented by the matrix

$$A = \begin{pmatrix} 3 & 0 \\ 0 & 2 \end{pmatrix},$$

the *one-dimensional vector spaces* containing the sets of vectors of the forms $(k \ \ 0)^T$ and $(0 \ \ k)^T$ are mapped *onto* themselves, respectively, and are called **invariant vector spaces**. The invariant vector spaces help characterize or describe a particular transformation of the plane.

EXAMPLE 1 / Determine the invariant vector spaces under a shear parallel to the x-axis represented by the matrix A where

$$A = \begin{pmatrix} 1 & 2 \\ 0 & 1 \end{pmatrix}.$$

The eigenvalues of A are $\lambda_1 = 1$ and $\lambda_2 = 1$. Associated with each eigenvalue is the set of eigenvectors of the form $(k \ \ 0)^T$, where k is any nonzero scalar. Then

$$\begin{pmatrix} 1 & 2 \\ 0 & 1 \end{pmatrix} \begin{pmatrix} k \\ 0 \end{pmatrix} = 1 \begin{pmatrix} k \\ 0 \end{pmatrix},$$

and the one-dimensional vector space containing the set of vectors of the form $(k \ \ 0)^T$ is an invariant vector space. Furthermore, since $\lambda_1 = \lambda_2 = 1$, each vector in the vector space is its own image.

EXAMPLE 2 / Determine the invariant vector spaces under a projection of the plane represented by the matrix A where

$$A = \begin{pmatrix} 1 & 0 \\ 1 & 0 \end{pmatrix}.$$

The eigenvalues of A are $\lambda_1 = 1$ and $\lambda_2 = 0$. Associated with the eigenvalue $\lambda_1 = 1$ is the set of eigenvectors of the form $(k \ \ k)^T$, where k is any nonzero scalar; associated with the eigenvalue $\lambda_2 = 0$ is the set of eigenvectors of the form $(0 \ \ k)^T$, where k is any nonzero scalar. Then

$$\begin{pmatrix} 1 & 0 \\ 1 & 0 \end{pmatrix} \begin{pmatrix} k \\ k \end{pmatrix} = 1 \begin{pmatrix} k \\ k \end{pmatrix}$$

and

$$\begin{pmatrix} 1 & 0 \\ 1 & 0 \end{pmatrix} \begin{pmatrix} 0 \\ k \end{pmatrix} = 0 \begin{pmatrix} 0 \\ k \end{pmatrix} = \begin{pmatrix} 0 \\ 0 \end{pmatrix}.$$

Since the vectors of the form $(0 \ \ k)^T$ are mapped onto the null vector, the one-dimensional vector space containing these vectors is mapped *into* but not *onto* itself, and the space is not considered an invariant vector space. However, the one-dimensional vector space containing the set of vectors of the form $(k \ \ k)^T$ is an invariant vector space. Note that these vectors lie along the line $y = x$ and that the plane is mapped onto this line under the projection of the plane represented by matrix A.

EXERCISES

In Exercises 1 through 6 determine the invariant vector spaces under the transformation of the plane represented by the given matrix.

1. $\begin{pmatrix} 1 & 0 \\ 0 & -1 \end{pmatrix}$.
2. $\begin{pmatrix} 1 & 0 \\ 0 & 2 \end{pmatrix}$.
3. $\begin{pmatrix} 1 & 0 \\ 3 & 1 \end{pmatrix}$.
4. $\begin{pmatrix} 0 & 1 \\ 1 & 0 \end{pmatrix}$.
5. $\begin{pmatrix} 1 & 0 \\ 0 & 0 \end{pmatrix}$.
6. $\begin{pmatrix} 2 & 0 \\ 0 & 2 \end{pmatrix}$.

7. Prove that, in general, no invariant vector spaces exist under a rotation of the plane about the origin.

7-3 / SOME THEOREMS

In this section several theorems concerning the eigenvalues and eigenvectors of matrices in general and of symmetric matrices in particular will be proved. These theorems are important for an understanding of the remaining sections of this text.

Notice that in Example 5 of §7-1 the eigenvectors associated with the distinct eigenvalues of matrix A are linearly independent; that is,

$$k_1 \begin{pmatrix} 1 \\ -2 \end{pmatrix} + k_2 \begin{pmatrix} 1 \\ 1 \end{pmatrix} = \begin{pmatrix} 0 \\ 0 \end{pmatrix}$$

implies $k_1 = k_2 = 0$. This is not a coincidence. The following theorem states a sufficient condition for eigenvectors associated with the eigenvalues of a matrix to be linearly independent.

THEOREM 7.1 / *If the eigenvalues of a matrix are distinct, then the associated eigenvectors are linearly independent.*

PROOF / Let A be a square matrix of order n with distinct eigenvalues $\lambda_1, \lambda_2, \ldots, \lambda_n$ and associated eigenvectors $X_1, X_2, \ldots, X_n$, respectively. Assume that the set of eigenvectors are linearly dependent. Then there exist scalars $k_1, k_2, \ldots, k_n$, not all zero, such that

$$k_1 X_1 + k_2 X_2 + \cdots + k_n X_n = 0. \tag{7-6}$$

Consider premultiplying both sides of (7-6) by

$$(A - \lambda_2 I)(A - \lambda_3 I) \cdots (A - \lambda_n I).$$

By use of equation (7-5), we obtain

$$k_1 (A - \lambda_2 I)(A - \lambda_3 I) \cdots (A - \lambda_n I) X_1 = 0. \tag{7-7}$$

Since $(A - \lambda_1 I) X_1 = 0$, then $AX_1 = \lambda_1 X_1$. Hence, equation (7-7) may be written as

$$k_1 (\lambda_1 - \lambda_2)(\lambda_1 - \lambda_3) \cdots (\lambda_1 - \lambda_n) X_1 = 0,$$

which implies $k_1 = 0$. Similarly, it can be shown that $k_2 = k_3 = \cdots = k_n = 0$, which is contrary to the hypothesis. Therefore, the eigenvectors are linearly independent.

It should be noted that if the eigenvalues of a matrix are not distinct, the associated eigenvectors may or may not be linearly independent. For example, consider the matrices

$$A = \begin{pmatrix} 3 & 0 \\ 0 & 3 \end{pmatrix} \quad \text{and} \quad B = \begin{pmatrix} 3 & 1 \\ 0 & 3 \end{pmatrix}.$$

Both matrices have $\lambda_1 = \lambda_2 = 3$; that is, an eigenvalue of multiplicity two. Any nonzero vector of the form $(x_1 \ \ x_2)^T$ is an eigenvector of A for λ_1 and λ_2. Hence, it is possible to choose any two linearly independent vectors, such as $(1 \ \ 0)^T$ and $(0 \ \ 1)^T$, as eigenvectors of A that are associated with λ_1 and λ_2, respectively. However, only a vector of the form $(x_1 \ \ 0)^T$ is an eigenvector of B for λ_1 and λ_2. Any two vectors of this form are linearly dependent; that is, one is a linear function of the other.

THEOREM 7.2 / *If A is a Hermitian matrix, then the eigenvalues of A are real.*

PROOF / Let A be a Hermitian matrix, λ_i be any eigenvalue of A, and X_i be an eigenvector associated with λ_i. Then

$$(A - \lambda_i I)X_i = 0,$$
$$AX_i - \lambda_i X_i = 0,$$
$$X_i^* A X_i - \lambda_i X_i^* X_i = 0.$$

Since every eigenvector is a nonzero vector, $X_i^* X_i$ is a nonzero real number and

$$\lambda_i = \frac{X_i^* A X_i}{X_i^* X_i}.$$

Furthermore,

$$X_i^* A X_i = X_i^* A^* X_i \quad \text{since } A = A^*$$
$$= (X_i^* A X_i)^* \quad \text{by Example 3 of §4-5}$$
$$= \overline{X_i^* A X_i} \quad \text{since } X_i^* A X_i \text{ is a matrix with one element;}$$

that is, $X_i^* A X_i$ equals its own conjugate, and hence is real. Therefore, λ_i is equal to the quotient of two real numbers, and is real.

THEOREM 7.3 / *If A is a real symmetric matrix, then the eigenvalues of A are real.*

PROOF / Since every real symmetric matrix is a Hermitian matrix, the proof follows from Theorem 7.2.

Before presenting the next theorem, it is necessary to consider the following definition: two complex eigenvectors X_1 and X_2 are defined as **orthogonal** if, and only if, $X_1^* X_2 = 0$. For example, if $X_1 = (-i \ \ 2)^T$ and $X_2 = (2i \ \ 1)^T$, then $X_1^* X_2 = (i \ \ 2)(2i \ \ 1)^T = 0$. Hence, X_1 and X_2 are orthogonal.

THEOREM 7.4 / *If A is a Hermitian matrix, then the eigenvectors of A associated with distinct eigenvalues are mutually orthogonal vectors.*

PROOF / Let A be a Hermitian matrix, and let X_1 and X_2 be eigenvectors associated with any two distinct eigenvalues λ_1 and λ_2, respectively. Then

$$(A - \lambda_1 I)X_1 = 0 \quad \text{and} \quad (A - \lambda_2 I)X_2 = 0;$$

that is,

$$AX_1 = \lambda_1 X_1 \quad \text{and} \quad AX_2 = \lambda_2 X_2. \tag{7-8}$$

Multiplying both sides of the first equation of (7-8) by X_2^*, we have

$$\begin{aligned}
\lambda_1 X_2^* X_1 &= X_2^* A X_1 \\
&= X_2^* A^* X_1 && \text{since } A = A^* \\
&= (AX_2)^* X_1 && \text{by Example 3 of §4-5} \\
&= \lambda_2 X_2^* X_1 && \text{by the second equation of (7-8).}
\end{aligned}$$

Then

$$X_2^* X_1 = \frac{\lambda_2}{\lambda_1} X_2^* X_1.$$

Since λ_1 and λ_2 are real and unequal, $X_2^* X_1 = 0$. Hence, X_1 and X_2 are orthogonal eigenvectors.

THEOREM 7.5 / *If A is a real symmetric matrix, then the eigenvectors of A associated with distinct eigenvalues are mutually orthogonal vectors.*

PROOF / Since every real symmetric matrix is a Hermitian matrix, the proof follows from Theorem 7.4.

EXERCISES

1. Verify Theorem 7.1 for the matrix

$$\begin{pmatrix} 3 & 5 \\ 4 & 4 \end{pmatrix}.$$

2. Verify Theorems 7.2 and 7.4 for the matrix

$$\begin{pmatrix} 0 & 1+i \\ 1-i & 1 \end{pmatrix}.$$

3. Verify Theorems 7.3 and 7.5 for the matrix

$$\begin{pmatrix} 2 & 2 & 0 \\ 2 & 2 & 0 \\ 0 & 0 & 1 \end{pmatrix}.$$

4. Prove that if X_i is a unit eigenvector associated with the eigenvalue λ_i of A, then $X_i^T A X_i = (\lambda_i)$.

5. Prove that the eigenvalues of A^* are the conjugates of the eigenvalues of A. (*Hint:* Use the results of Exercise 4.)

7-4 / DIAGONALIZATION OF MATRICES

It has been noted that an eigenvector X_i such that $(A - \lambda_i I)X_i = 0$, for $i = 1, 2, \ldots, n$, may be associated with each eigenvalue λ_i. This relationship may be expressed in the alternate form

$$AX_i = \lambda_i X_i, \quad \text{for } i = 1, 2, \ldots, n. \tag{7-9}$$

If a square matrix of order n whose columns are eigenvectors X_i of A is constructed and denoted by X, then the equations of (7-9) may be expressed in the form

$$AX = X\Lambda, \tag{7-10}$$

where Λ is a diagonal matrix whose diagonal elements are the eigenvalues of A; that is,

$$\Lambda = \begin{pmatrix} \lambda_1 & 0 & \cdots & 0 \\ 0 & \lambda_2 & \cdots & 0 \\ \cdots & \cdots & \cdots & \cdots \\ 0 & 0 & \cdots & \lambda_n \end{pmatrix}. \tag{7-11}$$

It has been proved that the eigenvectors associated with distinct eigenvalues are linearly independent (Theorem 7.1). Hence, the matrix X will be nonsingular if the λ_i's are distinct. If both sides of equation (7-10) are multiplied by X^{-1}, the result is

$$X^{-1}AX = \Lambda. \tag{7-12}$$

Thus, by use of a matrix of eigenvectors and its inverse, it is possible to transform any matrix A with distinct eigenvalues to a diagonal matrix whose diagonal elements are the eigenvalues of A. The transformation expressed by (7-12) is referred to as the **diagonalization** of matrix A. If the eigenvalues are not distinct, the diagonalization of matrix A may not be possible. For example, the matrix

$$A = \begin{pmatrix} 3 & 1 \\ 0 & 3 \end{pmatrix}$$

cannot be diagonalized as in (7-12).

A matrix such as matrix A in equation (7-12) sometimes is spoken of as being *similar* to the diagonal matrix. In general, if there exists a nonsingular matrix C such that $C^{-1}AC = B$ for two square matrices A and B of the same order, then A and B are called **similar matrices**, and the transformation of A to B is called a **similarity transformation**. Furthermore, if B is a diagonal matrix whose diagonal elements are the eigenvalues of A, then B is called the **classical canonical form** of matrix A. It is a unique matrix except for the order in which the eigenvalues appear along the principal diagonal.

The matrix X of (7-12) whose columns are eigenvectors of matrix A often is called a **modal matrix** of A. Recall that each eigenvector may be multiplied by any nonzero scalar. Hence, a modal matrix of A is not unique.

EXAMPLE 1 / Determine if

$$A = \begin{pmatrix} 6 & 2 \\ -2 & 1 \end{pmatrix} \quad \text{and} \quad B = \begin{pmatrix} 8 & 6 \\ -3 & -1 \end{pmatrix}$$

are similar matrices.

If A and B are similar matrices, there exists a nonsingular square matrix C of order two such that $C^{-1}AC = B$; that is, $AC = CB$. Let

$$C = \begin{pmatrix} a & b \\ c & d \end{pmatrix}$$

where $ad - bc \neq 0$. Then

$$\begin{pmatrix} 6 & 2 \\ -2 & 1 \end{pmatrix}\begin{pmatrix} a & b \\ c & d \end{pmatrix} = \begin{pmatrix} a & b \\ c & d \end{pmatrix}\begin{pmatrix} 8 & 6 \\ -3 & -1 \end{pmatrix},$$

$$\begin{pmatrix} 6a + 2c & 6b + 2d \\ -2a + c & -2b + d \end{pmatrix} = \begin{pmatrix} 8a - 3b & 6a - b \\ 8c - 3d & 6c - d \end{pmatrix}.$$

This single matrix equation leads to the system of homogeneous equations

$$\begin{cases} 2a - 3b - 2c & = 0 \\ 2a & + 7c - 3d = 0 \\ 6a - 7b & - 2d = 0 \\ & 2b + 6c - 2d = 0. \end{cases}$$

The system of homogeneous equations has infinitely many solutions of the form $a = 3t - 7s$, $b = 2t - 6s$, $c = 2s$, and $d = 2t$, where s and t are arbitrary real scalars. Therefore, a nonsingular matrix

$$C = \begin{pmatrix} 3t - 7s & 2t - 6s \\ 2s & 2t \end{pmatrix}$$

exists, where s and t are arbitrary real scalars, provided that $\det C \neq 0$; that is, provided that

$$6t^2 - 18st + 12s^2 \neq 0,$$
$$(6t - 12s)(t - s) \neq 0,$$
$$t \neq 2s \quad \text{and} \quad t \neq s.$$

Hence, A and B are similar matrices.

As an illustration that A and B are similar matrices, let $s = 0$ and $t = 1$. Then

$$C = \begin{pmatrix} 3 & 2 \\ 0 & 2 \end{pmatrix}, \quad C^{-1} = \begin{pmatrix} \frac{1}{3} & -\frac{1}{3} \\ 0 & \frac{1}{2} \end{pmatrix},$$

and

$$C^{-1}AC = \begin{pmatrix} \frac{1}{3} & -\frac{1}{3} \\ 0 & \frac{1}{2} \end{pmatrix} \begin{pmatrix} 6 & 2 \\ -2 & 1 \end{pmatrix} \begin{pmatrix} 3 & 2 \\ 0 & 2 \end{pmatrix} = \begin{pmatrix} 8 & 6 \\ -3 & -1 \end{pmatrix} = B.$$

EXAMPLE 2 / Prove that similar matrices have equal determinants and equal eigenvalues.

Let A and B be similar matrices. Then a nonsingular square matrix C of the same order as A and B such that $C^{-1}AC = B$ exists. Since the determinant of the product of two matrices is equal to the product of their determinants, it follows that

$$\det B = \det C^{-1} \det A \det C$$
$$= \det C^{-1} \det C \det A$$
$$= \det (C^{-1}C) \det A$$
$$= \det I \det A$$
$$= \det A.$$

Then

$$\det(A - \lambda I) = \det [C^{-1}(A - \lambda I)C]$$
$$= \det (C^{-1}AC - \lambda C^{-1}IC)$$
$$= \det(B - \lambda I);$$

that is, A and B have the same characteristic function. An immediate consequence of this is that A and B have the same characteristic equation and eigenvalues.

Careful note should be made that the converse of the statement of Example 2 is not necessarily true. For example, the matrices

$$A = \begin{pmatrix} 1 & 0 \\ 0 & 1 \end{pmatrix} \quad \text{and} \quad B = \begin{pmatrix} 1 & 2 \\ 0 & 1 \end{pmatrix}$$

have the same eigenvalues $\lambda_1 = \lambda_2 = 1$ and $\det A = \det B$, but $C^{-1}AC = I$ for any nonsingular square matrix C of order two and $I \neq B$. Hence, A and B cannot be similar matrices.

EXAMPLE 3 / Prove that $A^n X_i = \lambda_i^n X_i$ for all natural numbers n.

This useful relationship may be proved by mathematical induction.

$$AX_i = \lambda_i X_i \quad \text{by (7-9)}.$$

Next, assume that $A^k X_i = \lambda_i^k X_i$, where k is any natural number. Then

$$A^{k+1} X_i = A \lambda_i^k X_i$$
$$= \lambda_i^k A X_i$$
$$= \lambda_i^{k+1} X_i \quad \text{by (7-9)}.$$

Hence,

$$A^n X_i = \lambda_i^n X_i \tag{7-13}$$

for all natural numbers n.

It has been shown that if A is a real symmetric matrix of order n with n distinct real eigenvalues, the associated eigenvectors are mutually orthogonal (Theorem 7.5). A matrix of eigenvectors can be made proper orthogonal if each eigenvector is normalized by an appropriate scalar multiple. A similarity transformation employing an orthogonal modal matrix is called an **orthogonal transformation**; that is, an orthogonal transformation of matrix A is of the form $C^T A C$ where C is an orthogonal matrix.

If a real symmetric matrix of order n has multiple eigenvalues, it is always possible to determine n mutually orthogonal unit eigenvectors. It is possible to show that r linearly independent eigenvectors that are orthogonal to the other eigenvectors may be associated with an eigenvalue of multiplicity r. Furthermore, it is always possible to choose these vectors orthogonal to each other. These properties of symmetric matrices will be postulated for our purposes here and their proofs left for more advanced texts in linear algebra.

THEOREM 7.6 / *Every real symmetric matrix can be orthogonally transformed to the classical canonical form.*

Theorem 7.6 is sometimes called the **Principal Axes Theorem**. An application of this theorem to analytic geometry will be considered in later sections of this chapter.

EXAMPLE 4 / Determine a proper orthogonal modal matrix that transforms the matrix A to the classical canonical form where

$$A = \begin{pmatrix} 3 & 1 \\ 1 & 3 \end{pmatrix}.$$

The characteristic equation of A is $\lambda^2 - 6\lambda + 8 = 0$; then the eigenvalues of A are $\lambda_1 = 2$ and $\lambda_2 = 4$. Associated with the eigenvalue $\lambda_1 = 2$ is the unit eigenvector

$$\left(\frac{1}{\sqrt{2}} \quad \frac{-1}{\sqrt{2}}\right)^T.$$

Associated with the eigenvalue $\lambda_2 = 4$ is the unit eigenvector

$$\left(\frac{1}{\sqrt{2}} \quad \frac{1}{\sqrt{2}}\right)^T.$$

Therefore, a proper orthogonal modal matrix that transforms the matrix A to canonical form is

$$\begin{pmatrix} \frac{1}{\sqrt{2}} & \frac{1}{\sqrt{2}} \\ \frac{-1}{\sqrt{2}} & \frac{1}{\sqrt{2}} \end{pmatrix};$$

that is,

$$\begin{pmatrix} \frac{1}{\sqrt{2}} & \frac{-1}{\sqrt{2}} \\ \frac{1}{\sqrt{2}} & \frac{1}{\sqrt{2}} \end{pmatrix} \begin{pmatrix} 3 & 1 \\ 1 & 3 \end{pmatrix} \begin{pmatrix} \frac{1}{\sqrt{2}} & \frac{1}{\sqrt{2}} \\ \frac{-1}{\sqrt{2}} & \frac{1}{\sqrt{2}} \end{pmatrix} = \begin{pmatrix} 2 & 0 \\ 0 & 4 \end{pmatrix}.$$

EXERCISES

In Exercises 1 and 2 determine whether or not the matrices of each pair are similar matrices.

1. $\begin{pmatrix} -2 & -1 \\ 0 & 11 \end{pmatrix}$ and $\begin{pmatrix} 0 & 1 \\ 2 & 3 \end{pmatrix}$.

2. $\begin{pmatrix} 2 & 2 \\ 1 & 3 \end{pmatrix}$ and $\begin{pmatrix} 2 & 1 \\ 0 & 2 \end{pmatrix}$.

3. Verify the results of Example 2 where

$$A = \begin{pmatrix} 2 & 0 \\ 1 & 1 \end{pmatrix} \quad \text{and} \quad B = \begin{pmatrix} 4 & 3 \\ -2 & -1 \end{pmatrix}.$$

4. Show that

$$A = \begin{pmatrix} 2 & 0 \\ 0 & 2 \end{pmatrix} \quad \text{and} \quad B = \begin{pmatrix} 2 & 3 \\ 0 & 2 \end{pmatrix}$$

have the same eigenvalues but are not similar matrices.

5. Verify the results of Example 3 for each eigenvalue of A where

$$A = \begin{pmatrix} 2 & 3 \\ 0 & -1 \end{pmatrix}$$

and $n = 3$.

In Exercises 6 and 7 determine a modal matrix which transforms the matrix to the classical canonical form. Perform the transformation. Check the results.

6. $\begin{pmatrix} 2 & 1 \\ 0 & 3 \end{pmatrix}.$
7. $\begin{pmatrix} 5 & 4 \\ 12 & 7 \end{pmatrix}.$

In Exercises 8 and 9 determine a proper orthogonal modal matrix that transforms the matrix to the classical canonical form. Perform the orthogonal transformation. Check the results.

8. $\begin{pmatrix} 4 & 6 \\ 6 & -1 \end{pmatrix}.$
9. $\begin{pmatrix} 2 & 2 & 0 \\ 2 & 2 & 0 \\ 0 & 0 & 1 \end{pmatrix}.$

10. Prove that any matrix A similar to a diagonal matrix is similar to A^T.
11. Prove that if A is similar to the scalar matrix kI, then $A = kI$.

7-5 / THE HAMILTON-CAYLEY THEOREM

An important and interesting theorem of the theory of matrices is the **Hamilton-Cayley Theorem**:

THEOREM 7.7 / *Every square matrix A satisfies its own characteristic equation* $|A - \lambda I| = 0.$

More precisely, if λ is replaced by the matrix A of order n and each real number c_n is replaced by the scalar multiple $c_n I$ where I is the identity matrix of order n, then the characteristic equation of matrix A becomes a valid matrix equation; that is,

$$c_0 A^n + c_1 A^{n-1} + \cdots + c_{n-1} A + c_n I = 0. \qquad (7\text{-}14)$$

A heuristic argument may be used to prove the Hamilton-Cayley Theorem for a matrix A with distinct eigenvalues. Replace the variable λ by the square matrix A and c_n by $c_n I$ in the expression for the characteristic function of A and obtain

$$f(A) = c_0 A^n + c_1 A^{n-1} + \cdots + c_{n-1} A + c_n I. \tag{7-15}$$

Postmultiply both sides of equation (7-15) by an eigenvector X_i of A associated with λ_i and obtain

$$f(A)X_i = (c_0 \lambda_i^n + c_1 \lambda_i^{n-1} + \cdots + c_{n-1}\lambda_i + c_n)X_i$$

since $A^k X_i = \lambda_i^k X_i$ by Example 3 of §7-4. Since

$$c_0 \lambda_i^n + c_1 \lambda_i^{n-1} + \cdots + c_{n-1}\lambda_i + c_n = 0 \quad \text{for } i = 1, 2, \ldots, n,$$

then

$$f(A)X_i = 0 \quad \text{for } i = 1, 2, \ldots, n.$$

Hence,

$$f(A)X = 0, \tag{7-16}$$

where X is a matrix of eigenvectors. Since the eigenvectors are linearly independent by Theorem 7.1, the matrix of eigenvectors has a unique inverse X^{-1}. If both sides of equation (7-16) are postmultiplied by X^{-1}, the result is $f(A) = 0$, and the theorem is proved.

Proofs of the Hamilton-Cayley Theorem for the general case, without restrictions on the eigenvalues of A, may be found in most advanced texts on linear algebra.

EXAMPLE 1 / Show that the matrix A where

$$A = \begin{pmatrix} 3 & -2 \\ 1 & 2 \end{pmatrix}$$

satisfies its own characteristic equation.

The characteristic function of A is

$$f(\lambda) = |A - \lambda I| = \begin{vmatrix} 3 - \lambda & -2 \\ 1 & 2 - \lambda \end{vmatrix};$$

that is, $f(\lambda) = \lambda^2 - 5\lambda + 8$. Replace λ by A and 8 by $8I$ where I is the identity matrix of order two and obtain

$$f(A) = \begin{pmatrix} 3 & -2 \\ 1 & 2 \end{pmatrix}^2 - 5\begin{pmatrix} 3 & -2 \\ 1 & 2 \end{pmatrix} + 8\begin{pmatrix} 1 & 0 \\ 0 & 1 \end{pmatrix}$$

$$= \begin{pmatrix} 7 & -10 \\ 5 & 2 \end{pmatrix} + \begin{pmatrix} -15 & 10 \\ -5 & -10 \end{pmatrix} + \begin{pmatrix} 8 & 0 \\ 0 & 8 \end{pmatrix} = \begin{pmatrix} 0 & 0 \\ 0 & 0 \end{pmatrix}.$$

Hence, $f(A) = 0$ and the Hamilton-Cayley Theorem has been verified for matrix A.

The Hamilton-Cayley Theorem may be applied to the problem of determining the inverse of a nonsingular matrix A. Let

$$c_0\lambda^n + c_1\lambda^{n-1} + \cdots + c_{n-1}\lambda + c_n = 0$$

be the characteristic equation of A. Note that since A is a nonsingular matrix, $\lambda_i \neq 0$; that is, every eigenvalue is nonzero, and $c_n \neq 0$. By the Hamilton-Cayley Theorem,

$$c_0A^n + c_1A^{n-1} + \cdots + c_{n-1}A + c_nI = 0,$$

and

$$I = -\frac{1}{c_n}(c_0A^n + c_1A^{n-1} + \cdots + c_{n-1}A). \tag{7-17}$$

If both sides of equation (7-17) are multiplied by A^{-1}, the result is

$$A^{-1} = -\frac{1}{c_n}(c_0A^{n-1} + c_1A^{n-2} + \cdots + c_{n-1}I). \tag{7-18}$$

Note that the calculation of an inverse by use of equation (7-18) is quite adaptable to high-speed digital computers and is not difficult to compute manually for small values of n. In calculating the powers of matrix A necessary in equation (7-18), the necessary information concerning $tr(A^k)$ for calculating the c_i's is also obtained.

EXAMPLE 2 / Use the Hamilton-Cayley Theorem to find the inverse of A where

$$A = \begin{pmatrix} 1 & 0 & 1 \\ -1 & 1 & -3 \\ 2 & 2 & 4 \end{pmatrix}.$$

The characteristic equation of A is

$$\begin{vmatrix} 1-\lambda & 0 & 1 \\ -1 & 1-\lambda & -3 \\ 2 & 2 & 4-\lambda \end{vmatrix} = 0;$$

that is, $\lambda^3 - 6\lambda^2 + 13\lambda - 6 = 0$. By the Hamilton-Cayley Theorem,

$$A^3 - 6A^2 + 13A - 6I = 0,$$
$$I = \tfrac{1}{6}(A^3 - 6A^2 + 13A),$$

and

$$A^{-1} = \tfrac{1}{6}(A^2 - 6A + 13I).$$

Therefore,

$$A^{-1} = \tfrac{1}{6}\left[\begin{pmatrix} 1 & 0 & 1 \\ -1 & 1 & -3 \\ 2 & 2 & 4 \end{pmatrix}^2 - 6\begin{pmatrix} 1 & 0 & 1 \\ -1 & 1 & -3 \\ 2 & 2 & 4 \end{pmatrix} + 13\begin{pmatrix} 1 & 0 & 0 \\ 0 & 1 & 0 \\ 0 & 0 & 1 \end{pmatrix}\right]$$

$$= \tfrac{1}{6}\left[\begin{pmatrix} 3 & 2 & 5 \\ -8 & -5 & -16 \\ 8 & 10 & 12 \end{pmatrix} + \begin{pmatrix} -6 & 0 & -6 \\ 6 & -6 & 18 \\ -12 & -12 & -24 \end{pmatrix} + \begin{pmatrix} 13 & 0 & 0 \\ 0 & 13 & 0 \\ 0 & 0 & 13 \end{pmatrix}\right]$$

$$= \begin{pmatrix} \tfrac{5}{3} & \tfrac{1}{3} & -\tfrac{1}{6} \\ -\tfrac{1}{3} & \tfrac{1}{3} & \tfrac{1}{3} \\ -\tfrac{2}{3} & -\tfrac{1}{3} & \tfrac{1}{6} \end{pmatrix}.$$

A^{-k}, where k is a positive integer, is defined to be equal to $(A^{-1})^k$. By use of equation (7-18), it is now possible to express any negative integral power of a nonsingular matrix A of order n in terms of a linear function of the first $(n-1)$ powers of A.

EXAMPLE 3 / Find A^{-3} for the nonsingular matrix A where

$$A = \begin{pmatrix} 2 & 4 \\ 1 & 1 \end{pmatrix}.$$

Verify that A^{-3} is the inverse of A^3.

The characteristic equation of A is $\lambda^2 - 3\lambda - 2 = 0$; $c_0 = 1$, $c_1 = -3$, and $c_2 = -2$. By use of equation (7-18),

$$A^{-1} = \tfrac{1}{2}(A - 3I).$$

Then

$$A^{-2} = A^{-1}A^{-1} = \tfrac{1}{2}(I - 3A^{-1}) = \tfrac{11}{4}I - \tfrac{3}{4}A$$

and

$$A^{-3} = A^{-1}A^{-2} = \tfrac{11}{4}A^{-1} - \tfrac{3}{4}I = \tfrac{11}{8}A - \tfrac{39}{8}I.$$

Therefore,

$$A^{-3} = \tfrac{11}{8}\begin{pmatrix} 2 & 4 \\ 1 & 1 \end{pmatrix} - \tfrac{39}{8}\begin{pmatrix} 1 & 0 \\ 0 & 1 \end{pmatrix} = \begin{pmatrix} -\tfrac{17}{8} & \tfrac{11}{2} \\ \tfrac{11}{8} & -\tfrac{7}{2} \end{pmatrix}.$$

Verify this result by noting that

$$A^3 = \begin{pmatrix} 28 & 44 \\ 11 & 17 \end{pmatrix}, \quad \text{and} \quad \begin{pmatrix} 28 & 44 \\ 11 & 17 \end{pmatrix}\begin{pmatrix} -\tfrac{17}{8} & \tfrac{11}{2} \\ \tfrac{11}{8} & -\tfrac{7}{2} \end{pmatrix} = \begin{pmatrix} 1 & 0 \\ 0 & 1 \end{pmatrix}.$$

EXERCISES

In Exercises 1 through 4 verify the Hamilton-Cayley Theorem for the given matrix A.

1. $\begin{pmatrix} 2 & -1 \\ 3 & 4 \end{pmatrix}.$

2. $\begin{pmatrix} 7 & 9 \\ 2 & 1 \end{pmatrix}.$

3. $\begin{pmatrix} 5 & 0 \\ 0 & 2 \end{pmatrix}.$

4. $\begin{pmatrix} 1 & 1 & -2 \\ 1 & 0 & 3 \\ -2 & 3 & 2 \end{pmatrix}.$

In Exercises 5 and 6 use the Hamilton-Cayley Theorem to find the inverse of the given matrix A.

5. $\begin{pmatrix} 5 & 2 \\ 2 & 1 \end{pmatrix}.$

6. $\begin{pmatrix} 1 & 2 & 3 \\ 1 & 3 & 5 \\ 1 & 5 & 12 \end{pmatrix}.$

7. Find (a) A^{-2}; (b) A^{-3} for the nonsingular matrix A where

$$A = \begin{pmatrix} 7 & 2 \\ 5 & 1 \end{pmatrix}.$$

Verify that A^{-2} and A^{-3} are the inverses of A^2 and A^3, respectively.

8. Prove the Hamilton-Cayley Theorem for any square matrix of order two.

7-6 / QUADRATIC FORMS

Prior to a discussion of the simplification of the general quadratic equation in two variables by means of the rigid motion transformations, it is necessary to introduce the concept of real quadratic forms. A **quadratic form** is a homogeneous polynomial of the second degree in n variables $w_1, w_2, \ldots, w_n$; that is, a polynomial of the form

$$\sum_{i,j=1}^{n} a_{ij} w_i w_j. \tag{7-19}$$

Every quadratic form may be expressed as a matrix product $W^T A W$, where $W = (w_1 \ w_2 \ \cdots \ w_n)^T$ and $A = ((a_{ij}))$ is a unique real symmetric matrix. Therefore, every real quadratic form may be orthogonally transformed to the form $\lambda_1 z_1^2 + \lambda_2 z_2^2 + \cdots + \lambda_n z_n^2$, where the λ_i's are the eigenvalues of A, by an appropriate choice of a proper orthogonal matrix C such that $C^T A C$ is a diagonal matrix; that is,

$$W^T A W = Z^T C^T A C Z = Z^T \Lambda Z, \tag{7-20}$$

where $W = CZ$ and $Z = (z_1 \ z_2 \ \cdots \ z_n)^T$. Of course, the appropriate orthogonal matrix C which transforms a real symmetric matrix A into a diagonal matrix is a matrix of unit eigenvectors of A.

If (7-19) is a homogeneous polynomial of the second degree in two variables, the modal matrix C of the orthogonal transformation in (7-20) may be considered a rotation matrix. For example, the problem of reducing the quadratic form $ax^2 + 2bxy + cy^2$ to the canonical form $a'x'^2 + c'y'^2$ can be accomplished by means of a rotation of the plane about the origin. When the quadratic form is written in matrix notation, the result is

$$(x \ y) \begin{pmatrix} a & b \\ b & c \end{pmatrix} \begin{pmatrix} x \\ y \end{pmatrix}.$$

A rotation of the plane about the origin through an angle θ may be expressed by the matrix equation

$$\begin{pmatrix} x \\ y \end{pmatrix} = \begin{pmatrix} \cos\theta & \sin\theta \\ -\sin\theta & \cos\theta \end{pmatrix} \begin{pmatrix} x' \\ y' \end{pmatrix}.$$

Since the transpose of a product of two matrices is equal to the product of the transpose of the two matrices taken in reverse order, then

$$(x' \ y') \begin{pmatrix} \cos\theta & -\sin\theta \\ \sin\theta & \cos\theta \end{pmatrix} \begin{pmatrix} a & b \\ b & c \end{pmatrix} \begin{pmatrix} \cos\theta & \sin\theta \\ -\sin\theta & \cos\theta \end{pmatrix} \begin{pmatrix} x' \\ y' \end{pmatrix}$$

represents the quadratic form after a rotation of the plane about the origin through an angle θ. It is desired that θ, the angle of rotation, be such that

$$\begin{pmatrix} \cos\theta & -\sin\theta \\ \sin\theta & \cos\theta \end{pmatrix} \begin{pmatrix} a & b \\ b & c \end{pmatrix} \begin{pmatrix} \cos\theta & \sin\theta \\ -\sin\theta & \cos\theta \end{pmatrix} = \begin{pmatrix} a' & 0 \\ 0 & c' \end{pmatrix}.$$

This is clearly the problem of transforming a symmetric matrix to a diagonal matrix. Previously, it has been shown that in such a case the diagonal elements of the diagonal matrix are the eigenvalues of the symmetric matrix. Therefore, to accomplish the proper rotation, it is necessary only to solve the characteristic equation of the quadratic form:

$$\begin{vmatrix} a - \lambda & b \\ b & c - \lambda \end{vmatrix} = 0;$$

that is,

$$\lambda^2 - (a + c)\lambda + (ac - b^2) = 0.$$

The eigenvalues λ_1 and λ_2 become the coefficients of x'^2 and y'^2; that is,

$$(x \ y) \begin{pmatrix} a & b \\ b & c \end{pmatrix} \begin{pmatrix} x \\ y \end{pmatrix} = (x' \ y') \begin{pmatrix} \lambda_1 & 0 \\ 0 & \lambda_2 \end{pmatrix} \begin{pmatrix} x' \\ y' \end{pmatrix} = (\lambda_1 x'^2 + \lambda_2 y'^2).$$

The angle of rotation θ may be determined by considering the matrix equation

$$\begin{pmatrix} \cos\theta & -\sin\theta \\ \sin\theta & \cos\theta \end{pmatrix} \begin{pmatrix} a & b \\ b & c \end{pmatrix} \begin{pmatrix} \cos\theta & \sin\theta \\ -\sin\theta & \cos\theta \end{pmatrix} = \begin{pmatrix} \lambda_1 & 0 \\ 0 & \lambda_2 \end{pmatrix};$$

that is,

$$\begin{cases} a\cos^2\theta - b\sin\theta\cos\theta - b\sin\theta\cos\theta + c\sin^2\theta = \lambda_1 \\ a\sin\theta\cos\theta + b\cos^2\theta - b\sin^2\theta - c\sin\theta\cos\theta = 0 \\ a\sin\theta\cos\theta - b\sin^2\theta + b\cos^2\theta - c\sin\theta\cos\theta = 0 \\ a\sin^2\theta + b\sin\theta\cos\theta + b\sin\theta\cos\theta + c\cos^2\theta = \lambda_2. \end{cases} \quad (7\text{-}21)$$

If the second and third equations of (7-21) are added, the result is

$$(a - c)\sin\theta\cos\theta + b(\cos^2\theta - \sin^2\theta) = 0.$$

Therefore,

$$\frac{2b}{c-a} = \frac{\sin 2\theta}{\cos 2\theta} = \tan 2\theta.$$

Hence,

$$\theta = \tfrac{1}{2}\arctan\frac{2b}{c-a}. \quad (7\text{-}22)$$

Note that the unit vectors along the coordinate axes are the images of the unit eigenvectors $(\cos\theta \quad -\sin\theta)^T$ and $(\sin\theta \quad \cos\theta)^T$ associated with the eigenvalues of the symmetric matrix of the quadratic form. These unit eigenvectors lie along the principal axes of the locus of $ax^2 + 2bxy + y^2 = 0$.

EXERCISES

In Exercises 1 through 3 express each quadratic form in terms of matrices.

1. $3x^2 + 10xy + 3y^2$.   **2.** $x^2 - 2xy + y^2$.   **3.** $2x^2 + 2\sqrt{2}\,xy + y^2$.

In Exercises 4 through 6 use a proper orthogonal matrix to transform each quadratic form in the specified exercise to canonical form.

4. Exercise 1.   **5.** Exercise 2.   **6.** Exercise 3.

7-7 / CLASSIFICATION OF THE CONICS

The most general equation of the second degree in two variables x and y may be written in the form

$$f(x, y) = ax^2 + 2bxy + cy^2 + 2dx + 2ey + f = 0, \quad (7\text{-}23)$$

where a, b, c, d, e, and f are real numbers. The locus of equation (7-23) on the coordinate Euclidean plane is called a **conic section**, or simply a **conic**. By means of a rotation of the plane about the origin, a translation of the plane, or both, it is possible to represent every conic in a simplified standard, or canonical, form. These plane figures can be studied more readily in their canonical forms. There are nine classes of the conics; that is, equation (7-23) represents one of nine types of plane figures called conics. Two of these plane figures are imaginary in that there are no real points which satisfy equation (7-23).

If the general second degree function $f(x,y)$ of equation (7-23) is written in the form

$$f(x,y) = axx + bxy + dx \\ + byx + cyy + ey \\ + dx + ey + f, \qquad (7\text{-}24)$$

another valuable form of $f(x,y)$ is suggested. The function may be expressed as the product of three matrices:

$$(x \quad y \quad 1) \begin{pmatrix} a & b & d \\ b & c & e \\ d & e & f \end{pmatrix} \begin{pmatrix} x \\ y \\ 1 \end{pmatrix}. \qquad (7\text{-}25)$$

Note that the matrix of central importance

$$\Delta = \begin{pmatrix} a & b & d \\ b & c & e \\ d & e & f \end{pmatrix}, \qquad (7\text{-}26)$$

which defines the particular conic being studied, is a real symmetric matrix. The matrix Δ is called the **matrix of the conic section**.

In addition to Δ, the matrix whose determinant is the minor of f in Δ, the real symmetric matrix

$$F = \begin{pmatrix} a & b \\ b & c \end{pmatrix}, \qquad (7\text{-}27)$$

is of fundamental importance in the analysis of $f(x, y)$. The result of a premultiplication of matrix F by row vector $(x \quad y)$ and a postmultiplication of matrix F by column vector $(x \quad y)^T$ represents the quadratic form portion of the general second degree function; that is,

$$(x \ y)\begin{pmatrix} a & b \\ b & c \end{pmatrix}\begin{pmatrix} x \\ y \end{pmatrix} = ax^2 + 2bxy + cy^2.$$

After a rotation of the plane about the origin through an angle θ defined by the matrix equation

$$\begin{pmatrix} x \\ y \\ 1 \end{pmatrix} = \begin{pmatrix} \cos\theta & \sin\theta & 0 \\ -\sin\theta & \cos\theta & 0 \\ 0 & 0 & 1 \end{pmatrix}\begin{pmatrix} x' \\ y' \\ 1 \end{pmatrix},$$

where θ is $\frac{1}{2}$ arc tan $2b/(c-a)$, the general equation of a conic may be expressed in the form

$$\lambda_1 x'^2 + \lambda_2 y'^2 + 2\alpha x' + 2\beta y' + f = 0; \qquad (7\text{-}28)$$

that is,

$$(x' \ y' \ 1)\begin{pmatrix} \lambda_1 & 0 & \alpha \\ 0 & \lambda_2 & \beta \\ \alpha & \beta & f \end{pmatrix}\begin{pmatrix} x' \\ y' \\ 1 \end{pmatrix} = 0, \qquad (7\text{-}29)$$

where the λ_i's are the eigenvalues of matrix F and

$$\begin{cases} \alpha = d\cos\theta - e\sin\theta \\ \beta = d\sin\theta + e\cos\theta. \end{cases} \qquad (7\text{-}30)$$

The linear terms in x' and y' of (7-28) may be removed by a translation of the plane defined by the matrix equation

$$\begin{pmatrix} x' \\ y' \\ 1 \end{pmatrix} = \begin{pmatrix} 1 & 0 & -\dfrac{\alpha}{\lambda_1} \\ 0 & 1 & -\dfrac{\beta}{\lambda_2} \\ 0 & 0 & 1 \end{pmatrix}\begin{pmatrix} x'' \\ y'' \\ 1 \end{pmatrix}. \qquad (7\text{-}31)$$

If at least one λ_i equals zero, a translation of the plane that would remove all the linear terms of equation (7-28) does not exist. Hence, a geometric center for the conic does not exist and the conic is called a **noncentral conic**. If some λ_i equals zero, matrix F is necessarily singular. When F is nonsingular, a geometric center for the conic exists and the conic is called a **central conic**. After a translation of the plane described by (7-31), the equation of a central conic may be expressed as

$$\lambda_1 x''^2 + \lambda_2 y''^2 + f' = 0; \qquad (7\text{-}32)$$

that is,

$$(x'' \ y'' \ 1)\begin{pmatrix} \lambda_1 & 0 & 0 \\ 0 & \lambda_2 & 0 \\ 0 & 0 & f' \end{pmatrix}\begin{pmatrix} x'' \\ y'' \\ 1 \end{pmatrix} = 0, \qquad (7\text{-}33)$$

where

$$f' = f - \frac{\alpha^2}{\lambda_1} - \frac{\beta^2}{\lambda_2}. \qquad (7\text{-}34)$$

If one and only one λ_i is zero, then a translation of the plane exists which would remove one of the linear terms of equation (7-28). For example, consider $\lambda_2 = 0$. The translation of the plane defined by the matrix equation

$$\begin{pmatrix} x' \\ y' \\ 1 \end{pmatrix} = \begin{pmatrix} 1 & 0 & -\frac{\alpha}{\lambda_1} \\ 0 & 1 & 0 \\ 0 & 0 & 1 \end{pmatrix}\begin{pmatrix} x'' \\ y'' \\ 1 \end{pmatrix} \qquad (7\text{-}35)$$

transforms equation (7-28) to the form

$$\lambda_1 x''^2 + 2\beta y'' + f' = 0; \qquad (7\text{-}36)$$

that is,

$$(x'' \ y'' \ 1)\begin{pmatrix} \lambda_1 & 0 & 0 \\ 0 & 0 & \beta \\ 0 & \beta & f' \end{pmatrix}\begin{pmatrix} x'' \\ y'' \\ 1 \end{pmatrix} = 0, \qquad (7\text{-}37)$$

where

$$f' = f - \frac{\alpha^2}{\lambda_1}. \qquad (7\text{-}38)$$

EXAMPLE 1 / Transform the equation of the conic $5x^2 + 6xy + 5y^2 - 4x + 4y - 4 = 0$ to canonical form (Figure 7-2).

The equation of the conic may be written in the form

$$(x \ y \ 1)\begin{pmatrix} 5 & 3 & -2 \\ 3 & 5 & 2 \\ -2 & 2 & -4 \end{pmatrix}\begin{pmatrix} x \\ y \\ 1 \end{pmatrix} = 0.$$

7-7 / CLASSIFICATION OF THE CONICS

The characteristic equation of

$$F = \begin{pmatrix} 5 & 3 \\ 3 & 5 \end{pmatrix}$$

is $\lambda^2 - 10\lambda + 16 = 0$. Hence, the eigenvalues of F are $\lambda_1 = 8$ and $\lambda_2 = 2$ with associated unit eigenvectors

$$\left(\frac{1}{\sqrt{2}} \quad \frac{1}{\sqrt{2}}\right)^T \quad \text{and} \quad \left(-\frac{1}{\sqrt{2}} \quad \frac{1}{\sqrt{2}}\right)^T,$$

respectively. Therefore, a rotation of the plane defined by the matrix equation

$$\begin{pmatrix} x \\ y \\ 1 \end{pmatrix} = \begin{pmatrix} \frac{1}{\sqrt{2}} & -\frac{1}{\sqrt{2}} & 0 \\ \frac{1}{\sqrt{2}} & \frac{1}{\sqrt{2}} & 0 \\ 0 & 0 & 1 \end{pmatrix} \begin{pmatrix} x' \\ y' \\ 1 \end{pmatrix}$$

will transform the equation of the conic to the form of (7-28):

$$(x' \quad y' \quad 1) \begin{pmatrix} \frac{1}{\sqrt{2}} & \frac{1}{\sqrt{2}} & 0 \\ -\frac{1}{\sqrt{2}} & \frac{1}{\sqrt{2}} & 0 \\ 0 & 0 & 1 \end{pmatrix} \begin{pmatrix} 5 & 3 & -2 \\ 3 & 5 & 2 \\ -2 & 2 & -4 \end{pmatrix}$$

$$\begin{pmatrix} \frac{1}{\sqrt{2}} & -\frac{1}{\sqrt{2}} & 0 \\ \frac{1}{\sqrt{2}} & \frac{1}{\sqrt{2}} & 0 \\ 0 & 0 & 1 \end{pmatrix} \begin{pmatrix} x' \\ y' \\ 1 \end{pmatrix} = 0,$$

$$(x' \quad y' \quad 1) \begin{pmatrix} 8 & 0 & 0 \\ 0 & 2 & 2\sqrt{2} \\ 0 & 2\sqrt{2} & -4 \end{pmatrix} \begin{pmatrix} x' \\ y' \\ 1 \end{pmatrix} = 0;$$

that is,

$$8x'^2 + 2y'^2 + 4\sqrt{2}\,y' - 4 = 0.$$

A translation of the plane defined by (7-31) where $\alpha = 0$, $\beta = 2\sqrt{2}$, $\lambda_1 = 8$, and $\lambda_2 = 2$ will transform the equation of the conic to the form of (7-32):

$$(x'' \; y'' \; 1) \begin{pmatrix} 1 & 0 & 0 \\ 0 & 1 & 0 \\ 0 & -\sqrt{2} & 1 \end{pmatrix} \begin{pmatrix} 8 & 0 & 0 \\ 0 & 2 & 2\sqrt{2} \\ 0 & 2\sqrt{2} & -4 \end{pmatrix}$$

$$\begin{pmatrix} 1 & 0 & 0 \\ 0 & 1 & -\sqrt{2} \\ 0 & 0 & 1 \end{pmatrix} \begin{pmatrix} x'' \\ y'' \\ 1 \end{pmatrix} = 0,$$

$$(x'' \; y'' \; 1) \begin{pmatrix} 8 & 0 & 0 \\ 0 & 2 & 0 \\ 0 & 0 & -8 \end{pmatrix} \begin{pmatrix} x'' \\ y'' \\ 1 \end{pmatrix} = 0,$$

$$8x''^2 + 2y''^2 - 8 = 0,$$
$$4x''^2 + y''^2 - 4 = 0;$$

that is,

$$4x^2 + y^2 - 4 = 0.$$

Hence, the conic is a real ellipse. Note that the unit eigenvectors are parallel to the principal axes of the original conic.

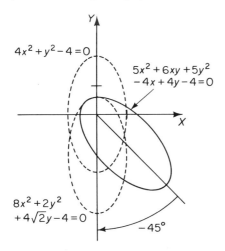

FIGURE 7-2

It should be evident from the discussion so far that the class of conic represented by equation (7-23) can be determined by an investigation of the algebraic values of the eigenvalues of matrix F. The conics will now be classified according to the eigenvalues of F.

If λ_1 and λ_2 are nonzero and have the same algebraic sign, then equation

7-7 / CLASSIFICATION OF THE CONICS

(7-32) represents an imaginary ellipse, a real ellipse, or a real point depending upon whether f' has the same sign as the λ_i's, differs in sign from the λ_i's, or is zero, respectively. Furthermore, if the eigenvalues are equal in the case of the real ellipse, the conic is a circle.

If λ_1 and λ_2 are nonzero and have opposite algebraic signs, then equation (7-32) represents a hyperbola or two intersecting lines depending upon whether f' is nonzero or zero, respectively.

For the case when only one eigenvalue is zero, consider λ_2 equal to zero; there is no loss in generality in doing so. Then equation (7-23) may be transformed to the form of (7-36). If β is nonzero, then equation (7-36) represents a parabola; if β is equal to zero, then equation (7-36) represents two parallel lines if λ_1 and f' have opposite signs, two imaginary parallel lines if λ_1 and f' have the same signs, and two coincident lines if f' is equal to zero.

EXAMPLE 2 / Identify the class of conic represented by the equation $x^2 - 4xy + 4y^2 - 4 = 0$.

The matrix of the conic section is

$$\begin{pmatrix} 1 & -2 & 0 \\ -2 & 4 & 0 \\ 0 & 0 & -4 \end{pmatrix}.$$

The characteristic equation of matrix F is

$$\begin{vmatrix} 1 - \lambda & -2 \\ -2 & 4 - \lambda \end{vmatrix} = 0;$$

that is, $\lambda^2 - 5\lambda = 0$. Hence, the eigenvalues of F are $\lambda_1 = 5$ and $\lambda_2 = 0$ with associated unit eigenvectors

$$\left(\frac{1}{\sqrt{5}}\ -\frac{2}{\sqrt{5}}\right)^T \text{ and } \left(\frac{2}{\sqrt{5}}\ \frac{1}{\sqrt{5}}\right)^T.$$

Since the inverse of the rotation matrix and the proper orthogonal matrix of eigenvectors are identical,

$$\begin{pmatrix} \cos\theta & \sin\theta \\ -\sin\theta & \cos\theta \end{pmatrix} = \begin{pmatrix} \frac{1}{\sqrt{5}} & \frac{2}{\sqrt{5}} \\ -\frac{2}{\sqrt{5}} & \frac{1}{\sqrt{5}} \end{pmatrix},$$

$\cos \theta = 1/\sqrt{5}$, and $\sin \theta = 2/\sqrt{5}$. Furthermore, $d = e = 0$; $f = -4$. Then $\alpha = \beta = 0$ by (7-30), and $f' = -4$ by (7-38). Since one eigenvalue is zero and the remaining eigenvalue and f' have opposite signs, the equation of the conic represents two parallel lines.

Note that the equation of the conic may be expressed in factored form as $(x - 2y + 2)(x - 2y - 2) = 0$. Hence, the equations of the two parallel lines are $x - 2y + 2 = 0$ and $x - 2y - 2 = 0$.

EXERCISES

In Exercises 1 through 4 transform the equation of the conic to canonical form and identify the conic.

1. $x^2 + xy + y^2 + 2x - 3y + 5 = 0$.
2. $x^2 - 2xy + y^2 + 8x + 8y = 0$.
3. $7x^2 - 48xy - 7y^2 + 60x - 170y + 225 = 0$.
4. $x^2 - 2xy + y^2 - 8 = 0$.

In Exercises 5 through 8 identify the class of conic represented by the equation.

5. $5x^2 + 4xy + 8y^2 - 36 = 0$.
6. $2x^2 + \sqrt{3}\,xy + y^2 + 14 = 0$.
7. $x^2 - 12xy - 4y^2 = 0$.
8. $3x^2 + 3xy + 3y^2 - 18x + 15y + 91 = 0$.

9. Show that the eigenvalues of matrix F in Example 1 are inversely proportional to the squares of the semi-axes.

10. Show that the geometric center (x,y) of the conic in Example 1 satisfies the system of equations

$$\begin{cases} ax + by + d = 0 \\ bx + cy + e = 0. \end{cases}$$

7-8 / INVARIANTS FOR CONICS

The rank of the matrix of a conic section is called the **rank of the conic** Since the rank of a matrix is an invariant under the elementary row transformations and since every rotation matrix and translation matrix may be expressed as a product of elementary row transformations, the rank of Δ is an invariant under the rigid motion transformations represented by these matrices.

If λ_1 and λ_2 are nonzero, then the transformed matrix of the conic section is represented in equation (7-33) by

$$\Delta_1 = \begin{pmatrix} \lambda_1 & 0 & 0 \\ 0 & \lambda_2 & 0 \\ 0 & 0 & f' \end{pmatrix}.$$

If f' is nonzero, Δ_1 is of rank three. Hence, from previous discussion of the classification of the conics, a real ellipse, an imaginary ellipse, and a hyperbola are conics of rank three. If f' is equal to zero, Δ_1 is of rank two. Hence, a real point and a pair of intersecting lines are conics of rank two.

If one and only one λ_i is zero (consider $\lambda_2 = 0$), then the transformed matrix of the conic section is represented in equation (7-37) by

$$\Delta_2 = \begin{pmatrix} \lambda_1 & 0 & 0 \\ 0 & 0 & \beta \\ 0 & \beta & f' \end{pmatrix}.$$

If β is nonzero, Δ_2 is of rank three. Hence, a parabola is a conic of rank three. If β is equal to zero, Δ_2 is of rank one or two depending upon whether f' is zero or nonzero, respectively. Hence, a pair of coincident lines is a conic of rank one; a pair of real parallel lines and a pair of imaginary parallel lines are conics of rank two.

The conics of rank three sometimes are called the **proper conics**; that is, the ellipses, the hyperbola, and the parabola are proper conics. The conics of rank less than three sometimes are called the **degenerate conics**.

Since a rotation of the plane about the origin transforms matrix F to a similar matrix, then by Example 2 of §7-4 the characteristic function of F is an invariant under the rotation transformation. Furthermore, since the elements of F are the coefficients of the second degree terms of $f(x,y)$ in (7-23) and since these coefficients are unchanged by a translation of the plane, the characteristic function of F is an invariant under the translation transformation. The characteristic function of F is

$$\lambda_2 - (a + c)\lambda + (ac - b^2). \tag{7-39}$$

Therefore, the following quantities are invariants for conics under the rotation and translation transformations:

$$a + c; \quad (7\text{-}40)$$

$$ac - b^2. \quad (7\text{-}41)$$

EXAMPLE 1 / Find the rank of the conic in Example 2 of §7-7.

The matrix of the conic section in Example 2 of §7-7 is

$$\Delta = \begin{pmatrix} 1 & -2 & 0 \\ -2 & 4 & 0 \\ 0 & 0 & -4 \end{pmatrix}.$$

Since det $\Delta = 0$, the rank of Δ cannot be three. The rank of Δ is two since

$$\begin{vmatrix} 4 & 0 \\ 0 & -4 \end{vmatrix} \neq 0.$$

Hence, the rank of the conic is two.

EXAMPLE 2 / Verify the invariance of (a) $a + c$; (b) $ac - b^2$ for the conic of Example 1 of §7-7 under the rotation and translation transformations.

In Example 1 of §7-7, $a = 5$, $b = 3$, and $c = 5$. After the rotation and translation transformations, the coefficients of x^2, $2xy$, and y^2 are 8, 0, and 2, respectively. Therefore,

(a) $a + c = 5 + 5 = 8 + 2$;
(b) $ac - b^2 = (5)(5) - (3)^2 = (8)(2) - (0)^2$.

EXERCISES

Find the rank of each conic in the specified exercises of §7-7.

1. Exercise 1. **2.** Exercise 2.
3. Exercise 3. **4.** Exercise 4.

Verify the invariance of (a) $a + c$; (b) $ac - b^2$ *for each conic in the specified exercises of §7-7.*

5. Exercise 1. **6.** Exercise 2.
7. Exercise 3. **8.** Exercise 4.

9. Show that (a) $a + c = \lambda_1 + \lambda_2$; (b) $ac - b^2 = \lambda_1\lambda_2$; (c) $\det F = ac - b^2$.

10. Discuss the classification of the conics in terms of $\det \Delta$ and $\det F$. (See Exercise 9.)

11. Prove that $a + c + f$ is an invariant for conics under the rotation transformation.

12. Prove that $d^2 + e^2$ of (7-23) is an invariant for conics under the rotation transformation.

ANSWERS

TO ODD-NUMBERED EXERCISES

ONE

ELEMENTARY VECTOR OPERATIONS

1-1 / SCALARS AND VECTORS

1. Scalar quantity.
3. Scalar quantity.
5. Vector quantity.

1-2 / EQUALITY OF VECTORS

1. Not equal.
3. Equal.
5. Not equal.
7.
9. →P→

11. ⟶____ P

1-3 / VECTOR ADDITION AND SUBTRACTION

1. (a) $\overrightarrow{AD}$; (b) $\vec{0}$.   **3.**

5. $(\vec{a} + \vec{b}) + \vec{c} = \vec{a} + (\vec{b} + \vec{c})$ (Theorem 1.2)
$= \vec{a} + (\vec{c} + \vec{b})$ (Theorem 1.1)
$= (\vec{a} + \vec{c}) + \vec{b}$ (Theorem 1.2)
$= (\vec{c} + \vec{a}) + \vec{b}$. (Theorem 1.1).

7. $\overrightarrow{AB} = \overrightarrow{DC}$; that is, two opposite sides of the quadrilateral are equal and parallel. Hence, $ABCD$ is a parallelogram.

9. (a) $\vec{a}, \vec{b}, \vec{c}, \vec{a} + \vec{b}, \vec{a} + \vec{c}, \vec{b} + \vec{c}, \vec{a} + \vec{b} + \vec{c}$;
(b) $\vec{a} + (\vec{b} + \vec{c}) + (\vec{a} + \vec{b} + \vec{c}) = \vec{b} + \vec{c} + (\vec{a} + \vec{b}) + (\vec{a} + \vec{c})$.

There are also other correct answers.

1-4 / MULTIPLICATION OF A VECTOR BY A SCALAR

1. (a) $3\overrightarrow{AB}$; (b) $2\overrightarrow{BA}$.   **3.**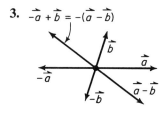

5. Let $\vec{c}$ and $\vec{d}$ represent the adjacent sides of the parallelogram. Assume $|\vec{a}| \geq |\vec{b}|$. Then $\vec{a} = \vec{c} + \vec{d}$ and $\vec{b} = \vec{c} - \vec{d}$. Therefore, $2\vec{c} = \vec{a} + \vec{b}$ and $\vec{c} = \frac{1}{2}(\vec{a} + \vec{b})$; $2\vec{d} = \vec{a} - \vec{b}$ and $\vec{d} = \frac{1}{2}(\vec{a} - \vec{b})$. Construct a parallelogram with sides $\vec{c}$ and $\vec{d}$.

7. Let $ABCD$ be any trapezoid with bases AB and DC; that is, $\overrightarrow{AB}$ is parallel to $\overrightarrow{DC}$. Let E and F be the midpoints of sides AD and BC, respectively; that is, line segment EF is the median of the trapezoid. Then

$\overrightarrow{EF} = \overrightarrow{EA} + \overrightarrow{AB} + \overrightarrow{BF}$
$= \frac{1}{2}\overrightarrow{DA} + \overrightarrow{AB} + \frac{1}{2}\overrightarrow{BC}$
$= \frac{1}{2}\overrightarrow{AB} + \frac{1}{2}(\overrightarrow{DA} + \overrightarrow{AB} + \overrightarrow{BC})$
$= \frac{1}{2}\overrightarrow{AB} + \frac{1}{2}\overrightarrow{DC}$
$= \frac{1}{2}(\overrightarrow{AB} + \overrightarrow{DC})$.

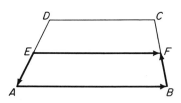

Since $\overrightarrow{AB}$ is parallel to $\overrightarrow{DC}$, then $\overrightarrow{EF}$ is parallel to both $\overrightarrow{AB}$ and $\overrightarrow{DC}$. Furthermore, $|\overrightarrow{EF}| = \frac{1}{2}|\overrightarrow{AB} + \overrightarrow{DC}| = \frac{1}{2}|\overrightarrow{AB}| + \frac{1}{2}|\overrightarrow{DC}|$. Therefore, line segment EF is parallel to the bases and equal to one-half their sum.

9. Let M, N, R, and S divide sides AB, BC, CD, and DA of a parallelogram in the same ratio k to $1 - k$. Then $\overrightarrow{MN} = \overrightarrow{MB} + \overrightarrow{BN} = (1 - k)\overrightarrow{AB} + k\overrightarrow{BC}$ and $\overrightarrow{SR} = \overrightarrow{SD} + \overrightarrow{DR} = k\overrightarrow{AD} + (1 - k)\overrightarrow{DC} = k\overrightarrow{BC} + (1 - k)\overrightarrow{AB}$. Therefore, $\overrightarrow{MN} = \overrightarrow{SR}$; that is, two opposite sides of the quadrilateral $MNRS$ are equal and parallel. Hence, $MNRS$ is a parallelogram.

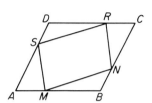

1-5 / LINEAR DEPENDENCE OF VECTORS

1. (a) $\overrightarrow{OA}$; (b) $-\frac{1}{2}\overrightarrow{OC}$; (c) $\overrightarrow{OA}$.
3. (a) $-2\overrightarrow{OA}$; (b) $\overrightarrow{OC}$; (c) $-2\overrightarrow{OA}$.
5. (a) $-\overrightarrow{OA} - \overrightarrow{OB}$; (b) $-\overrightarrow{OB} + \frac{1}{2}\overrightarrow{OC}$; (c) $5\overrightarrow{OA} - 2\overrightarrow{OD}$.
7. (a) $-\overrightarrow{OA} + \overrightarrow{OB}$; (b) $\overrightarrow{OB} + \frac{1}{2}\overrightarrow{OC}$; (c) $-7\overrightarrow{OA} + 2\overrightarrow{OD}$.
9. $\overrightarrow{OM} = \frac{1}{2}\overrightarrow{OA} + \frac{1}{2}\overrightarrow{OB}$. By Theorem 1.9, M is the midpoint of line segment AB.
11. $\overrightarrow{AD} = \overrightarrow{AB} + \frac{2}{5}\overrightarrow{BC} = \overrightarrow{AB} + \frac{2}{5}(\overrightarrow{AC} - \overrightarrow{AB}) = \frac{3}{5}\overrightarrow{AB} + \frac{2}{5}\overrightarrow{AC}$; $\frac{3}{5} + \frac{2}{5} = 1$. Then D, B, and C are collinear by Theorem 1.9.
13. Any two nonzero, nonparallel vectors of the form $m\vec{a} + n\vec{b}$ may be used; for example, $\vec{a} - \vec{b}$ and $\vec{a} + \vec{b}$.
15. Let $\vec{b}$ be a linear function of $\vec{a_1}, \vec{a_2}, \ldots, \vec{a_n}$. Assume that

$$\vec{b} = m_1\vec{a_1} + m_2\vec{a_2} + \cdots + m_n\vec{a_n} = k_1\vec{a_1} + k_2\vec{a_2} + \cdots + k_n\vec{a_n}.$$

Then

$$(m_1 - k_1)\vec{a_1} + (m_2 - k_2)\vec{a_2} + \cdots + (m_n - k_n)\vec{a_n} = \vec{0}.$$

Since $\vec{a_1}, \vec{a_2}, \ldots, \vec{a_n}$ are linearly independent vectors, then $m_1 - k_1 = m_2 - k_2 = \cdots = m_n - k_n = 0$ and $m_i = k_i$ for $i = 1, 2, \ldots, n$. Hence, the two representations of $\vec{b}$ are identical.

1-6 / APPLICATIONS OF LINEAR DEPENDENCE

1. Let $ABCD$ be any parallelogram. Let M be a point which divides side BC in the ratio 1 to n and P be the point of intersection of line segment AM and diagonal BD. Then, by Theorem 1.9,

$$\overrightarrow{AM} = \frac{n}{n+1}\overrightarrow{AB} + \frac{1}{n+1}\overrightarrow{AC}$$
$$= \frac{n}{n+1}\overrightarrow{AB} + \frac{1}{n+1}(\overrightarrow{AD} + \overrightarrow{AB})$$
$$= \overrightarrow{AB} + \frac{1}{n+1}\overrightarrow{AD},$$

and

$$\overrightarrow{AP} = k\overrightarrow{AM} = k\overrightarrow{AB} + \frac{k}{n+1}\overrightarrow{AD}.$$

Since B, P, and D are collinear, $k + [k/(n+1)] = 1$, and $k = (n+1)/(n+2)$. Hence, $\overrightarrow{AP} = [(n+1)/(n+2)]\overrightarrow{AB} + [1/(n+2)]\overrightarrow{AD}$; that is, P divides diagonal BD in the ratio 1 to $n+1$. If we let M be a point that divides side CB in the ratio 1 to n, then, in a similar manner, it follows that P divides diagonal BD in the ratio n to $n+1$.

3. Let ABC be any triangle with centroid at P and M be the midpoint of side BC. Let O be any reference point in space. Then, by Theorem 1.9,

$$\overrightarrow{OP} = \tfrac{1}{3}\overrightarrow{OA} + \tfrac{2}{3}\overrightarrow{OM}$$
$$= \tfrac{1}{3}\overrightarrow{OA} + \tfrac{2}{3}(\tfrac{1}{2}\overrightarrow{OB} + \tfrac{1}{2}\overrightarrow{OC})$$
$$= \tfrac{1}{3}(\overrightarrow{OA} + \overrightarrow{OB} + \overrightarrow{OC}).$$

5. Let $ABCD$ be any quadrilateral with M, N, R, and S midpoints of sides AB, BC, CD, and DA, respectively. Let O be any reference point. Then $\overrightarrow{OM} = \tfrac{1}{2}\overrightarrow{OA} + \tfrac{1}{2}\overrightarrow{OB}$; $\overrightarrow{ON} = \tfrac{1}{2}\overrightarrow{OB} + \tfrac{1}{2}\overrightarrow{OC}$; $\overrightarrow{OR} = \tfrac{1}{2}\overrightarrow{OC} + \tfrac{1}{2}\overrightarrow{OD}$; and $\overrightarrow{OS} = \tfrac{1}{2}\overrightarrow{OD} + \tfrac{1}{2}\overrightarrow{OA}$. Now,

$$\overrightarrow{MN} = \overrightarrow{ON} - \overrightarrow{OM} = \tfrac{1}{2}(\overrightarrow{OC} - \overrightarrow{OA}) = \tfrac{1}{2}\overrightarrow{AC};$$
$$\overrightarrow{SR} = \overrightarrow{OR} - \overrightarrow{OS} = \tfrac{1}{2}(\overrightarrow{OC} - \overrightarrow{OA}) = \tfrac{1}{2}\overrightarrow{AC}.$$

Hence $\overrightarrow{MN} = \overrightarrow{SR}$; that is, $\overrightarrow{MN}$ is parallel to $\overrightarrow{SR}$ and $|\overrightarrow{MN}| = |\overrightarrow{SR}|$. Since two opposite sides, MN and SR, of the quadrilateral $MNRS$ are equal in length and parallel, the quadrilateral is a parallelogram.

7. Let $ABCD$ be any tetrahedron. Let M, N, R, and S be the midpoints of edges AB, CD, BD, and AC, respectively. Let P and P' be the midpoints of line segments MN and RS, respectively. If O is any reference point, then

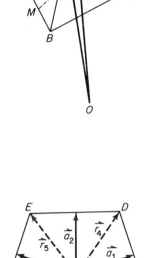

$$\overrightarrow{OP} = \tfrac{1}{2}\overrightarrow{OM} + \tfrac{1}{2}\overrightarrow{ON}$$
$$= \tfrac{1}{2}(\tfrac{1}{2}\overrightarrow{OA} + \tfrac{1}{2}\overrightarrow{OB}) + \tfrac{1}{2}(\tfrac{1}{2}\overrightarrow{OC} + \tfrac{1}{2}\overrightarrow{OD})$$
$$= \tfrac{1}{4}(\overrightarrow{OA} + \overrightarrow{OB} + \overrightarrow{OC} + \overrightarrow{OD});$$
$$\overrightarrow{OP'} = \tfrac{1}{2}\overrightarrow{OR} + \tfrac{1}{2}\overrightarrow{OS}$$
$$= \tfrac{1}{2}(\tfrac{1}{2}\overrightarrow{OB} + \tfrac{1}{2}\overrightarrow{OD}) + \tfrac{1}{2}(\tfrac{1}{2}\overrightarrow{OA} + \tfrac{1}{2}\overrightarrow{OC})$$
$$= \tfrac{1}{4}(\overrightarrow{OA} + \overrightarrow{OB} + \overrightarrow{OC} + \overrightarrow{OD}).$$

Since $\overrightarrow{OP} = \overrightarrow{OP'}$, the points P and P' coincide; that is, the line segments MN and RS bisect each other.

9. Let $ABCDE$ be any regular pentagon with center at O. Let $\vec{r}_1, \vec{r}_2, \ldots, \vec{r}_5$ be vectors drawn from the center to vertices $A, B, \ldots, E$, respectively. Let $\vec{a}_1, \vec{a}_2, \ldots, \vec{a}_5$ be vectors drawn from the center to the midpoints of sides CD, DE, $\ldots$, BC, respectively. Then

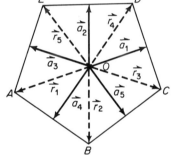

$$\vec{a}_1 = \tfrac{1}{2}\vec{r}_3 + \tfrac{1}{2}\vec{r}_4,$$
$$\vec{a}_2 = \tfrac{1}{2}\vec{r}_4 + \tfrac{1}{2}\vec{r}_5,$$
$$\vdots$$
$$\vec{a}_5 = \tfrac{1}{2}\vec{r}_2 + \tfrac{1}{2}\vec{r}_3.$$

Therefore, $\vec{a}_1 + \vec{a}_2 + \cdots + \vec{a}_5 = \vec{r}_1 + \vec{r}_2 + \cdots + \vec{r}_5$. Now, $\vec{a}_1 = m\vec{r}_1$, $\vec{a}_2 = m\vec{r}_2, \ldots, \vec{a}_5 = m\vec{r}_5$. Hence, $m(\vec{r}_1 + \vec{r}_2 + \cdots + \vec{r}_5) = \vec{r}_1 + \vec{r}_2 + \cdots + \vec{r}_5$; $(m-1)(\vec{r}_1 + \vec{r}_2 + \cdots + \vec{r}_5) = \vec{0}$. Since $m \neq 1$, then $\vec{r}_1 + \vec{r}_2 + \cdots + \vec{r}_5 = \vec{0}$.

1-7 / POSITION VECTORS

1. (a) $2\vec{i} + 2\vec{j} + \vec{k}$; (b) 3; (c) $\frac{2}{3}\vec{i} + \frac{2}{3}\vec{j} + \frac{1}{3}\vec{k}$.
3. (a) $3\vec{i}$; (b) 3; (c) $\vec{i}$.   **5.** yz-plane.
7. A plane parallel to and three units from the yz-plane along the positive x-axis.
9. A unit sphere with center at the origin.
11. (a) $|z|$; (b) $\sqrt{y^2 + z^2}$.
13. M, N, and P are not collinear.
15. (a) $(\frac{11}{2}, 4, \frac{19}{2})$; (b) (5, 3, 11); (c) (13, 19, −13).

TWO

PRODUCTS OF VECTORS

2-1 / THE SCALAR PRODUCT

1. 3; $\frac{2}{3}\vec{i} + \frac{2}{3}\vec{j} + \frac{1}{3}\vec{k}$.
3. $\frac{13}{5}$.
5. (a) Since $|\cos{(\vec{r_1}, \vec{r_2})}| \leq 1$, $|\vec{r_1} \cdot \vec{r_2}| \leq |\vec{r_1}||\vec{r_2}|$; (b) either $\vec{r_1} = \vec{0}$, $\vec{r_2} = \vec{0}$, or $\cos{(\vec{r_1}, \vec{r_2})} = 1$, that is, $\vec{r_1}$ has the same direction as $\vec{r_2}$; (c) either $\vec{r_1} = \vec{0}$, $\vec{r_2} = \vec{0}$, or $\cos{(\vec{r_1}, \vec{r_2})} = -1$, that is, $\vec{r_1}$ has an opposite direction to $\vec{r_2}$.
7. $\frac{\sqrt{3}}{3}$.
9. By Theorem 2.4, $\angle ABC$ is a right angle since $\overrightarrow{BA} \cdot \overrightarrow{BC} = 0$; $|\overrightarrow{BA}| = |\overrightarrow{BC}|$; hence, $\triangle ABC$ is a right isosceles triangle.
11. By Theorem 2.2, $\vec{a} \cdot \vec{a} = |\vec{a}|^2$. If $\vec{a} \cdot \vec{a} = 0$, then $|\vec{a}| = 0$. If $|\vec{a}| = 0$, then $\vec{a} \cdot \vec{a} = 0^2 = 0$.
13. $\overrightarrow{AB} = -\vec{i} + \vec{j}$; $\overrightarrow{AC} = -\vec{i} + \vec{k}$; $\dfrac{\overrightarrow{AB} \cdot \overrightarrow{AC}}{|\overrightarrow{AC}|} = \dfrac{\sqrt{2}}{2}$.

2-2 / APPLICATIONS OF THE SCALAR PRODUCT

1. Let $ABCD$ be a rhombus. Let $\vec{a}$ and $\vec{b}$ be associated with the adjacent sides AB and BC, respectively. Then $\vec{a} + \vec{b}$ and $\vec{a} - \vec{b}$ are vectors associated with the diagonals. By Theorems 2.5, 2.1, and 2.2,

$$(\vec{a} + \vec{b}) \cdot (\vec{a} - \vec{b}) = (\vec{a} + \vec{b}) \cdot \vec{a} - (\vec{a} + \vec{b}) \cdot \vec{b}$$
$$= \vec{a} \cdot \vec{a} + \vec{b} \cdot \vec{a} - \vec{a} \cdot \vec{b} - \vec{b} \cdot \vec{b}$$
$$= \vec{a} \cdot \vec{a} - \vec{b} \cdot \vec{b}$$
$$= |\vec{a}|^2 - |\vec{b}|^2$$
$$= 0 \quad (\text{since } |\vec{a}| = |\vec{b}|).$$

Since $|\vec{a} + \vec{b}| \neq 0$ and $|\vec{a} - \vec{b}| \neq 0$, then by Theorem 2.4 $(\vec{a} + \vec{b}) \perp (\vec{a} - \vec{b})$; that is, the diagonals of a rhombus are perpendicular.

3. Let ABC be any right triangle with right angle at C. If M is the midpoint of the hypotenuse AB, then $\overrightarrow{CM} = \frac{1}{2}(\overrightarrow{CA} + \overrightarrow{CB})$, and

$$\overrightarrow{CM} \cdot \overrightarrow{CM} = \frac{1}{2}(\overrightarrow{CA} + \overrightarrow{CB}) \cdot \frac{1}{2}(\overrightarrow{CA} + \overrightarrow{CB})$$
$$= \frac{1}{4}(\overrightarrow{CA} \cdot \overrightarrow{CA} + \overrightarrow{CB} \cdot \overrightarrow{CA} + \overrightarrow{CA} \cdot \overrightarrow{CB} + \overrightarrow{CB} \cdot \overrightarrow{CB})$$
$$= \frac{1}{4}(\overrightarrow{CA} \cdot \overrightarrow{CA} + \overrightarrow{CB} \cdot \overrightarrow{CB})$$
$$(\text{since } \overrightarrow{CB} \cdot \overrightarrow{CA} = \overrightarrow{CA} \cdot \overrightarrow{CB} = 0)$$
$$= \frac{1}{4}(|\overrightarrow{CA}|^2 + |\overrightarrow{CB}|^2)$$
$$= \frac{1}{4}|\overrightarrow{AB}|^2 \quad (\text{by the Pythagorean Theorem}).$$

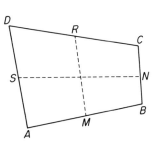

Hence, $|\overrightarrow{CM}|^2 = \frac{1}{4}|\overrightarrow{AB}|^2$ and $|\overrightarrow{CM}| = \frac{1}{2}|\overrightarrow{AB}|$; that is, the median to the hypotenuse is equal to one-half the length of the hypotenuse.

5. Let $ABCD$ be any quadrilateral with M, N, R, and S midpoints of sides AB, BC, CD, and DA, respectively. Since

$$\overrightarrow{MR} = \overrightarrow{MB} + \overrightarrow{BC} + \overrightarrow{CR}$$
$$= \frac{1}{2}\overrightarrow{AB} + \overrightarrow{BC} + \frac{1}{2}\overrightarrow{CD}$$

and

$$\overrightarrow{SN} = \overrightarrow{SA} + \overrightarrow{AB} + \overrightarrow{BN}$$
$$= \frac{1}{2}\overrightarrow{DA} + \overrightarrow{AB} + \frac{1}{2}\overrightarrow{BC},$$

then

$$2\overrightarrow{MR} = \overrightarrow{AB} + 2\overrightarrow{BC} + \overrightarrow{CD}$$
$$= (\overrightarrow{AB} + \overrightarrow{BC}) + (\overrightarrow{BC} + \overrightarrow{CD})$$
$$= \overrightarrow{AC} + \overrightarrow{BD}$$

and

$$2\overrightarrow{SN} = \overrightarrow{DA} + 2\overrightarrow{AB} + \overrightarrow{BC}$$
$$= (\overrightarrow{DA} + \overrightarrow{AB}) + (\overrightarrow{AB} + \overrightarrow{BC})$$
$$= \overrightarrow{DB} + \overrightarrow{AC}$$
$$= \overrightarrow{AC} - \overrightarrow{BD}.$$

Now,

$$4|\overrightarrow{MR}|^2 = 2\overrightarrow{MR} \cdot 2\overrightarrow{MR} = (\overrightarrow{AC} + \overrightarrow{BD}) \cdot (\overrightarrow{AC} + \overrightarrow{BD})$$
$$= |\overrightarrow{AC}|^2 + |\overrightarrow{BD}|^2 + 2\overrightarrow{AC} \cdot \overrightarrow{BD}$$

and

$$4|\overrightarrow{SN}|^2 = 2\overrightarrow{SN} \cdot 2\overrightarrow{SN} = (\overrightarrow{AC} - \overrightarrow{BD}) \cdot (\overrightarrow{AC} - \overrightarrow{BD})$$
$$= |\overrightarrow{AC}|^2 + |\overrightarrow{BD}|^2 - 2\overrightarrow{AC} \cdot \overrightarrow{BD}.$$

Hence,

$$4|\overrightarrow{MR}|^2 + 4|\overrightarrow{SN}|^2 = 2|\overrightarrow{AC}|^2 + 2|\overrightarrow{BD}|^2;$$

that is,

$$2(|\overrightarrow{MR}|^2 + |\overrightarrow{SN}|^2) = |\overrightarrow{AC}|^2 + |\overrightarrow{BD}|^2.$$

7. Let $\vec{a}$ and $\vec{b}$ be unit position vectors on a rectangular Cartesian coordinate plane forming angles θ and $-\phi$, respectively, with the positive half of the x-axis. Then $\vec{a} = \cos\theta \vec{i} + \sin\theta \vec{j}$ and $\vec{b} = \cos(-\phi)\vec{i} + \sin(-\phi)\vec{j} = \cos\phi \vec{i} - \sin\phi \vec{j}$. By Definition 2.1, $\vec{a} \cdot \vec{b} = |\vec{a}||\vec{b}|\cos(\vec{a}, \vec{b}) = |\vec{a}||\vec{b}|\cos[-(\theta + \phi)] = |\vec{a}||\vec{b}|\cos(\theta + \phi) = \cos(\theta + \phi)$. By the properties of the scalar product of the unit position vectors,

$$\vec{a}\cdot\vec{b} = \cos\theta\cos\phi - \sin\theta\sin\phi.$$

Hence, $\cos(\theta + \phi) = \cos\theta\cos\phi - \sin\theta\sin\phi$.

9. Let ABC be any triangle. Then

$$\overrightarrow{AC} = \overrightarrow{AB} + \overrightarrow{BC},$$
$$\overrightarrow{AB}\cdot\overrightarrow{AC} = \overrightarrow{AB}\cdot(\overrightarrow{AB} + \overrightarrow{BC})$$
$$= \overrightarrow{AB}\cdot\overrightarrow{AB} + \overrightarrow{AB}\cdot\overrightarrow{BC};$$
$$|\overrightarrow{AB}||\overrightarrow{AC}|\cos A = |\overrightarrow{AB}||\overrightarrow{AB}| + |\overrightarrow{AB}||\overrightarrow{BC}|\cos(180° - B),$$
$$|\overrightarrow{AC}|\cos A = |\overrightarrow{AB}| - |\overrightarrow{BC}|\cos B.$$

Hence,
$$|\overrightarrow{AB}| = |\overrightarrow{AC}|\cos A + |\overrightarrow{BC}|\cos B.$$

2-3 / CIRCLES AND LINES ON A COORDINATE PLANE

1. $x^2 + y^2 - 2x + 6y - 6 = 0$. 3. 2 units.
5. $\vec{r} = m\vec{b}$. 7. $\vec{r} = \vec{a} + m\vec{b}$.
9. $r = \dfrac{|c|}{\sqrt{a^2 + b^2}}$.

2-4 / ORTHOGONAL BASES

1. -1 and -2.
3. $\dfrac{\sqrt{2}}{2}\vec{i} + \dfrac{\sqrt{2}}{2}\vec{k}$ and $-\dfrac{\sqrt{3}}{3}\vec{i} + \dfrac{\sqrt{3}}{3}\vec{j} + \dfrac{\sqrt{3}}{3}\vec{k}$.

There are also other correct answers.

5. $2\vec{i} - \vec{j} + 2\vec{k}$ and $\vec{i} + 2\vec{j}$.

2-5 / THE VECTOR PRODUCT

1. $2\vec{i} - 3\vec{j} - 5\vec{k}$.
3. $3\vec{i} \times (\vec{i} + 2\vec{j}) = 6\vec{k}$ and $3\vec{i} \times 2\vec{j} = 6\vec{k}$.
5. $\overrightarrow{AB} \times \overrightarrow{CD} = (3\vec{i} - 3\vec{j} - 2\vec{k}) \times (2\vec{i} + 2\vec{j} - \vec{k})$
$= 7\vec{i} - \vec{j} + 12\vec{k}$.

7. $(\vec{a} + \vec{b}) \times (\vec{c} + \vec{d}) = (\vec{a} + \vec{b}) \times \vec{c} + (\vec{a} + \vec{b}) \times \vec{d}$
$= -\vec{c} \times (\vec{a} + \vec{b}) - \vec{d} \times (\vec{a} + \vec{b})$
$= (-\vec{c} \times \vec{a}) - (\vec{c} \times \vec{b}) - (\vec{d} \times \vec{a}) - (\vec{d} \times \vec{b})$
$= (\vec{a} \times \vec{c}) + (\vec{b} \times \vec{c}) + (\vec{a} \times \vec{d}) + (\vec{b} \times \vec{d})$
$= (\vec{a} \times \vec{c}) + (\vec{a} \times \vec{d}) + (\vec{b} \times \vec{c}) + (\vec{b} \times \vec{d}).$

9. If $\vec{a} \times \vec{b} = \vec{0}$ is given, then $\vec{a}$ is parallel to $\vec{b}$ or $\vec{a} = \vec{0}$ or $\vec{b} = \vec{0}$. If $\vec{a}$ is parallel to $\vec{b}$, then $\vec{a} = m\vec{b}$ and $\vec{a}$ and $\vec{b}$ are linearly dependent. If either $\vec{a}$ or $\vec{b}$ is the null vector, then $\vec{a}$ and $\vec{b}$ are linearly dependent. Thus, if $\vec{a} \times \vec{b} = \vec{0}$, the vectors are linearly dependent. Conversely, if $\vec{a}$ and $\vec{b}$ are given as linearly dependent vectors, then either $\vec{a}$ or $\vec{b}$ is the null vector in which case $\vec{a} \times \vec{b} = \vec{0}$, or $\vec{a}$ is parallel to $\vec{b}$ in which case $\vec{a} = m\vec{b}$ and $\vec{a} \times \vec{b} = m\vec{b} \times \vec{b} = \vec{0}$.

2-6 / APPLICATIONS OF THE VECTOR PRODUCT

1. One square unit.

3. By Theorem 2.15, $\vec{b} \times \vec{a} = (\sin \theta \cos \phi + \cos \theta \sin \phi)\vec{k}$. By Definition 2.2, $\vec{b} \times \vec{a} = |\vec{b}||\vec{a}| \sin [\theta - (-\phi)]\vec{k} = \sin (\theta + \phi)\vec{k}$. Hence, $\sin (\theta + \phi) = \sin \theta \cos \phi + \cos \theta \sin \phi$.

5. The law of reflection of light states that when a ray of light strikes a plane mirror and is reflected, the incident ray and the reflected ray each form equal angles with the surface of the mirror. Therefore, $\vec{a} \times \vec{n} = \vec{b} \times \vec{n}$ represents a vector form of the law of reflection of light since $\sin (\vec{a}, \vec{n}) = \sin (\vec{b}, \vec{n})$.

7. The area of triangle ABC equals $\frac{1}{2}|\overrightarrow{AB} \times \overrightarrow{AC}|$ and the area of triangle HJG equals $\frac{1}{2}|\overrightarrow{JG} \times \overrightarrow{JH}|$. Now,

$\overrightarrow{JG} \times \overrightarrow{JH} = \frac{3}{7}\overrightarrow{AD} \times \frac{3}{7}\overrightarrow{FC}$
$= \frac{3}{7}(\overrightarrow{AB} + \overrightarrow{BD}) \times \frac{3}{7}(\overrightarrow{FA} + \overrightarrow{AC})$
$= \frac{3}{7}(\overrightarrow{AB} + \frac{1}{3}\overrightarrow{BC}) \times \frac{3}{7}(-\frac{1}{3}\overrightarrow{AB} + \overrightarrow{AC})$
$= (\frac{3}{7}\overrightarrow{AB} + \frac{1}{7}\overrightarrow{BC}) \times (-\frac{1}{7}\overrightarrow{AB} + \frac{3}{7}\overrightarrow{AC})$
$= (\frac{3}{7}\overrightarrow{AB} + \frac{1}{7}\overrightarrow{AC} - \frac{1}{7}\overrightarrow{AB}) \times (-\frac{1}{7}\overrightarrow{AB} + \frac{3}{7}\overrightarrow{AC})$
$= (\frac{2}{7}\overrightarrow{AB} + \frac{1}{7}\overrightarrow{AC}) \times (-\frac{1}{7}\overrightarrow{AB} + \frac{3}{7}\overrightarrow{AC})$
$= \frac{6}{49}(\overrightarrow{AB} \times \overrightarrow{AC}) - \frac{1}{49}(\overrightarrow{AC} \times \overrightarrow{AB})$
$= \frac{6}{49}(\overrightarrow{AB} \times \overrightarrow{AC}) + \frac{1}{49}(\overrightarrow{AB} \times \overrightarrow{AC})$
$= \frac{1}{7}(\overrightarrow{AB} \times \overrightarrow{AC}).$

Hence, the area of triangle HJG is one-seventh the area of triangle ABC.

2-7 / THE SCALAR TRIPLE PRODUCT

1. (a) 1; (b) -1; (c) 0.
3. 2 cubic units.
5. 6 cubic units.
7. $\vec{a}\cdot(\vec{b}\times\vec{c}) = (2\vec{i} + 3\vec{j} - 4\vec{k})\cdot(-3\vec{i} - \vec{j} + 2\vec{k}) = -17$; $(\vec{a}\times\vec{b})\cdot\vec{c}$
   $= (-\vec{i} - 6\vec{j} - 5\vec{k})\cdot(\vec{i} + \vec{j} + 2\vec{k}) = -17$.
9. No.

2-8 / THE VECTOR TRIPLE PRODUCT

1. (a) $\vec{0}$; (b) $\vec{j}$; (c) $\vec{j}$.
3. By (2-43), $\vec{a}\times(\vec{b}\times\vec{c}) = (\vec{a}\cdot\vec{c})\vec{b} - (\vec{a}\cdot\vec{b})\vec{c}$;
   $\vec{b}\times(\vec{c}\times\vec{a}) = (\vec{b}\cdot\vec{a})\vec{c} - (\vec{b}\cdot\vec{c})\vec{a}$;
   $\vec{c}\times(\vec{a}\times\vec{b}) = (\vec{c}\cdot\vec{b})\vec{a} - (\vec{c}\cdot\vec{a})\vec{b}$.
   Hence, $\vec{a}\times(\vec{b}\times\vec{c}) + \vec{b}\times(\vec{c}\times\vec{a}) + \vec{c}\times(\vec{a}\times\vec{b}) = \vec{0}$.

2-9 / QUADRUPLE PRODUCTS

1. By (2-46), $(\vec{a}\times\vec{b})\cdot(\vec{b}\times\vec{c})\times(\vec{c}\times\vec{a})$
   $= [\vec{a}\cdot(\vec{b}\times\vec{c})][\vec{b}\cdot(\vec{c}\times\vec{a})] - [\vec{a}\cdot(\vec{c}\times\vec{a})][\vec{b}\cdot(\vec{b}\times\vec{c})]$
   $= (\vec{a}\vec{b}\vec{c})(\vec{a}\vec{b}\vec{c}) - (0)(0) = (\vec{a}\vec{b}\vec{c})^2$.
3. $(\vec{a}\times\vec{b})\cdot(\vec{c}\times\vec{d})\times(\vec{e}\times\vec{f}) = (\vec{c}\times\vec{d})\cdot(\vec{e}\times\vec{f})\times(\vec{a}\times\vec{b})$. Then, by (2-46), $(\vec{c}\times\vec{d})\cdot(\vec{e}\times\vec{f})\times(\vec{a}\times\vec{b})$
   $= [\vec{c}\cdot(\vec{e}\times\vec{f})][\vec{d}\cdot(\vec{a}\times\vec{b})] - [\vec{c}\cdot(\vec{a}\times\vec{b})][\vec{d}\cdot(\vec{e}\times\vec{f})]$
   $= (\vec{c}\vec{e}\vec{f})(\vec{d}\vec{a}\vec{b}) - (\vec{c}\vec{a}\vec{b})(\vec{d}\vec{e}\vec{f})$
   $= (\vec{c}\vec{e}\vec{f})(\vec{a}\vec{b}\vec{d}) - (\vec{a}\vec{b}\vec{c})(\vec{d}\vec{e}\vec{f})$
   $= (\vec{a}\vec{b}\vec{d})(\vec{c}\vec{e}\vec{f}) - (\vec{a}\vec{b}\vec{c})(\vec{d}\vec{e}\vec{f})$.
5. $(\vec{a}\times\vec{b})\cdot(\vec{c}\times\vec{d}) = (4\vec{i} + 8\vec{j} + 4\vec{k})\cdot(-3\vec{i} + \vec{j} - 2\vec{k}) = -12$;
   $(\vec{a}\cdot\vec{c})(\vec{b}\cdot\vec{d}) - (\vec{a}\cdot\vec{d})(\vec{b}\cdot\vec{c}) = (-3)(4) - (0)(2) = -12$.

2-10 / QUATERNIONS

1. (a) (2, 5, 0, 2); (b) (7, 13, -3, 9); (c) (6, 3, 0, -9).
3. $\dfrac{a - bi - cj - dk}{a^2 + b^2 + c^2 + d^2}$.

5. If $a + bi \longleftrightarrow a + bi + 0j + 0k$ and
$c + di \longleftrightarrow c + di + 0j + 0k$, then $(a + bi) + (c + di)$
$= (a + c) + (b + d)i \longleftrightarrow (a + bi + 0j + 0k) + (c + di + 0j + 0k)$
$= (a + c) + (b + d)i + 0j + 0k$, and $(a + bi)(c + di)$
$= (ac - bd) + (ad + bc)i \longleftrightarrow (a + bi + 0j + 0k)(c + di + 0j + 0k)$
$= (ac - bd) + (ad + bc)i + 0j + 0k$.

7. Let $x = a + bi + cj + dk$ and $y = e + fi + gj + hk$. Then
$xy = (ae - bf - cg - dh) + (af + be + ch - dg)i$
$\quad + (ag + ce + df - bh)j + (ah + de + bg - cf)k;$
$(xy)^* = (ae - bf - cg - dh) - (af + be + ch - dg)i$
$\quad - (ag + ce + df - bh)j - (ah + de + bg - cf)k;$
$y^* = e - fi - gj - hk;\quad x^* = a - bi - cj - dk;$
$y^*x^* = (ae - bf - cg - dh) - (af + be + ch - dg)i$
$\quad -(ag + ce + df - bh)j - (ah + de + bg - cf)k;$
$(xy)^* = y^*x^*$.

2-11 / REAL VECTOR SPACES

1. By the definition of the sum of two vectors stated in §2-10, the set V is closed under the addition of vectors (Property 1). The addition of vectors is commutative (Property 2) and associative (Property 3) since the addition of real numbers is commutative and associative. Property 4 is satisfied because $(0, 0, \ldots, 0)$ is the additive identity element for the set V. Since for each $(a_1, a_2, \ldots, a_n) \in V$ there exists a vector $(-a_1, -a_2, \ldots, -a_n)$ such that

$$(a_1, a_2, \ldots, a_n) + (-a_1, -a_2, \ldots, -a_n) = (0, 0, \ldots, 0),$$

Property 5 is satisfied. Let $(a_1, a_2, \ldots, a_n)$ and $(b_1, b_2, \ldots, b_n)$ be any two vectors of V; let m and n be any two real numbers. Then, by the definition of the multiplication of a vector by a real number stated in §2-10 and the properties of real numbers,

$1(a_1, a_2, \ldots, a_n)$
$\quad = (1 \cdot a_1, 1 \cdot a_2, \ldots, 1 \cdot a_n)$
$\quad = (a_1, a_2, \ldots, a_n) \quad$ (Property 6);

$(mn)(a_1, a_2, \ldots, a_n)$
$\quad = ((mn)a_1, (mn)a_2, \ldots, (mn)a_n)$
$\quad = (m(na_1), m(na_2), \ldots, m(na_n))$

$$= m(na_1, na_2, \ldots, na_n)$$
$$= m[n(a_1, a_2, \ldots, a_n)] \quad \text{(Property 7)};$$

$$m(a_1 + b_1, a_2 + b_2, \ldots, a_n + b_n)$$
$$= (m(a_1 + b_1), m(a_2 + b_2), \ldots, m(a_n + b_n))$$
$$= (ma_1 + mb_1, ma_2 + mb_2, \ldots, ma_n + mb_n)$$
$$= (ma_1, ma_2, \ldots, ma_n) + (mb_1, mb_2, \ldots, mb_n)$$
$$= m(a_1, a_2, \ldots, a_n) + m(b_1, b_2, \ldots, b_n) \quad \text{(Property 8)};$$

$$(m + n)(a_1, a_2, \ldots, a_n)$$
$$= ((m + n)a_1, (m + n)a_2, \ldots, (m + n)a_n)$$
$$= (ma_1 + na_1, ma_2 + na_2, \ldots, ma_n + na_n)$$
$$= (ma_1, ma_2, \ldots, ma_n) + (na_1, na_2, \ldots, na_n)$$
$$= m(a_1, a_2, \ldots, a_n) + n(a_1, a_2, \ldots, a_n) \quad \text{(Property 9)}.$$

Therefore, the set V of ordered n-tuples of real numbers is a real vector space.

3. By the definition of the sum of two quaternions stated in §2-10, the set Q is closed under the addition of quaternions (Property 1). The addition of quaternions is commutative (Property 2) and associative (Property 3) since the addition of real numbers is commutative and associative. Property 4 is satisfied because $(0, 0, 0, 0)$ is the additive identity element for the set Q. Since for each $(a, b, c, d) \in Q$ there exists a quaternion $(-a, -b, -c, -d)$ such that

$$(a, b, c, d) + (-a, -b, -c, -d) = (0, 0, 0, 0),$$

Property 5 is satisfied. Let (a, b, c, d) and (e, f, g, h) be any two quaternions of Q; let m and n be any two real numbers. Then, by the definition of the multiplication of a quaternion by a real number stated in §2-10 and the properties of real numbers,

$$1(a, b, c, d) = (1 \cdot a, 1 \cdot b, 1 \cdot c, 1 \cdot d)$$
$$= (a, b, c, d) \quad \text{(Property 6)};$$
$$(mn)(a, b, c, d) = ((mn)a, (mn)b, (mn)c, (mn)d)$$
$$= (m(na), m(nb), m(nc), m(nd))$$
$$= m(na, nb, nc, nd)$$
$$= m[n(a, b, c, d)] \quad \text{(Property 7)};$$

$$m(a + e, b + f, c + g, d + h)$$
$$= (m(a + e), m(b + f), m(c + g), m(d + h))$$
$$= (ma + me, mb + mf, mc + mg, md + mh)$$
$$= (ma, mb, mc, md) + (me, mf, mg, mh)$$
$$= m(a, b, c, d) + m(e, f, g, h) \quad \text{(Property 8)};$$

$$(m + n)(a, b, c, d) = ((m + n)a, (m + n)b, (m + n)c, (m + n)d)$$

$$= (ma + na, mb + nb, mc + nc, md + nd)$$
$$= (ma, mb, mc, md) + (na, nb, nc, nd)$$
$$= m(a, b, c, d) + n(a, b, c, d) \quad \text{(Property 9).}$$

Therefore, the set Q of quaternions is a real vector space.

5. Let p, q, and w be any three polynomial functions of P such that

$$p(x) = a_0 + a_1 x + a_2 x^2 + \cdots + a_r x^r,$$
$$q(x) = b_0 + b_1 x + b_2 x^2 + \cdots + b_s x^s,$$

and

$$w(x) = c_0 + c_1 x + c_2 x^2 + \cdots + c_t x^t.$$

There is no loss of generality in assuming $r \geq s \geq t$. By the definition of the sum of two functions stated in Example 4 and the properties of real numbers,

$$(p + q)x = p(x) + q(x)$$
$$= a_0 + a_1 x + a_2 x^2 + \cdots + a_r x^r + b_0 + b_1 x$$
$$\quad + b_2 x^2 + \cdots + b_s x^s$$
$$= (a_0 + b_0) + (a_1 + b_1)x + (a_2 + b_2)x^2 + \cdots$$
$$\quad + (a_s + b_s)x^s + a_{s+1} x^{s+1} + \cdots + a_r x^r;$$

that is, $p + q \in P$ and the set P is closed under the addition of functions (Property 1). Note that if $r = s$, then $a_{s+1} = a_{s+2} = \cdots = a_r = 0$. Since the addition of real numbers is commutative and associative, then

$$p(x) + q(x) = q(x) + p(x)$$

and

$$[p(x) + q(x)] + w(x) = p(x) + [q(x) + w(x)];$$

that is, the addition of polynomial functions is commutative and associative (Properties 2 and 3):

$$p + q = q + p$$

and

$$(p + q) + w = p + (q + w).$$

Property 4 is satisfied because the polynomial function 0 is the additive identity element for the set P. Since for each $p \in F$ there exists a polynomial function $-p$, where $(-p)(x) = -p(x) = -a_0 - a_1 x - a_2 x^2 - \cdots - a_r x^r$, such that $p + (-p) = 0$, Property 5 is satisfied. Let m and n be any two real numbers. If mp is defined as that function of P represented by

$$(mp)(x) = ma_0 + ma_1x + ma_2x^2 + \cdots + ma_rx^r,$$

then using the properties of real numbers, we obtain

$$1(a_0 + a_1x + a_2x^2 + \cdots + a_rx^r) = 1 \cdot a_0 + 1 \cdot a_1x + 1 \cdot a_2x^2 + \cdots + 1 \cdot a_rx^r$$
$$= a_0 + a_1x + a_2x^2 + \cdots + a_rx^r,$$

that is, $1 \cdot p = p$ (Property 6);

$$(mn)(a_0 + a_1x + a_2x^2 + \cdots + a_rx^r)$$
$$= (mn)a_0 + (mn)a_1x + (mn)a_2x^2 + \cdots + (mn)a_rx^r$$
$$= m(na_0) + m(na_1)x + m(na_2)x^2 + \cdots + m(na_r)x^r$$
$$= m(na_0 + na_1x + na_2x^2 + \cdots + na_rx^r)$$
$$= m[n(a_0 + a_1x + a_2x^2 + \cdots + a_rx^r)],$$

that is, $(mn)p = m(np)$ (Property 7);

$$m[(a_0 + b_0) + (a_1 + b_1)x + (a_2 + b_2)x^2 + \cdots + (a_s + b_s)x^s$$
$$+ a_{s+1}x^{s+1} + \cdots + a_rx^r]$$
$$= m(a_0 + b_0) + m(a_1 + b_1)x + m(a_2 + b_2)x^2 + \cdots$$
$$+ m(a_s + b_s)x^s + ma_{s+1}x^{s+1} + \cdots + ma_rx^r$$
$$= (ma_0 + mb_0) + (ma_1 + mb_1)x + (ma_2 + mb_2)x^2 + \cdots$$
$$+ (ma_s + mb_s)x^s + ma_{s+1}x^{s+1} + \cdots + ma_rx^r$$
$$= (ma_0 + ma_1x + ma_2x^2 + \cdots + ma_rx^r)$$
$$+ (mb_0 + mb_1x + mb_2x^2 + \cdots + mb_sx^s)$$
$$= m(a_0 + a_1x + a_2x^2 + \cdots + a_rx^r)$$
$$+ m(b_0 + b_1x + b_2x^2 + \cdots + b_sx^s),$$

that is, $m(p + q) = mp + mq$ (Property 8);

$$(m + n)(a_0 + a_1x + a_2x^2 + \cdots + a_rx^r)$$
$$= (m + n)a_0 + (m + n)a_1x + (m + n)a_2x^2$$
$$+ \cdots + (m + n)a_rx^r$$
$$= (ma_0 + na_0) + (ma_1 + na_1)x$$
$$+ (ma_2 + na_2)x^2 + \cdots + (ma_r + na_r)x^r$$
$$= (ma_0 + ma_1x + ma_2x^2 + \cdots + ma_rx^r)$$
$$+ (na_0 + na_1x + na_2x^2 + \cdots + na_rx^r)$$
$$= m(a_0 + a_1x + a_2x^2 + \cdots + a_rx^r)$$
$$+ n(a_0 + a_1x + a_2x^2 + \cdots + a_rx^r),$$

that is, $(m + n)p = mp + np$ (Property 9).

Therefore, the set P of real polynomial functions over R is a real vector space.

7. A real vector space.

9. Since $a = a + 0$, then
$$ma = m(a + 0) = ma + m0$$
for each $m \in R$ and $a \in V$. Since $ma \in V$ and $ma = ma + 0$, then $m0 = 0$.

11. Since $0 = m0 = m[a + (-a)] = ma + m(-a)$ for each $m \in R$ and $a \in V$, then $m(-a)$ must be the additive inverse of ma; that is,
$$-(ma) = m(-a).$$
Since $m(-a) = m[(-1)a] = [m(-1)]a = (-m)a$, it follows that
$$-(ma) = (-m)a = m(-a).$$

13. Not a finite-dimensional vector space.

15. Let $a + bi$ be any complex number. Consider
$$a + bi = r(1 + i) + s(3 - 5i),$$
where r and s are real numbers. Then
$$a + bi = (r + 3s) + (r - 5s)i,$$
whereby
$$\begin{cases} a = r + 3s \\ b = r - 5s. \end{cases}$$
Hence,
$$r = \frac{5a + 3b}{8} \quad \text{and} \quad s = \frac{a - b}{8}.$$
Since the real numbers r and s exist, then any complex number $a + bi$ can be expressed as a linear function of $1 + i$ and $3 - 5i$. Therefore, $1 + i$ and $3 - 5i$ form a basis for the real vector space C.

17. (a) A subspace of V; (b) a subspace of V; (c) not a subspace of V; (d) not a subspace of V; (e) a subspace of V; (f) a subspace of V.

THREE

PLANES AND LINES
IN SPACE

3-1 / DIRECTION COSINES AND NUMBERS

1. $(\frac{3}{13} : \frac{12}{13} : -\frac{4}{13})$.
3. $(3k : 12k : -4k)$, where k is any real number such that $k > 0$.
5. (b).
7. (a) 45° or 135°; (b) 45°, 135°, 225°, or 315°.
9. $\left(\frac{\sqrt{10}}{10} : 0 : -\frac{3\sqrt{10}}{10}\right)$ or $\left(-\frac{\sqrt{10}}{10} : 0 : \frac{3\sqrt{10}}{10}\right)$.
11. Any three points of the form $(2 + 3k, 1 - k, -4 + k)$, where k is any real number such that $k \neq 0$.
13. 60°.
15. $\frac{\sqrt{3}}{3}$.

3-2 / EQUATION OF A PLANE

1. $3x - 2y + z - 14 = 0$.
3. $3x + 6y + z = 0$.
5. $x = 1 - 4m + 4n$; $y = 2 - n$; $z = 4m - n$.
7. (a) $2b + d = 0$ where $bd \neq 0$; (b) $b = c$ where $bc \neq 0$; (c) $a = b = c$ where $abc \neq 0$; (d) $a = 2t$, $b = -t$, and $c = t$ for some real number $t \neq 0$; (e) $a = t$, $b = 3t$, and $c = 5t$ for some real number $t \neq 0$; (f) $d = 0$; (g) $2a - b + 4c + d = 0$; (h) $b = 0$; (i) $b = c = 0$; (j) $c = d = 0$.
9. $2x + 3y - 7z + 28 = 0$.
11. $\overrightarrow{OP} = \overrightarrow{OC} + m\overrightarrow{OA} + n\overrightarrow{OB}$ or $\overrightarrow{CP} \cdot \overrightarrow{OA} \times \overrightarrow{OB} = 0$.

3-3 / EQUATION OF A SPHERE

1. $x^2 + y^2 + z^2 - 2x + 4y - 8z + 12 = 0$.
3. $x + y + 2z - 18 = 0$.   5. $x - 2 = 0$.

7. $x^2 + y^2 + z^2 - 4z - 1 = 0$.   **9.** $x - 1 = 0$.

3-4 / ANGLE BETWEEN TWO PLANES

1. 45° and 135°.

3. The planes are parallel since the nonzero vectors perpendicular to the planes are parallel: $(3\vec{i} - 2\vec{j} + \vec{k}) \times (9\vec{i} - 6\vec{j} + 3\vec{k}) = \vec{0}$. An alternative proof may be given by showing that the same direction numbers $(3: -2: 1)$ may be used for vectors perpendicular to each plane.

5. 45°.

3-5 / DISTANCE BETWEEN A POINT AND A PLANE

1. $2\sqrt{6}$ units.   **3.** 3 or -15.

5. 2 units.   **7.** -2.

9. $x^2 + y^2 + z^2 - 4x - 4y - 2z + 5 = 0$.

3-6 / EQUATION OF A LINE

1. $x = 4 + 2m$, $y = 5 - 3m$, $z = 2 + m$.

3. $\dfrac{x-1}{3} = \dfrac{y-2}{1} = \dfrac{z}{-2}$.

5. $\dfrac{x}{3} = \dfrac{y}{5} = \dfrac{z}{-1}$.

7. $x = 4 - 3m$, $y = 1$, $z = 3 - m$.

9. $\dfrac{x-3}{3} = \dfrac{y+1}{5} = \dfrac{z-4}{2}$.

11. $(\tfrac{2}{3}: -\tfrac{2}{3}: \tfrac{1}{3})$ or $(-\tfrac{2}{3}: \tfrac{2}{3}: -\tfrac{1}{3})$.

13. $\left(0: \dfrac{\sqrt{2}}{2}: \dfrac{\sqrt{2}}{2}\right)$ or $\left(0: -\dfrac{\sqrt{2}}{2}: -\dfrac{\sqrt{2}}{2}\right)$.

15. $(\tfrac{46}{5}, -\tfrac{13}{5}, 0)$.

17. $[(3\vec{i} - \vec{j}) \times (4\vec{i} - \vec{k})] \cdot [(\vec{j} + \vec{k}) \times (\vec{i} - \vec{j})]$
$= (\vec{i} + 3\vec{j} + 4\vec{k}) \cdot (\vec{i} + \vec{j} - \vec{k}) = 0$.

19. $5x - 2y + 11z = 0$.

FOUR

MATRICES

4-1 / DEFINITIONS AND ELEMENTARY PROPERTIES

1.
$$\begin{pmatrix} 0 & 2 & 4 \\ 3 & 5 & 7 \\ 8 & 10 & 12 \end{pmatrix}.$$

3. (a) The elements are in row two; (b) the elements are in column one; (c) the elements are those in the upper left-hand corner and the lower right-hand corner.

5.
$$(A + B) + C = \begin{pmatrix} 4 & -2 & 1 \\ -2 & 6 & 4 \end{pmatrix} + \begin{pmatrix} 2 & 7 & -1 \\ -2 & 1 & 3 \end{pmatrix} = \begin{pmatrix} 6 & 5 & 0 \\ -4 & 7 & 7 \end{pmatrix};$$

$$A + (B + C) = \begin{pmatrix} 3 & 1 & 1 \\ -2 & 5 & 0 \end{pmatrix} + \begin{pmatrix} 3 & 4 & -1 \\ -2 & 2 & 7 \end{pmatrix} = \begin{pmatrix} 6 & 5 & 0 \\ -4 & 7 & 7 \end{pmatrix};$$

$$(A + B) + C = A + (B + C).$$

7.
$$\begin{pmatrix} a_{11} & a_{12} \\ a_{21} & a_{22} \end{pmatrix} = \begin{pmatrix} -6 & 1 \\ 5 & 0 \end{pmatrix}.$$

9. $\begin{pmatrix} 1 & 0 \\ 0 & 0 \end{pmatrix}, \begin{pmatrix} 0 & 1 \\ 0 & 0 \end{pmatrix}, \begin{pmatrix} 0 & 0 \\ 1 & 0 \end{pmatrix},$ and $\begin{pmatrix} 0 & 0 \\ 0 & 1 \end{pmatrix}.$

Other bases exist.

4-2 / MATRIX MULTIPLICATION

1. $AB = \begin{pmatrix} 18 & 1 & 26 \\ -8 & 7 & 6 \end{pmatrix};$ BA does not exist.

3.
$$A(BC) = \begin{pmatrix} 2 & 1 \\ -1 & 0 \end{pmatrix} \begin{pmatrix} 2 & 4 \\ -4 & 0 \end{pmatrix} = \begin{pmatrix} 0 & 8 \\ -2 & -4 \end{pmatrix};$$

$$(AB)C = \begin{pmatrix} 8 & 0 \\ -3 & -1 \end{pmatrix} \begin{pmatrix} 0 & 1 \\ 2 & 1 \end{pmatrix} = \begin{pmatrix} 0 & 8 \\ -2 & -4 \end{pmatrix};$$

$$A(BC) = (AB)C.$$

5.
$$\begin{pmatrix} 1 & 0 \\ 0 & 1 \end{pmatrix}.$$

7.
$$(A+B)(A-B) = \begin{pmatrix} 3 & 4 \\ -3 & 2 \end{pmatrix}\begin{pmatrix} -1 & 2 \\ -1 & -2 \end{pmatrix} = \begin{pmatrix} -7 & -2 \\ 1 & -10 \end{pmatrix};$$

$$A^2 - B^2 = \begin{pmatrix} -5 & 3 \\ -2 & -6 \end{pmatrix} - \begin{pmatrix} 3 & 4 \\ -4 & 3 \end{pmatrix} = \begin{pmatrix} -8 & -1 \\ 2 & -9 \end{pmatrix};$$

$(A+B)(A-B) \neq A^2 - B^2$.

9. (a) No matrices exist;

(b) $\begin{pmatrix} r & s \\ 3s & r \end{pmatrix},$

where r and s are arbitrary scalars.

11. $\begin{pmatrix} 20 & 21 \\ 7 & 6 \end{pmatrix}.$

13.

$C(AB + BA) = C(AB) + C(BA)$	distributive property
$= (CA)B + (CB)A$	associative property of multiplication
$= (AC)B + (BC)A$	$AC = CA$ and $BC = CB$
$= A(CB) + B(CA)$	associative property of multiplication
$= A(BC) + B(AC)$	$AC = CA$ and $BC = CB$
$= (AB)C + (BA)C$	associative property of multiplication
$= (AB + BA)C$	distributive property.

4-3 / DIAGONAL MATRICES

1. $\begin{pmatrix} 4 & 3 & 3 & 3 \\ 6 & 10 & 6 & 6 \\ 9 & 9 & 18 & 9 \end{pmatrix}.$

3. $\begin{pmatrix} -1 & -1 & -1 \\ 0 & 1 & 0 \\ 0 & 0 & 1 \end{pmatrix}\begin{pmatrix} -1 & -1 & -1 \\ 0 & 1 & 0 \\ 0 & 0 & 1 \end{pmatrix}$

$= \begin{pmatrix} 1+0+0 & 1-1+0 & 1+0-1 \\ 0+0+0 & 0+1+0 & 0+0+0 \\ 0+0+0 & 0+0+0 & 0+0+1 \end{pmatrix} = \begin{pmatrix} 1 & 0 & 0 \\ 0 & 1 & 0 \\ 0 & 0 & 1 \end{pmatrix} = I.$

4-4 / SPECIAL REAL MATRICES

1. (a) C, D, and J;   (b) F and J.

3. $n^2 - n + 1$.

5.
$$EE^T = \begin{pmatrix} 5 & 2 & 1 \\ 4 & 2 & 4 \\ 1 & 2 & 3 \end{pmatrix} \begin{pmatrix} 5 & 4 & 1 \\ 2 & 2 & 2 \\ 1 & 4 & 3 \end{pmatrix} = \begin{pmatrix} 30 & 28 & 12 \\ 28 & 36 & 20 \\ 12 & 20 & 14 \end{pmatrix} = (EE^T)^T.$$

7.
$$E - E^T = \begin{pmatrix} 5 & 2 & 1 \\ 4 & 2 & 4 \\ 1 & 2 & 3 \end{pmatrix} - \begin{pmatrix} 5 & 4 & 1 \\ 2 & 2 & 2 \\ 1 & 4 & 3 \end{pmatrix} = \begin{pmatrix} 0 & -2 & 0 \\ 2 & 0 & 2 \\ 0 & -2 & 0 \end{pmatrix};$$

$$(E - E^T)^T = \begin{pmatrix} 0 & 2 & 0 \\ -2 & 0 & -2 \\ 0 & 2 & 0 \end{pmatrix}; (E - E^T)^T = -(E - E^T).$$

9. Let A be any skew-symmetric matrix. Then $A = -A^T$ and $A^2 = (-A^T)(-A^T) = A^T A^T = (AA)^T = (A^2)^T$. Hence, A^2 is a symmetric matrix.

11. Let A and B be skew-symmetric matrices of the same order such that AB is a symmetric matrix. Then $AB = (AB)^T = B^T A^T = (-B)(-A) = BA$. To prove the converse, let A and B be skew-symmetric matrices of the same order such that $AB = BA$. Then $AB = BA = (-B^T)(-A^T) = B^T A^T = (AB)^T$; that is, AB is a symmetric matrix.

4-5 / SPECIAL COMPLEX MATRICES

1. (a) $\begin{pmatrix} 1 & 3 - 2i \\ i & 2 + i \end{pmatrix}$;   (b) $\begin{pmatrix} 1 & i \\ 3 - 2i & 2 + i \end{pmatrix}$.

3. Let $A = ((a_{ij} + b_{ij}i))$ be a skew-Hermitian matrix. Then $A = -A^* = -(\bar{A})^T$. Since the diagonal elements of A and $-(\bar{A})^T$ are equal, $a_{ii} + b_{ii}i = -(a_{ii} - b_{ii}i)$; $a_{ii} = -a_{ii}$; and $a_{ii} = 0$. Therefore, the diagonal elements of A are of the form $b_{ii}i$. Hence, the diagonal elements of a skew-Hermitian matrix are either zeros or pure imaginary numbers.

5. $(A + B)^* = \overline{(A + B)}^T = (\bar{A} + \bar{B})^T = (\bar{A})^T + (\bar{B})^T = A^* + B^*$.

7. $(AA^*)^* = (A^*)^* A^* = AA^*$.

9. $(A - A^*)^* = A^* - (A^*)^* = A^* - A; A - A^* = -(A - A^*)^*$.

11. Let $((h_{ij}))$ be any Hermitian matrix where $h_{ij} = a_{ij} + b_{ij}i$. Since $((h_{ij})) = ((\overline{h_{ij}}))^T$, then $a_{ij} + b_{ij}i = a_{ji} - b_{ji}i$ for all pairs (i, j); $a_{ij} = a_{ji}$ and

$b_{ij} = -b_{ji}$. Hence, $((h_{ij})) = ((a_{ij})) + ((b_{ij}))i$ where $((a_{ij}))$ is a real symmetric matrix and $((b_{ij}))$ is a real skew-symmetric matrix.

FIVE

INVERSES AND
SYSTEMS OF MATRICES

5-1 / DETERMINANTS

1. 14. **3.** -2. **5.** 27.

7. By Theorem 5.5, the elements of one of the two identical rows can be made zero by adding to each element the product of (-1) and the corresponding element of the other row. Then, by Theorem 5.3, the value of the determinant is zero.

9. $k^4 m$.

11.
$$AB = \begin{pmatrix} 7 & 36 \\ 7 & 39 \end{pmatrix};$$
det $AB = 21 = (-7)(-3) = $ det A det B.

13.
$$\begin{vmatrix} 5-\lambda & 1 \\ 2 & 3-\lambda \end{vmatrix} = \lambda^2 - 8\lambda + 13 = 0;$$

$$\begin{pmatrix} 5 & 1 \\ 2 & 3 \end{pmatrix}\begin{pmatrix} 5 & 1 \\ 2 & 3 \end{pmatrix} - 8\begin{pmatrix} 5 & 1 \\ 2 & 3 \end{pmatrix} + 13\begin{pmatrix} 1 & 0 \\ 0 & 1 \end{pmatrix}$$

$$= \begin{pmatrix} 27 & 8 \\ 16 & 11 \end{pmatrix} + \begin{pmatrix} -40 & -8 \\ -16 & -24 \end{pmatrix} + \begin{pmatrix} 13 & 0 \\ 0 & 13 \end{pmatrix} = \begin{pmatrix} 0 & 0 \\ 0 & 0 \end{pmatrix}.$$

15.
$$\begin{vmatrix} a+b & a & a \\ a & a+b & a \\ a & a & a+b \end{vmatrix} = \begin{vmatrix} b & 0 & -b \\ a & a+b & a \\ a & a & a+b \end{vmatrix} = \begin{vmatrix} b & 0 & 0 \\ a & a+b & 2a \\ a & a & 2a+b \end{vmatrix}$$

$$= b\begin{vmatrix} a+b & 2a \\ a & 2a+b \end{vmatrix} = b(3ab + b^2) = b^2(3a + b).$$

17. $\vec{a} \times \vec{b} = (y_1 z_2 - y_2 z_1)\vec{i} + (z_1 x_2 - z_2 x_1)\vec{j} + (x_1 y_2 - x_2 y_1)\vec{k}$
$= \vec{i}(y_1 z_2 - y_2 z_1) - \vec{j}(x_1 z_2 - x_2 z_1) + \vec{k}(x_1 y_2 - x_2 y_1)$
$= \vec{i}\begin{vmatrix} y_1 & z_1 \\ y_2 & z_2 \end{vmatrix} - \vec{j}\begin{vmatrix} x_1 & z_1 \\ x_2 & z_2 \end{vmatrix} + \vec{k}\begin{vmatrix} x_1 & y_1 \\ x_2 & y_2 \end{vmatrix}$
$= \begin{vmatrix} \vec{i} & \vec{j} & \vec{k} \\ x_1 & y_1 & z_1 \\ x_2 & y_2 & z_2 \end{vmatrix}.$

19. $\vec{a} \cdot (\vec{b} \times \vec{c}) = x_1(y_2 z_3 - y_3 z_2) + y_1(z_2 x_3 - z_3 x_2) + z_1(x_2 y_3 - x_3 y_2)$
$= x_1(y_2 z_3 - y_3 z_2) - y_1(x_2 z_3 - x_3 z_2) + z_1(x_2 y_3 - x_3 y_2)$
$= x_1 \begin{vmatrix} y_2 & z_2 \\ y_3 & z_3 \end{vmatrix} - y_1 \begin{vmatrix} x_2 & z_2 \\ x_3 & z_3 \end{vmatrix} + z_1 \begin{vmatrix} x_2 & y_2 \\ x_3 & y_3 \end{vmatrix}$
$= \begin{vmatrix} x_1 & y_1 & z_1 \\ x_2 & y_2 & z_2 \\ x_3 & y_3 & z_3 \end{vmatrix}.$

21. $\begin{vmatrix} x - x_1 & y - y_1 & z - z_1 \\ x_2 - x_1 & y_2 - y_1 & z_2 - z_1 \\ x_3 - x_1 & y_3 - y_1 & z_3 - z_1 \end{vmatrix} = 0.$

5-2 / INVERSE OF A MATRIX

1. $\begin{pmatrix} 2 & -1 \\ -\frac{5}{2} & \frac{3}{2} \end{pmatrix}.$

3. $\begin{pmatrix} \cos\theta & \sin\theta \\ -\sin\theta & \cos\theta \end{pmatrix}.$

5. $\begin{pmatrix} 1 & -3 & 2 \\ -3 & 3 & -1 \\ 2 & -1 & 0 \end{pmatrix}.$

7. $\begin{pmatrix} x \\ y \end{pmatrix} = \begin{pmatrix} \frac{4}{5} & -\frac{1}{5} \\ -\frac{3}{5} & \frac{2}{5} \end{pmatrix}\begin{pmatrix} 4 \\ 1 \end{pmatrix} = \begin{pmatrix} 3 \\ -2 \end{pmatrix};$
that is, $x = 3$ and $y = -2.$

9. A left multiplicative inverse is of the form
$$\begin{pmatrix} 4 & -1 & r \\ -3 & 1 & s \end{pmatrix},$$
where r and s are arbitrary scalars. A right multiplicative inverse does not exist since
$$\begin{pmatrix} 1 & 1 \\ 3 & 4 \\ 0 & 0 \end{pmatrix}\begin{pmatrix} a & b & c \\ d & e & f \end{pmatrix} = \begin{pmatrix} a+d & b+e & c+f \\ 3a+4d & 3b+4e & 3c+4f \\ 0 & 0 & 0 \end{pmatrix} \neq \begin{pmatrix} 1 & 0 & 0 \\ 0 & 1 & 0 \\ 0 & 0 & 1 \end{pmatrix}$$
for any values of $a, b, c, d, e,$ and f.

11. Assume A is a nonsingular matrix. Then A^{-1} exists and $A^{-1}(AB) = A^{-1}0$; $(A^{-1}A)B = 0$; $IB = 0$; $B = 0$, which is contrary to the hypothesis. Hence, A is a singular matrix.

13.
$$AB = \begin{pmatrix} 4 & 4 \\ 3 & 4 \end{pmatrix}; \quad (AB)^{-1} = \begin{pmatrix} 1 & -1 \\ -\frac{3}{4} & 1 \end{pmatrix};$$

$$B^{-1}A^{-1} = \begin{pmatrix} 1 & 0 \\ \frac{1}{2} & \frac{1}{2} \end{pmatrix}\begin{pmatrix} 1 & -1 \\ -\frac{5}{2} & 3 \end{pmatrix} = \begin{pmatrix} 1 & -1 \\ -\frac{3}{4} & 1 \end{pmatrix}; \quad (AB)^{-1} = B^{-1}A^{-1}.$$

15. Let A be any nonsingular symmetric matrix. Then $A^{-1}A = AA^{-1} = I = (AA^{-1})^T = (A^{-1})^TA^T = (A^{-1})^TA$. Hence, $A^{-1} = (A^{-1})^T$; that is, A^{-1} is a symmetric matrix.

17. Let A be any nonsingular matrix. Then $AA^{-1} = I = (AA^{-1})^T = (A^{-1})^TA^T$. Hence, $(A^T)^{-1} = (A^{-1})^T$.

5-3 / SYSTEMS OF MATRICES

1. Not a ring. **3.** A ring. **5.** A ring.

7. $\begin{pmatrix} 5 & -3 \\ 3 & 5 \end{pmatrix} + \begin{pmatrix} 2 & -4 \\ 4 & 2 \end{pmatrix} = \begin{pmatrix} 7 & -7 \\ 7 & 7 \end{pmatrix} \longleftrightarrow 7 - 7i = (5 - 3i) + (2 - 4i)$.

9. $\begin{pmatrix} 0 & 1 \\ -1 & 0 \end{pmatrix}\begin{pmatrix} 0 & 1 \\ -1 & 0 \end{pmatrix} = \begin{pmatrix} -1 & 0 \\ 0 & -1 \end{pmatrix} \longleftrightarrow -1 = (i) \times (i)$.

11. $A \odot B = AB - BA = -(BA - AB) = -B \odot A$.

13. $\begin{pmatrix} 5i & 1+i \\ -1+i & -5i \end{pmatrix}\begin{pmatrix} 3+i & 2+4i \\ -2+4i & 3-i \end{pmatrix} = \begin{pmatrix} -11+17i & -16+12i \\ 16+12i & -11-17i \end{pmatrix} \longleftrightarrow$
$-11 + 17i - 16j + 12k = (5i + j + k) \times (3 + i + 2j + 4k)$.

5-4 / LINEAR ALGEBRAS

1. Under the operations of addition and multiplication by a real number, the set R is a real vector space (Exercise 6 of §2-11). Hence, Property 1 of a linear algebra is satisfied. Furthermore, the product of two real numbers is a real number; the multiplication of real numbers is associative; the multiplication of real numbers is distributive with respect to the addition of real numbers. Hence, the set R is a ring, and Property 2 of a linear algebra is satisfied. Since the multiplication of real numbers is associative, Property 3 of a linear algebra is satisfied. Therefore, the set R of real numbers is a real linear algebra.

3. Under the definitions of equality, addition, and multiplication by a real number stated in §2-10, the set Q of quaternions is a real vector space (Exercise 3 of §2-11). Hence, Property 1 of a linear algebra is satisfied. Furthermore, the product of two quaternions is a quaternion by definition (§2-10); the multiplication of quaternions is associative and distributive with respect to addition since the set of quaternions is isomorphic to a set of square matrices and the multiplication of square matrices is associative and distributive with respect to addition. Hence, the set Q is a ring, and Property 2 of a linear algebra is satisfied. Now, let (a, b, c, d), $(e, f, g, h) \in Q$ and $m \in R$. Then, by the definitions of multiplication, multiplication by a real number, and the properties of real numbers,

$[m(a, b, c, d)] \times (e, f, g, h) = (ma, mb, mc, md) \times (e, f, g, h)$
$= ((ma)e - (mb)f - (mc)g - (md)h, (ma)f + (mb)e + (mc)h - (md)g,$
$\quad (ma)g + (mc)e + (md)f - (mb)h, (ma)h + (md)e + (mb)g - (mc)f)$
$= (a(me) - b(mf) - c(mg) - d(mh), a(mf) + b(me) + c(mh) - d(mg),$
$\quad a(mg) + c(me) + d(mf) - b(mh), a(mh) + d(me) + b(mg) - c(mf))$
$= (a, b, c, d) \times (me, mf, mg, mh)$
$= (a, b, c, d) \times [m(e, f, g, h)];$

$[m(a, b, c, d)] \times (e, f, g, h) = (m(ae) - m(bf) - m(cg) - m(dh),$
$\quad m(af) + m(be) + m(ch) - m(dg), m(ag) + m(ce) + m(df) - m(bh),$
$\quad m(ah) + m(de) + m(bg) - m(cf))$
$= (m(ae - bf - cg - dh), m(af + be + ch - dg),$
$\quad m(ag + ce + df - bh), m(ah + de + bg - cf))$
$= m(ae - bf - cg - dh, af + be + ch - dg,$
$\quad ag + ce + df - bh, ah + de + bg - cf)$
$= m[(a, b, c, d) \times (e, f, g, h)].$

Hence, Property 3 of a linear algebra is satisfied. Therefore, the set Q of quaternions is a real linear algebra.

5. A real linear algebra.

7. Not a real linear algebra.

5-5 / RANK OF A MATRIX

1. $t(2x + z) - 2t(x + y) + t(2y - z) = 0$ for any nonzero scalar t.

3. Three.   **5.** Two.   **7.** $\dfrac{n(n + 1)}{2}$.

9. $\begin{pmatrix} 1 & 0 \\ 0 & 3 \end{pmatrix} \begin{pmatrix} 1 & -1 \\ 0 & 1 \end{pmatrix} \begin{pmatrix} 1 & 0 \\ -2 & 1 \end{pmatrix} \begin{pmatrix} \frac{1}{6} & 0 \\ 0 & 1 \end{pmatrix} = \begin{pmatrix} \frac{1}{2} & -1 \\ -1 & 3 \end{pmatrix}.$

Other elementary row transformation matrices exist whose product is

$$\begin{pmatrix} \frac{1}{2} & -1 \\ -1 & 3 \end{pmatrix}.$$

11. $\begin{pmatrix} 1 & 0 & 2 \\ 0 & 1 & 0 \\ 0 & 0 & 1 \end{pmatrix} \begin{pmatrix} 1 & 0 & 0 \\ 0 & 1 & 1 \\ 0 & 0 & 1 \end{pmatrix} \begin{pmatrix} 1 & 0 & 0 \\ 0 & 1 & 0 \\ 0 & 0 & \frac{1}{2} \end{pmatrix} \begin{pmatrix} 1 & 0 & 0 \\ 0 & 1 & 0 \\ 0 & 1 & 1 \end{pmatrix} \begin{pmatrix} 1 & 2 & 0 \\ 0 & 1 & 0 \\ 0 & 0 & 1 \end{pmatrix}$

$\begin{pmatrix} 1 & 0 & 0 \\ 0 & -1 & 0 \\ 0 & 0 & 1 \end{pmatrix} \begin{pmatrix} 1 & 0 & 0 \\ 0 & 1 & 0 \\ 5 & 0 & 1 \end{pmatrix} \begin{pmatrix} 1 & 0 & 0 \\ 2 & 1 & 0 \\ 0 & 0 & 1 \end{pmatrix} = \begin{pmatrix} 0 & -3 & 1 \\ -\frac{1}{2} & -\frac{3}{2} & \frac{1}{2} \\ \frac{3}{2} & -\frac{1}{2} & \frac{1}{2} \end{pmatrix}.$

Other elementary row transformation matrices exist whose product is

$$\begin{pmatrix} 0 & -3 & 1 \\ -\frac{1}{2} & -\frac{3}{2} & \frac{1}{2} \\ \frac{3}{2} & -\frac{1}{2} & \frac{1}{2} \end{pmatrix}.$$

5-6 / SYSTEMS OF LINEAR EQUATIONS

1. Infinitely many solutions of the form $x = -\frac{15}{14}z - \frac{17}{7}$ and $y = \frac{5}{7}z - \frac{26}{7}$ exist.

3. $k = 18$.

5. A nontrivial solution does not exist.

SIX

TRANSFORMATIONS OF THE PLANE

6-1 / MAPPINGS

1. (a) T is a one-to-one mapping of R onto R; (b) T is not a mapping of R onto R; (c) T is a one-to-one mapping of R onto R; (d) T is a mapping of R onto R.

3. $n!$

5. A single-valued mapping T^{-1} does not exist such that $T^{-1}(b) = 1$ and $T^{-1}(b) = 3$.

6-2 / ROTATIONS

1. $(3, 2)$. 3. $(1, \sqrt{3})$. 5. $\left(-\dfrac{\sqrt{2}}{2}, \dfrac{3\sqrt{2}}{2}\right)$.

7. $3x^2 + y^2 = 32$. 9. $x^2 + y^2 = r^2$.

11. Let

$$\begin{pmatrix} \cos\theta & -\sin\theta \\ \sin\theta & \cos\theta \end{pmatrix} \quad \text{and} \quad \begin{pmatrix} \cos\phi & -\sin\phi \\ \sin\phi & \cos\phi \end{pmatrix}$$

be any two rotation matrices of the form (6-3).

(a) $\begin{pmatrix} \cos\theta & -\sin\theta \\ \sin\theta & \cos\theta \end{pmatrix}\begin{pmatrix} \cos\phi & -\sin\phi \\ \sin\phi & \cos\phi \end{pmatrix} = \begin{pmatrix} \cos(\theta+\phi) & -\sin(\theta+\phi) \\ \sin(\theta+\phi) & \cos(\theta+\phi) \end{pmatrix}$,

a rotation matrix of the form (6-3);

(b) $\begin{pmatrix} \cos(\theta+\phi) & -\sin(\theta+\phi) \\ \sin(\theta+\phi) & \cos(\theta+\phi) \end{pmatrix} = \begin{pmatrix} \cos(\phi+\theta) & -\sin(\phi+\theta) \\ \sin(\phi+\theta) & \cos(\phi+\theta) \end{pmatrix}$

$$= \begin{pmatrix} \cos\phi & -\sin\phi \\ \sin\phi & \cos\phi \end{pmatrix}\begin{pmatrix} \cos\theta & -\sin\theta \\ \sin\theta & \cos\theta \end{pmatrix}.$$

13. Let (x_1, y_1) and (x_2, y_2) be any two points on a coordinate plane. The distance d between these two points is equal to $\sqrt{(x_2 - x_1)^2 + (y_2 - y_1)^2}$. Under any rotation of the plane about the origin represented by

$$\begin{pmatrix} \cos\theta & -\sin\theta \\ \sin\theta & \cos\theta \end{pmatrix},$$

these points are mapped onto $(x_1 \cos\theta - y_1 \sin\theta, x_1 \sin\theta + y_1 \cos\theta)$ and $(x_2 \cos\theta - y_2 \sin\theta, x_2 \sin\theta + y_2 \cos\theta)$, respectively. The distance d' between these image points is equal to

$$\{[(x_2 \cos\theta - y_2 \sin\theta) - (x_1 \cos\theta - y_1 \sin\theta)]^2$$
$$+ [(x_2 \sin\theta + y_2 \cos\theta) - (x_1 \sin\theta + y_1 \cos\theta)]^2\}^{1/2}$$
$$= \sqrt{[(x_2 - x_1)\cos\theta - (y_2 - y_1)\sin\theta]^2 + [(x_2 - x_1)\sin\theta + (y_2 - y_1)\cos\theta]^2}$$
$$= \sqrt{(x_2 - x_1)^2(\cos^2\theta + \sin^2\theta) + (y_2 - y_1)^2(\sin^2\theta + \cos^2\theta)}$$
$$= \sqrt{(x_2 - x_1)^2 + (y_2 - y_1)^2}.$$

Hence, $d = d'$ and the distance between two points on a plane is invariant under a rotation of the plane about the origin.

6-3 / REFLECTIONS, DILATIONS, AND MAGNIFICATIONS

1. Each point (x, y) is mapped onto the point $(2x, 2y)$; that is, the matrix represents a dilation of the plane. The transformation is an example of a one-to-one mapping of the set of points on the plane onto itself.

3. Each point (x, y) is mapped onto the point $(2y, 2x)$; that is, the matrix represents the product of a dilation of the plane and a reflection of the plane with respect to the line $y = x$. The transformation is an example of a one-to-one mapping of the set of points on the plane onto itself.

5. Let (x_1, y_1) and (x_2, y_2) be any two points on a coordinate plane. The distance d between these two points is equal to $\sqrt{(x_2 - x_1)^2 + (y_2 - y_1)^2}$.

(a) Under a reflection of the plane with respect to the x-axis represented by

$$\begin{pmatrix} 1 & 0 \\ 0 & -1 \end{pmatrix},$$

these points are mapped onto $(x_1, -y_1)$ and $(x_2, -y_2)$, respectively. The distance d' between these image points is equal to

$$\sqrt{(x_2 - x_1)^2 + (-y_2 + y_1)^2} = \sqrt{(x_2 - x_1)^2 + (y_2 - y_1)^2}.$$

Hence, $d = d'$ and the distance between two points on a plane is invariant under a reflection of the plane with respect to the x-axis.

(b) Under a reflection of the plane with respect to the y-axis represented by

$$\begin{pmatrix} -1 & 0 \\ 0 & 1 \end{pmatrix},$$

these points are mapped onto $(-x_1, y_1)$ and $(-x_2, y_2)$, respectively. The distance d'' between these image points is equal to

$$\sqrt{(-x_2 + x_1)^2 + (y_2 - y_1)^2} = \sqrt{(x_2 - x_1)^2 + (y_2 - y_1)^2}.$$

Hence, $d = d''$ and the distance between two points on a plane is invariant under a reflection of the plane with respect to the y-axis.

7. The multiplication of any matrix A and a conformable scalar matrix is commutative. Hence, the multiplication of any dilation matrix (a scalar matrix) and any rotation matrix of the form (6-3) is commutative.

9. (a) $\begin{pmatrix} 4 & 0 \\ 0 & 4 \end{pmatrix}$;   (b) $\begin{pmatrix} \frac{1}{2} & 0 \\ 0 & \frac{1}{2} \end{pmatrix}$;   (c) $\begin{pmatrix} r & 0 \\ 0 & r \end{pmatrix}$.

11. The line $2x + 5y = 10$ is mapped onto the line $x + y = 1$.

6-4 / OTHER TRANSFORMATIONS

1. Each point (x, y) is mapped onto a point $(0, y)$; that is, the matrix represents a vertical projection of the points on the plane onto the y-axis. The transformation is a mapping of the set of points on the plane into itself.

3. Each point (x, y) is mapped onto a point $(x - 2y, y)$; that is, the matrix represents a shear parallel to the x-axis. The transformation is a one-to-one mapping of the set of points on the plane onto itself.

5. (a) The circle is mapped onto the ellipse $5x^2 - 4xy + y^2 = 1$.
(b) The rectangle is mapped onto a parallelogram with vertices at $(0, 0)$, $(2, 2k)$, $(2, 2k + 1)$, and $(0, 1)$.

7. $\begin{pmatrix} 2 & 3 \\ 1 & 2 \end{pmatrix}$.

6-5 / LINEAR HOMOGENEOUS TRANSFORMATIONS

In Exercises 1 and 3 let

$$T = \begin{pmatrix} a & b \\ c & d \end{pmatrix}$$

be a nonsingular matrix.

1. $\begin{pmatrix} a & b \\ c & d \end{pmatrix} \begin{pmatrix} 0 \\ 0 \end{pmatrix} = \begin{pmatrix} 0 \\ 0 \end{pmatrix}$.

3. Let $Ax + By + C = 0$ and $Ax + By + D = 0$ be the equations of parallel lines. By the results of Exercise 2, the images of the parallel lines under the transformation represented by T are the lines

$$\frac{Ad - Bc}{ad - bc} x + \frac{Ba - Ab}{ad - bc} y + C = 0$$

and

$$\frac{Ad - Bc}{ad - bc} x + \frac{Ba - Ab}{ad - bc} y + D = 0,$$

a pair of parallel lines. Hence, the images of parallel lines are parallel lines.

5. Under the *singular* homogeneous transformation represented by

$$\begin{pmatrix} 6 & 3 \\ 2 & 1 \end{pmatrix},$$

the image (x', y') of each point (x, y) is such that $x' = 6x + 3y$ and $y' = 2x + y$; that is, $x' = 3y'$, or $x = 3y$. Hence, the set of points on the plane are mapped onto the line $x = 3y$.

6-6 / ORTHOGONAL MATRICES

1. A proper orthogonal matrix.
3. Neither a proper orthogonal matrix nor an improper orthogonal matrix.
5.
$$(A^{-1})(A^{-1})^T = \begin{pmatrix} \frac{2}{3} & \frac{1}{3} & \frac{2}{3} \\ -\frac{2}{3} & \frac{2}{3} & \frac{1}{3} \\ \frac{1}{3} & \frac{2}{3} & -\frac{2}{3} \end{pmatrix} \begin{pmatrix} \frac{2}{3} & -\frac{2}{3} & \frac{1}{3} \\ \frac{1}{3} & \frac{2}{3} & \frac{2}{3} \\ \frac{2}{3} & \frac{1}{3} & -\frac{2}{3} \end{pmatrix} = I.$$

7.
$$\det\left[\begin{pmatrix} \frac{12}{13} & \frac{5}{13} \\ \frac{5}{13} & -\frac{12}{13} \end{pmatrix} + \begin{pmatrix} 1 & 0 \\ 0 & 1 \end{pmatrix}\right] = \det \begin{pmatrix} \frac{25}{13} & \frac{5}{13} \\ \frac{5}{13} & \frac{1}{13} \end{pmatrix} = \frac{25}{13} \cdot \frac{1}{13} - \frac{5}{13} \cdot \frac{5}{13} = 0.$$

9.
$$\begin{pmatrix} 1 & 0 \\ 0 & 1 \end{pmatrix} \text{ and } \begin{pmatrix} 0 & 1 \\ 1 & 0 \end{pmatrix}.$$

6-7 / TRANSLATIONS

1. In (a) through (d), k is any nonzero real number.
(a) $(k, -2k, k)$;   (b) $(3k, 0, k)$;
(c) $(0, 0, k)$;   (d) $(3k, 4k, k)$.

3. (a) $\begin{pmatrix} x' \\ y' \\ 1 \end{pmatrix} = \begin{pmatrix} 1 & 0 & -2 \\ 0 & 1 & 4 \\ 0 & 0 & 1 \end{pmatrix} \begin{pmatrix} x \\ y \\ 1 \end{pmatrix}$;   (b) $\begin{pmatrix} x' \\ y' \\ 1 \end{pmatrix} = \begin{pmatrix} 1 & 0 & 0 \\ 0 & 1 & 1 \\ 0 & 0 & 1 \end{pmatrix} \begin{pmatrix} x \\ y \\ 1 \end{pmatrix}.$

5. $(8, -2)$.   **7.** $(10, 4)$.   **9.** $3x^2 + 2y^2 - 18 = 0$.

11. Let (x_1, y_1) and (x_2, y_2) be any two points on a plane. The distance d between these two points is equal to $\sqrt{(x_2 - x_1)^2 + (y_2 - y_1)^2}$. Under any translation of the plane represented by

$$\begin{pmatrix} 1 & 0 & a \\ 0 & 1 & b \\ 0 & 0 & 1 \end{pmatrix},$$

the points are mapped onto $(x_1 + a, y_1 + b)$ and $(x_2 + a, y_2 + b)$, respectively. The distance d' between these image points is equal to

$$\sqrt{[(x_2 + a) - (x_1 + a)]^2 + [(y_2 + b) - (y_1 + b)]^2};$$

that is,

$$\sqrt{(x_2 - x_1)^2 + (y_2 - y_1)^2}.$$

Hence, $d = d'$ and the distance between two points on a plane is invariant under a translation of the plane.

13. $\left(\dfrac{\sqrt{3} - 6}{2}, \dfrac{1}{2}\right).$

6-8 / RIGID MOTION TRANSFORMATIONS

1. $$\mathscr{R} = \begin{pmatrix} \dfrac{1}{2} & -\dfrac{\sqrt{3}}{2} & 2 \\ \dfrac{\sqrt{3}}{2} & \dfrac{1}{2} & 0 \\ 0 & 0 & 1 \end{pmatrix}; \quad (2, 0, 1).$$

3. $a_{31} = a_{32} = 0, \quad a_{33} = 1, \quad \begin{vmatrix} a_{11} & a_{12} \\ a_{21} & a_{22} \end{vmatrix} = \begin{vmatrix} \dfrac{1}{2} & -\dfrac{\sqrt{3}}{2} \\ \dfrac{\sqrt{3}}{2} & \dfrac{1}{2} \end{vmatrix} = 1;$

$$\left(\dfrac{1}{2}\right)^2 + \left(\dfrac{\sqrt{3}}{2}\right)^2 = 1, \quad \left(-\dfrac{\sqrt{3}}{2}\right)^2 + \left(\dfrac{1}{2}\right)^2 = 1;$$

$$\left(\dfrac{1}{2}\right)\left(-\dfrac{\sqrt{3}}{2}\right) + \left(\dfrac{\sqrt{3}}{2}\right)\left(\dfrac{1}{2}\right) = 0.$$

5. (a) $\begin{pmatrix} -1 & 0 & 6 \\ 0 & 1 & 0 \\ 0 & 0 & 1 \end{pmatrix};$ \quad (b) $\begin{pmatrix} \dfrac{1}{2} & \dfrac{\sqrt{3}}{2} & -\dfrac{1}{2} \\ \dfrac{\sqrt{3}}{2} & -\dfrac{1}{2} & \dfrac{\sqrt{3}}{2} \\ 0 & 0 & 0 \end{pmatrix}.$

7. The matrices representing a rotation of the plane about the origin are of the form

$$R = \begin{pmatrix} \cos\theta & -\sin\theta & 0 \\ \sin\theta & \cos\theta & 0 \\ 0 & 0 & 1 \end{pmatrix}.$$

The reflection matrices of (6-6) are of the form

$$F = \begin{pmatrix} \mp 1 & 0 & 0 \\ 0 & \pm 1 & 0 \\ 0 & 0 & 1 \end{pmatrix}. \quad \text{Then} \quad RF = \begin{pmatrix} \mp \cos\theta & \mp \sin\theta & 0 \\ \mp \sin\theta & \pm \cos\theta & 0 \\ 0 & 0 & 1 \end{pmatrix},$$

where $a_{31} = a_{32} = 0$, $a_{33} = 1$, and

$$\begin{vmatrix} a_{11} & a_{12} \\ a_{21} & a_{22} \end{vmatrix} = \begin{vmatrix} \mp \cos\theta & \mp \sin\theta \\ \mp \sin\theta & \pm \cos\theta \end{vmatrix} = -1;$$

$(\mp \cos\theta)^2 + (\mp \sin\theta)^2 = 1$, $(\mp \sin\theta)^2 + (\pm \cos\theta)^2 = 1$; $(\mp \cos\theta)(\mp \sin\theta) + (\mp \sin\theta)(\pm \cos\theta) = 0$. Similarly,

$$FR = \begin{pmatrix} \mp \cos\theta & \pm \sin\theta & 0 \\ \pm \sin\theta & \pm \cos\theta & 0 \\ 0 & 0 & 1 \end{pmatrix},$$

where $a_{31} = a_{32} = 0$, $a_{33} = 1$, and

$$\begin{vmatrix} a_{11} & a_{12} \\ a_{21} & a_{22} \end{vmatrix} = \begin{vmatrix} \mp \cos\theta & \pm \sin\theta \\ \pm \sin\theta & \pm \cos\theta \end{vmatrix} = -1;$$

$(\mp \cos\theta)^2 + (\pm \sin\theta)^2 = 1$, $(\pm \sin\theta)^2 + (\pm \cos\theta)^2 = 1$; $(\mp \cos\theta)(\pm \sin\theta) + (\pm \sin\theta)(\pm \cos\theta) = 0$.

S E V E N

EIGENVALUES
AND EIGENVECTORS

7-1 / CHARACTERISTIC FUNCTIONS

1. $\lambda^2 - 9\lambda + 14 = 0$; $\lambda_1 = 2$ and $\lambda_2 = 7$; $(k \;\; -k)^T$ and $(3k \;\; 2k)^T$, where k is any nonzero scalar.

3. $\lambda^2 - 2\lambda = 0$; $\lambda_1 = 0$ and $\lambda_2 = 2$; $(0 \;\; k)^T$ and $(k \;\; 0)^T$, where k is any nonzero scalar.

5. $\lambda^3 - 2\lambda^2 - 5\lambda + 6 = 0$; $\lambda_1 = 1$, $\lambda_2 = -2$, and $\lambda_3 = 3$; $(k \ -k \ -k)^T$, $(11k \ k \ -14k)^T$, and $(k \ k \ k)^T$, where k is any nonzero scalar.

7. (a) Let

$$A = \begin{pmatrix} 5 & 3 \\ 2 & 4 \end{pmatrix}. \quad \text{Then} \quad A^2 = \begin{pmatrix} 31 & 27 \\ 18 & 22 \end{pmatrix},$$

$t_1 = 9$, and $t_2 = 53$. By (7-4), $c_0 = 1$, $c_1 = -9$, and $c_2 = 14$. Hence, $\lambda^2 - 9\lambda + 14 = 0$.

(b) Let

$$A = \begin{pmatrix} 2 & -2 & 3 \\ 1 & 1 & 1 \\ 1 & 3 & -1 \end{pmatrix}. \quad \text{Then} \quad A^2 = \begin{pmatrix} 5 & 3 & 1 \\ 4 & 2 & 3 \\ 4 & -2 & 7 \end{pmatrix},$$

$$A^3 = \begin{pmatrix} 14 & -4 & 17 \\ 13 & 3 & 11 \\ 13 & 11 & 3 \end{pmatrix},$$

$t_1 = 2$, $t_2 = 14$, and $t_3 = 20$. By (7-4), $c_0 = 1$, $c_1 = -2$, $c_2 = -5$, and $c_3 = 6$. Hence, $\lambda^3 - 2\lambda^2 - 5\lambda + 6 = 0$.

9. Every eigenvalue of A satisfies the characteristic equation $f(\lambda) = |A - \lambda I| = 0$. If $\lambda = 0$, then $|A - 0I| = |A| = 0$; that is, the value of the determinant of A is zero.

11. (a) $k\lambda_1$, $k\lambda_2$, and $k\lambda_3$. (b) $\lambda_1 - k$, $\lambda_2 - k$, and $\lambda_3 - k$.

13. Let $A = ((\delta_{ij}a_{ij}))$ be any diagonal matrix of order n. Then the characteristic equation of A is $|A - \lambda I| = (a_{11} - \lambda)(a_{22} - \lambda) \cdots (a_{nn} - \lambda) = 0$. Therefore, $\lambda_1 = a_{11}$, $\lambda_2 = a_{22}$, ..., $\lambda_n = a_{nn}$; that is, the eigenvalues of a diagonal matrix are equal to the diagonal elements.

7-2 / A GEOMETRIC INTERPRETATION OF EIGENVECTORS

1. The one-dimensional vector spaces containing the sets of vectors of the forms $(k \ 0)^T$ and $(0 \ k)^T$, respectively.

3. The one-dimensional vector space containing the set of vectors of the form $(0 \ k)^T$.

5. The one-dimensional vector space containing the set of vectors of the form $(k \ 0)^T$.

7. The characteristic equation of the matrix

$$\begin{pmatrix} \cos\theta & -\sin\theta \\ \sin\theta & \cos\theta \end{pmatrix}$$

representing a rotation of the plane about the origin through an angle θ is $\lambda^2 - 2\cos\theta\lambda + 1 = 0$. Real eigenvalues (and hence real eigenvectors) of A exist if, and only if, the discriminant of the characteristic equation is greater than or equal to zero; that is, if, and only if,

$$4\cos^2\theta - 4 \geq 0,$$
$$\cos^2\theta - 1 \geq 0,$$
$$-\sin^2\theta \geq 0,$$
$$\theta = 180°k,$$

where k is any integer. Therefore, invariant vector spaces exist only under rotations of the plane about the origin through angles which are integral multiples of $180°$.

7-3 / SOME THEOREMS

1. Associated with the distinct eigenvalues $\lambda_1 = -1$ and $\lambda_2 = 8$ are the eigenvectors $(5a \quad -4a)^T$ and $(b \quad b)^T$, respectively, where a and b are any nonzero scalars. Then $m_1(5a \quad -4a)^T + m_2(b \quad b)^T = 0$ if, and only if, $m_1 = m_2 = 0$. Hence, $(5a \quad -4a)^T$ and $(b \quad b)^T$ are linearly independent.

3. Associated with the distinct real eigenvalues $\lambda_1 = 0$, $\lambda_2 = 1$, and $\lambda_3 = 4$ are the eigenvectors $(a \quad -a \quad 0)^T$, $(0 \quad 0 \quad b)^T$, and $(c \quad c \quad 0)^T$, respectively, where a, b, and c are any nonzero scalars. The eigenvectors are mutually orthogonal since $(a \quad -a \quad 0)(0 \quad 0 \quad b)^T = 0$, $(0 \quad 0 \quad b)(c \quad c \quad 0)^T = 0$, and $(c \quad c \quad 0)(a \quad -a \quad 0)^T = 0$.

5. Let X_i be a unit eigenvector associated with the eigenvalue λ_i of A. Then

$$X_i^T A X_i = (\lambda_i),$$
$$(X_i^T A X_i)^* = (\lambda_i)^*,$$
$$(\bar{X}_i)^T A^* \bar{X}_i = (\bar{\lambda}_i);$$

that is, $\bar{\lambda}_i$ is an eigenvalue of A^*. Hence, the eigenvalues of A^* are the conjugates of the eigenvalues of A.

7-4 / DIAGONALIZATION OF MATRICES

1. The matrices are not similar matrices.

3. Matrices A and B are similar matrices since any nonsingular matrix of the form

$$\begin{pmatrix} 3c - 2d & 3c - 2d \\ c & d \end{pmatrix},$$

where c and d are arbitrary scalars, is such that

$$\begin{pmatrix} 3c - 2d & 3c - 2d \\ c & d \end{pmatrix}^{-1} \begin{pmatrix} 2 & 0 \\ 1 & 1 \end{pmatrix} \begin{pmatrix} 3c - 2d & 3c - 2d \\ c & d \end{pmatrix} = \begin{pmatrix} 4 & 3 \\ -2 & -1 \end{pmatrix};$$

$\det A = 2 = \det B$; the eigenvalues of both A and B are $\lambda_1 = 1$ and $\lambda_2 = 2$.

5. Associated with the eigenvalues $\lambda_1 = -1$ and $\lambda_2 = 2$ are the eigenvectors $(k \;\; -k)^T$ and $(k \;\; 0)^T$, respectively, where k is any nonzero scalar. Since

$$A^3 = \begin{pmatrix} 8 & 9 \\ 0 & -1 \end{pmatrix},$$

$\lambda_1^3 = -1$, and $\lambda_2^3 = 8$, then

$$A^3(k \;\; -k)^T = \begin{pmatrix} -k \\ k \end{pmatrix} = \lambda_1^3(k \;\; -k)^T$$

and

$$A^3(k \;\; 0)^T = \begin{pmatrix} 8k \\ 0 \end{pmatrix} = \lambda_2^3(k \;\; 0)^T.$$

7. Any matrix of the form

$$\begin{pmatrix} a & 2b \\ 2a & -3b \end{pmatrix},$$

where a and b are any scalars such that $ab \neq 0$;

$$\begin{pmatrix} \frac{3}{7a} & \frac{2}{7a} \\ \frac{2}{7b} & -\frac{1}{7b} \end{pmatrix} \begin{pmatrix} 5 & 4 \\ 12 & 7 \end{pmatrix} \begin{pmatrix} a & 2b \\ 2a & -3b \end{pmatrix} = \begin{pmatrix} 13 & 0 \\ 0 & -1 \end{pmatrix}.$$

9.
$$\begin{pmatrix} \frac{1}{\sqrt{2}} & \frac{1}{\sqrt{2}} & 0 \\ 0 & 0 & 1 \\ \frac{1}{\sqrt{2}} & -\frac{1}{\sqrt{2}} & 0 \end{pmatrix} \begin{pmatrix} 2 & 2 & 0 \\ 2 & 2 & 0 \\ 0 & 0 & 1 \end{pmatrix} \begin{pmatrix} \frac{1}{\sqrt{2}} & 0 & \frac{1}{\sqrt{2}} \\ \frac{1}{\sqrt{2}} & 0 & -\frac{1}{\sqrt{2}} \\ 0 & 1 & 0 \end{pmatrix} = \begin{pmatrix} 4 & 0 & 0 \\ 0 & 1 & 0 \\ 0 & 0 & 0 \end{pmatrix}.$$

Other orthogonal transformations exist.

11. If A is similar to kI, a nonsingular matrix C exists such that $A = C^{-1}(kI)C$. Then $A = C^{-1}k(IC) = C^{-1}kC = kC^{-1}C = kI$.

7-5 / THE HAMILTON-CAYLEY THEOREM

1. $f(\lambda) = \lambda^2 - 6\lambda + 11$;

$$f(A) = \begin{pmatrix} 2 & -1 \\ 3 & 4 \end{pmatrix}^2 - 6\begin{pmatrix} 2 & -1 \\ 3 & 4 \end{pmatrix} + 11\begin{pmatrix} 1 & 0 \\ 0 & 1 \end{pmatrix}$$

$$= \begin{pmatrix} 1 & -6 \\ 18 & 13 \end{pmatrix} + \begin{pmatrix} -12 & 6 \\ -18 & -24 \end{pmatrix} + \begin{pmatrix} 11 & 0 \\ 0 & 11 \end{pmatrix} = \begin{pmatrix} 0 & 0 \\ 0 & 0 \end{pmatrix}.$$

3. $f(\lambda) = \lambda^2 - 7\lambda + 10$;

$$f(A) = \begin{pmatrix} 5 & 0 \\ 0 & 2 \end{pmatrix}^2 - 7\begin{pmatrix} 5 & 0 \\ 0 & 2 \end{pmatrix} + 10\begin{pmatrix} 1 & 0 \\ 0 & 1 \end{pmatrix}$$

$$= \begin{pmatrix} 25 & 0 \\ 0 & 4 \end{pmatrix} + \begin{pmatrix} -35 & 0 \\ 0 & -14 \end{pmatrix} + \begin{pmatrix} 10 & 0 \\ 0 & 10 \end{pmatrix} = \begin{pmatrix} 0 & 0 \\ 0 & 0 \end{pmatrix}.$$

5. $f(\lambda) = \lambda^2 - 6\lambda + 1$; $f(A) = A^2 - 6A + I = 0$, $I = -A^2 + 6A$, and

$$A^{-1} = -A + 6I = -\begin{pmatrix} 5 & 2 \\ 2 & 1 \end{pmatrix} + \begin{pmatrix} 6 & 0 \\ 0 & 6 \end{pmatrix} = \begin{pmatrix} 1 & -2 \\ -2 & 5 \end{pmatrix}.$$

7. (a) $f(\lambda) = \lambda^2 - 8\lambda - 3$; $f(A) = A^2 - 8A - 3I = 0$, $I = \frac{1}{3}(A^2 - 8A)$,

$A^{-1} = \frac{1}{3}(A - 8I)$,

$A^{-2} = A^{-1}A^{-1} = \frac{1}{3}(I - 8A^{-1}) = \frac{67}{9}I - \frac{8}{9}A$

$$= \frac{67}{9}\begin{pmatrix} 1 & 0 \\ 0 & 1 \end{pmatrix} - \frac{8}{9}\begin{pmatrix} 7 & 2 \\ 5 & 1 \end{pmatrix} = \begin{pmatrix} \frac{11}{9} & -\frac{16}{9} \\ -\frac{40}{9} & \frac{59}{9} \end{pmatrix};$$

$$A^2 A^{-2} = \begin{pmatrix} 59 & 16 \\ 40 & 11 \end{pmatrix} \begin{pmatrix} \frac{11}{9} & -\frac{16}{9} \\ -\frac{40}{9} & \frac{59}{9} \end{pmatrix} = \begin{pmatrix} 1 & 0 \\ 0 & 1 \end{pmatrix}.$$

(b) $A^{-3} = A^{-1}A^{-2} = \frac{67}{9}A^{-1} - \frac{8}{9}I = \frac{67}{27}A - \frac{560}{27}I$

$= \frac{67}{27}\begin{pmatrix} 7 & 2 \\ 5 & 1 \end{pmatrix} - \frac{560}{27}\begin{pmatrix} 1 & 0 \\ 0 & 1 \end{pmatrix} = \begin{pmatrix} -\frac{91}{27} & \frac{134}{27} \\ \frac{335}{27} & -\frac{493}{27} \end{pmatrix};$

$A^3 A^{-3} = \begin{pmatrix} 493 & 134 \\ 335 & 91 \end{pmatrix}\begin{pmatrix} -\frac{91}{27} & \frac{134}{27} \\ \frac{335}{27} & -\frac{493}{27} \end{pmatrix} = \begin{pmatrix} 1 & 0 \\ 0 & 1 \end{pmatrix}.$

7-6 / QUADRATIC FORMS

1. $(x \ y)\begin{pmatrix} 3 & 5 \\ 5 & 3 \end{pmatrix}\begin{pmatrix} x \\ y \end{pmatrix}.$ **3.** $(x \ y)\begin{pmatrix} 2 & \sqrt{2} \\ \sqrt{2} & 1 \end{pmatrix}\begin{pmatrix} x \\ y \end{pmatrix}.$

5.

$(x' \ y')\begin{pmatrix} \frac{1}{\sqrt{2}} & -\frac{1}{\sqrt{2}} \\ \frac{1}{\sqrt{2}} & \frac{1}{\sqrt{2}} \end{pmatrix}\begin{pmatrix} 1 & -1 \\ -1 & 1 \end{pmatrix}\begin{pmatrix} \frac{1}{\sqrt{2}} & \frac{1}{\sqrt{2}} \\ -\frac{1}{\sqrt{2}} & \frac{1}{\sqrt{2}} \end{pmatrix}\begin{pmatrix} x' \\ y' \end{pmatrix}$

$= (x' \ y')\begin{pmatrix} 2 & 0 \\ 0 & 0 \end{pmatrix}\begin{pmatrix} x' \\ y' \end{pmatrix} = (2x'^2).$

Other orthogonal transformations exist.

7-7 / CLASSIFICATION OF THE CONICS

1. $3x''^2 + 9y''^2 = 8$, a real ellipse. **3.** $x''^2 - y''^2 = 4$, a hyperbola.

5. A real ellipse. **7.** Two intersecting lines.

9. $\lambda_1 = 8$ and $\lambda_2 = 2$; the squares of the semi-axes are $a^2 = 1$ and $b^2 = 4$; $\lambda_1 a^2 = \lambda_2 b^2$.

7-8 / INVARIANTS FOR CONICS

1. Three. **3.** Three.

5. (a) $a + c = 1 + 1 = \frac{1}{2} + \frac{3}{2}$; (b) $ac - b^2 = (1)(1) - (\frac{1}{2})^2 = (\frac{1}{2})(\frac{3}{2})$.

7. (a) $a + c = 7 + (-7) = 1 + (-1)$;
(b) $ac - b^2 = (7)(-7) - (-24)^2 = (25)(-25)$.

9. (a) The sum of the roots of equation (7-39) is equal to the opposite of the coefficient of λ; that is, $a + c = \lambda_1 + \lambda_2$;

(b) the product of the roots of equation (7-39) is equal to the constant term; that is, $ac - b^2 = \lambda_1\lambda_2$;

(c) $$\det F = \begin{vmatrix} a & b \\ b & c \end{vmatrix} = ac - b^2.$$

11. By the results of this section, $a + c$ is an invariant for conics under a rotation transformation. By (7-28), f is an invariant for conics under a rotation transformation. Hence, $a + c + f$ is an invariant for conics under the rotation transformation.

INDEX

Addition:
  of matrices, 114
  of quaternions, 75
  of vectors, 6, 28, 76, 77
Additive identity element:
  for matrices, 115
  for a ring, 154
Additive inverses:
  for matrices, 116
  for a ring, 154
Angle:
  between two planes, 96–97
  between two vectors, 33
Anti-symmetric matrix, 128
Associative property:
  for addition of matrices, 115
  for addition of vectors, 7, 77
  for multiplication of matrices, 123
  for a ring, 154
Augmented matrix, 169

Basis for a vector space, 14, 15, 16, 48, 49, 78
  normal orthogonal, 49
  orthogonal, 48, 49
  orthonormal, 49
Bound vectors, 4

Central conic, 240
Centroid, 20
Ceva, Theorem of, 26
Characteristic equation, 213
Characteristic function, 213
Characteristic values, 217 (see also Eigenvalues)
Characteristic vectors, 217–218 (see also Eigenvectors)
Circle, equation of a, 45
  coordinate form, 45
  vector form, 45

Classical canonical form, 226
Closure property for a ring, 154
Cofactor, 142
Cofactors, matrix of, 150
Collinear points, 17
Column index, 112
Column matrix, 120
Column vector, 120
Commutative property:
  for addition of matrices, 114–115
  for addition of vectors, 7, 77
  for a ring, 154
  for the scalar product, 34
Complex matrix, 133
Complex numbers as matrices, 156
Components of a vector, 28
Conformable matrices, 119
Conic, 239
  central, 240
  noncentral, 240
  rank of a, 245
Conics:
  degenerate, 246
  proper, 246
Conic section, 239
  matrix of a, 239
Conjugate:
  of a matrix, 133
  of a quaternion, 76
Consistent system of linear equations, 169
Coordinate axes, 26
Coordinate form, 87
Coordinate planes, 26
Coordinates of a point, 26
  homogeneous, 199
  nonhomogeneous, 199
Coordinate systems:
  left-handed, 27
  right-handed, 26–27

Coplanar points, 67–68
Cosine law:
  of plane trigonometry, 41
  of spherical trigonometry, 70–71
Cross product, 51 (see also Vector product)

Degenerate conics, 246
Degenerate parallelepiped, 66
Desargues' Theorem, 23
Determinant of a matrix:
  of order n, 139
  of order three, 138
  of order two, 137
  value of the, 137, 138, 139
Diagonal, main, 125
Diagonal, principal, 125
Diagonal elements, 125
Diagonalization of a matrix, 226
Diagonal matrix, 125
Difference:
  of two matrices, 116
  of two vectors, 8, 28
Dilation of the plane, 186
Dimension of a real vector space, 78
Directed line segments, 2
Direction angles, 83
Direction cosines, 83
Direction numbers, 84
Distance:
  between a point and a line, 46–47
  between a point and a plane, 99, 100
  between two parallel planes, 100–101
  between two points, 44
Distributive property:
  left-hand, 122

287

Distributive property (*cont.*):
  for matrices, 121–122
  for quaternions, 75
  right-hand, 122
  for a ring, 154
  for the scalar product, 35
  for the vector product, 53, 64, 67
Division of a line segment in a given ratio, 17–18
Dot product, 33 (*see also* Scalar product)

Echelon form, 165
Eigenvalues, 215
  of Hermitian matrices, 223
  of symmetric matrices, 223
Eigenvector, 216–217
Eigenvectors:
  of Hermitian matrices, 223
  of symmetric matrices, 224
Elementary row transformation matrices, 167
Elementary row transformations, 165
Equality:
  of quaternions, 75
  of vectors, 4, 76
Equal matrices, 113
Equal vectors, 5, 28
Equation of a circle, 45
  coordinate form, 45
  vector form, 45
Equation of a line, 29–30, 46, 47–48, 102–104, 106–107
  coordinate forms, 30, 46, 48, 103, 104, 106, 107
  general form, 107
  parametric forms, 30, 103, 104
  point-direction number form, 103
  symmetric form, 103, 106
  tangent to a circle, 47–48
  two-point form, 29–30, 104
  vector forms, 46, 47, 48, 102, 104
Equation of a plane, 86–90, 94, 95
  coordinate forms, 87, 90, 94, 95
  intercept form, 87–88
  parametric forms, 90
  point-direction number form, 87
  tangent to a sphere, 94, 95
  vector forms, 86–87, 89, 90, 94
Equation of a sphere, 92–93
  coordinate forms, 93, 94
  vector forms, 93

Fibonacci sequence, 125

Finite-dimensional vector space, 79
Fixed point, 181
Free vectors, 5

General linear transformations of the plane, 198–199, 208–209
Geometric vector, 2
Gibbs, J. W., 74
Gram–Schmidt process, 49

Hamilton, William Rowan, 74
Hamilton–Cayley Theorem, 231
Hermitian matrix, 133
  eigenvalues of a, 223
  eigenvectors of a, 223
Hero's formula, 60–61
Homogeneous coordinates, 199

Identity matrix, 127
Identity transformation, 181
Image, 175
Improper orthogonal matrix, 196
Inconsistent system of linear equations, 169
Index:
  column, 112
  row, 112
Initial point of a vector, 2
Inner product, 33 (*see also* Scalar product)
Invariant, under a transformation 183, 196, 208–211, 245–247
Invariants for conics, 245–247
Invariant vector spaces, 219–220
Inverse:
  additive, 116, 154
  mapping, 177
  multiplicative, 146, 149, 150, 167, 233
  left, 146
  right, 146
Isomorphic rings, 155
Isomorphism, 155

Kronecker delta, 127

Latent values, 217 (*see also* Eigenvalues)
Latent vectors, 217–218 (*see also* Eigenvectors)
Law of cosines:
  of plane trigonometry, 41
  of spherical trigonometry, 70–71
Law of sines:
  of plane trigonometry, 59–60, 63

Law of sines (*cont.*):
  of spherical trigonometry, 73
Law of vector addition, 6
Left-hand distributive property, 122
Left-handed coordinate system, 27
Left multiplicative inverse, 146
Line, equation of a, 29–30, 46, 47–48, 102–104, 106–107
  coordinate forms, 30, 46, 48, 103, 104, 106, 107
  general form, 107
  parametric forms, 30, 103, 104
  point-direction number form, 103
  symmetric form, 103, 106
  tangent to a circle, 47–48
  two-point form, 29–30, 104
  vector forms, 46, 47, 48, 102, 104
Linear algebra, real, 161
Linear equations, systems of, 169
  consistent, 169
  homogeneous, 173
  inconsistent, 169
Linear function of vectors, 14
Linearly dependent, 14, 163
Linearly independent, 14, 163
Linear transformations of the plane (*see also* Transformations):
  general, 198–199, 208–209
  homogeneous, 193, 209
  nonsingular, 193
  singular, 278
  rigid motion, 208, 209
Line vectors, 4

Magnification of the plane, 187
Magnitude of a vector, 2, 34
Main diagonal, 125
Mapping, single-valued, 175
  into, 175
  inverse, 177
  one-to-one, 176
  onto, 176
Matrices, 112
  addition of, 114
    associative property for, 115
    commutative property for, 114–115
  additive identity element for, 115
  conformable, 119
  diagonal, 125
  difference of, 116

Matrices (*cont.*):
  elementary row transformation, 167
  equal, 113
  multiplication of, 118–119
    associative property for, 123
    distributive properties for, 121–122
  notation for, 112, 113, 115, 127, 129, 133
  Pauli, 125
  product of, 119
  reflection, 183
  similar, 226
  sum of, 114
Matrix, 112
  additive inverse of a, 116
  anti-symmetric, 128
  augmented, 169
  characteristic equation of a, 213
  characteristic function of a, 213
  characteristic values of a, 217
  characteristic vectors of a, 217–218
  classical canonical form of a, 226
  of coefficients, 118, 169
  of cofactors, 150
  column, 120
  complex, 133
  of a conic section, 239
  conjugate of a, 133
  determinant of a, 137, 138, 139
    value of a, 137, 138, 139
  diagonal, 125
  diagonalization of a, 226
  echelon form of a, 165
  eigenvalues of a, 215
  eigenvector of a, 216–217
  element of a, 112
  Hermitian, 133
    eigenvalues of a, 223
    eigenvectors of a, 223
  identity, 127
  latent values of a, 217
  latent vectors of a, 217–218
  modal, 226
  multiplication, 118–119
  multiplicative inverse of a, 146, 149, 150, 167, 233
    left, 146
    right, 146
  nonsingular, 150
  null, 115
  order of a, 112, 113
  orthogonal, 195
    improper, 196
    proper, 196

Matrix (*cont.*):
  postmultiplication of a, 119
  premultiplication of a, 119
  proper values of a, 217
  proper vectors of a, 217–218
  rank of a, 164
  real, 113
  reflection, 183
  rotation, 179
  row, 120
  r-rowed minors of a, 164
  scalar, 125
  scalar multiple of a, 116
  singular, 150
  skew-Hermitian, 133
  skew-symmetric, 128
  square, 113
    order of a, 113
  symmetric, 128
    eigenvalues of a, 223
    eigenvectors of a, 224
  trace of a, 214
  translation, 200–201
  transpose of a, 129
  unit, 127
  upper triangular, 144
  zero, 115
Menelaus, Theorem of, 22
Minor:
  of an element, 142
  of a matrix, 164
Modal matrix, 226
Multiplication:
  of matrices, 118–119
  of a matrix by a scalar (real number), 115–116
  of a quaternion by a scalar (real number), 75
  of quaternions, 75
  of a vector by a scalar (real number), 9–10, 28, 76, 77
Multiplicative inverse of a matrix, 146, 149, 150, 167, 233
  left, 146
  right, 146

n-dimensional vector, 76
Noncentral conic, 240
Nonhomogeneous coordinates, 199
Nonsingular linear homogeneous transformation, 193
Nonsingular matrix, 150
Normal orthogonal basis, 49
Null matrix, 115
Null parallelepiped, 66
Null parallelogram, 52
Null space, 195

Null vector, 3

One-dimensional vector space, 14
One-to-one mapping, 176
Ordered n-tuple, 76
Ordered quadruples, 75
Order of a matrix, 112, 113
Origin, 26
  of a vector, 2
Origin point of a vector, 2
Orthogonal basis, 48, 49
  normal, 49
Orthogonal matrix, 195
  improper, 196
  proper, 196
Orthogonal transformation, 229
Orthogonal vectors, 48, 152
Orthonormal basis, 49
Outer product, 51 (*see also* Vector product)

Parallelogram law of vector addition, 7
Parallel planes:
  distance between, 100–101
  equations of, 100
Pauli matrices, 125
Piercing points, 107
Plane, equation of a, 86–90, 94, 95
  coordinate forms, 87, 90, 94, 95
  intercept form, 87–88
  parametric forms, 90
  point-direction number form, 87
  tangent to a sphere, 94, 95
  vector forms, 86–87, 89, 90, 94
Point transformation, 178
Position vector, 27
Postmultiplication of a matrix, 119
Premultiplication of a matrix, 119
Principal Axes Theorem, 229
Principal diagonal, 125
Product of matrices, 119
Product of quaternions, 75
Product of vectors:
  cross, 51
  dot, 33
  inner, 33
  outer, 51
  scalar, 33, 36, 76
    quadruple, 70
  triple, 64, 145
  vector, 51, 55, 145
    quadruple, 72
    triple, 69
Product transformation, 181

INDEX   289

Projections of the plane, 192
Proper conics, 246
Proper orthogonal matrix, 196
Proper values, 217 (*see also* Eigenvalues)
Proper vectors, 217–218 (*see also* Eigenvectors)
Pure quaternion, 75
Pythagorean Theorem, 44

Quadratic form, 236
Quadruple products:
 scalar, 70
 vector, 72
Quaternion, 75
 algebra, 75
 conjugate of a, 76
 pure, 75
Quaternions, 79, 158
 addition of, 75
 equality of, 75
 multiplication of, 75
 scalar multiplication of, 75

Rank:
 of a conic, 245
 of a matrix, 164
Real linear algebra, 161
Real matrix, 113
Real vector space, 77 (*see also* Vector space)
 basis for a, 14, 15, 16, 48, 49, 78
 dimension of a, 78
 finite-dimensional, 79
 subspace of a, 79
Reflection matrices, 183
Reflections of the plane, 183, 205
Right-hand distributive property, 122
Right-handed coordinate system, 26–27
Right multiplicative inverse, 146
Rigid motion transformations, 208, 209
Ring, 154
Rings, isomorphic, 155
Rotation matrix, 179
Rotation of the plane, 178, 205
Row index, 112
Row matrix, 120
Row transformation(s),
 elementary, 165
 matrices, 167
Row vector, 120
r-rowed minors, 164

Scalar matrix, 125

Scalar multiple:
 of a matrix, 116
 of a quaternion, 75
 of a vector, 9–10, 28, 76, 77
Scalar product, 33, 36, 76
Scalar quadruple product, 70
Scalar quantities, 1
Scalars, 1
Scalar triple product, 64, 145
Shear:
 parallel to the x-axis, 190
 parallel to the y-axis, 192
Similarity transformation, 226
Similar matrices, 226
Sine law:
 of plane trigonometry, 59–60, 63
 of spherical trigonometry, 73
Single-valued mapping, 175
 into, 175
 inverse, 177
 one-to-one, 176
 onto, 176
Singular linear homogeneous transformation, 278
Singular matrix, 150
Skew-commutative, 52
Skew-Hermitian matrix, 133
Skew-symmetric matrix, 128
Snell's law, 62–63
Spanned, a vector space, 49
Sphere, equation of a, 92–93
 coordinate forms, 93, 94
 vector forms, 93
Square matrix, 113
 order of a, 113
Square submatrices, 164
Subspace of a real vector space, 79
Subtraction of vectors, 8, 28
Sum:
 of matrices, 114
 of quaternions, 75
 of vectors, 6, 8, 28, 76, 77
Symmetric matrix, 128
 eigenvalues of a, 223
 eigenvectors of a, 224
Systems of linear equations, 169
 consistent, 169
 homogeneous, 173
 inconsistent, 169

Tangent line to a circle, 47–48
Tangent plane to a sphere, 94, 95
Terminal point of a vector, 2
Three-dimensional vector space, 16
Trace, 214

Transformation (*see also* Transformations of the plane):
 identity, 181
 invariant under a, 183, 196, 208–211, 245–247
 orthogonal, 229
 point, 178
 product, 181
 similarity, 226
Transformation(s), elementary
 row, 165
 matrices, 167
Transformations of the plane, 178
 dilations, 186
 general linear, 198–199, 208–209
 linear homogeneous, 193, 209
 nonsingular, 193
 singular, 278
 magnifications, 187
 null space of, 195
 projections, 192
 reflections, 183, 205
 rigid motion, 208, 209
 rotations, 178, 205
 shears:
  parallel to the x-axis, 190
  parallel to the y-axis, 192
 translations, 200
Translation matrix, 200–201
Translation of the plane, 200
Transpose of a matrix, 129
Transposition, 129
Triangle law of vector addition, 7
Triple product:
 scalar, 64, 145
 vector, 69
Two-dimensional vector space, 15

Uniqueness of representation of a vector, 17, 19
Unit matrix, 127
Unit vector, 13
Upper triangular matrix, 144

Vector, 2, 76
 column, 120
 components of a, 28
 direction angles of a, 83
 direction cosines of a, 83
 direction numbers of a, 84
 geometric, 2
 initial point of a, 2
 magnitude of a, 2, 34
 multiplication of by a scalar (real number), 9–10, 28, 76, 77

Vectors (*cont.*):
   n-dimensional, 76
   notation for a, 2, 76
   null, 3
   origin of a, 2
   origin point of a, 2
   position, 27
   row, 120
   terminal point of a, 2
   uniqueness of representation of
      a, 17, 19
   unit, 13
   zero, 3
Vector product, 51, 55, 145
Vector quadruple product, 72
Vector quantities, 1
Vectors, 1, 76, 77
   addition of, 6, 28, 76, 77
   angle between two, 33
   bound, 4

Vectors (*cont.*):
   cross product of, 51
   difference of, 8, 28
   dot product of, 33
   equal, 5, 28
   equality of, 4, 76
   free, 5
   inner product of, 33
   line, 4
   linear function of, 14
   linearly dependent, 14, 163
   linearly independent, 14, 163
   orthogonal, 48, 152
   outer product of, 51
   quadruple products of:
      scalar, 70
      vector, 72
   scalar product of, 33, 36, 76
   subtraction of, 8, 28
   sum of, 6, 8, 28, 76, 77

Vectors (*cont.*):
   triple products of:
      scalar, 64, 145
      vector, 69
   vector product of, 51, 55, 145
Vector space, 14, 15, 16, 77
   basis for a, 14, 15, 16, 48, 49, 78
   dimension of a, 78
   finite-dimensional, 79
   one-dimensional, 14
   subspace of a, 79
   three-dimensional, 16
   two-dimensional, 15
Vector spaces, invariant, 219–220
Vector triple product, 69

Zero divisors, 123
Zero matrix, 115
Zero vector, 3